Student Solutions Manual

Elementary Algebra

NINTH EDITION

Charles P. McKeague
Cuesta College

Ross Rueger
College of the Sequoias

BROOKS/COLE CENGAGE Learning

Australia • Brazil • Japan • Korea • Mexico • Singapore • Spain • United Kingdom • United States

© 2012 Brooks/Cole, Cengage Learning

ALL RIGHTS RESERVED. No part of this work covered by the copyright herein may be reproduced, transmitted, stored, or used in any form or by any means graphic, electronic, or mechanical, including but not limited to photocopying, recording, scanning, digitizing, taping, Web distribution, information networks, or information storage and retrieval systems, except as permitted under Section 107 or 108 of the 1976 United States Copyright Act, without the prior written permission of the publisher.

For product information and technology assistance, contact us at **Cengage Learning Customer & Sales Support, 1-800-354-9706**

For permission to use material from this text or product, submit all requests online at **www.cengage.com/permissions**
Further permissions questions can be emailed to **permissionrequest@cengage.com**

ISBN-13: 978-1-111-57180-1
ISBN-10: 1-111-57180-5

Brooks/Cole
20 Davis Drive
Belmont, CA 94002-3098
USA

Cengage Learning is a leading provider of customized learning solutions with office locations around the globe, including Singapore, the United Kingdom, Australia, Mexico, Brazil, and Japan. Locate your local office at: **www.cengage.com/global**

Cengage Learning products are represented in Canada by Nelson Education, Ltd.

To learn more about Brooks/Cole, visit **www.cengage.com/brookscole**

Purchase any of our products at your local college store or at our preferred online store **www.cengagebrain.com**

Printed in the United States of America
1 2 3 4 5 6 7 14 13 12 11 10

Contents

© 2012 Cengage Learning. All Rights Reserved. May not be scanned, copied or duplicated, or posted to a publicly accessible website, in whole or in part.

© 2012 Cengage Learning. All Rights Reserved. May not be scanned, copied or duplicated, or posted to a publicly accessible website, in whole or in part.

© 2012 Cengage Learning. All Rights Reserved. May not be scanned, copied or duplicated, or posted to a publicly accessible website, in whole or in part.

© 2012 Cengage Learning. All Rights Reserved. May not be scanned, copied or duplicated, or posted to a publicly accessible website, in whole or in part.

Preface

This *Student Solutions Manual* contains complete solutions to all odd-numbered exercises and all chapter test exercises of *Elementary Algebra* by Charles P. McKeague. Every attempt has been made to format solutions for readability and accuracy, and I apologize for any errors that you may encounter. If you have any comments, suggestions, error corrections, or alternative solutions please feel free to drop me a note or send an email (address below).

Please use this manual with some degree of caution. Be sure that you have attempted a solution, and re-attempted it, before you look it up in this manual. Algebra can only be learned by *doing*, and not by observing! As you use this manual, do not just read the solution but work it along with the manual, using my solution to check your work. If you use this manual in that fashion then it should be helpful to you as you do homework and study for tests.

I would like to thank a number of people for their assistance in preparing this manual. Thanks go to Shaun Williams at Cengage Learning for her valuable assistance and support. Also thanks go to Patrick McKeague for keeping me updated to manuscript corrections and prompt response to my emails.

I wish to express my appreciation to Pat McKeague for asking me to be involved with this textbook. This book provides a complete course in beginning algebra, and you will find the text very easy to read and understand. Good luck!

Ross Rueger
College of the Sequoias
rossrueger@gmail.com

© 2012 Cengage Learning. All Rights Reserved. May not be scanned, copied or duplicated, or posted to a publicly accessible website, in whole or in part.

Preface

[illegible]

[illegible]

[illegible]

[illegible]

[illegible]

Chapter 1
The Basics

Getting Ready for Chapter 1

1. Multiplying: $5 \cdot 5 \cdot 5 = 125$
2. Dividing: $12 \div 4 = 3$
3. Subtracting: $1.7 - 1.2 = 0.5$
4. Subtracting: $10 - 9.5 = 0.5$
5. Multiplying: $7(0.2) = 1.4$
6. Multiplying: $0.3(6) = 1.8$
7. Subtracting: $14 - 9 = 5$
8. Dividing: $39 \div 13 = 3$
9. Multiplying: $5 \cdot 9 = 45$
10. Multiplying: $2 \cdot 2 \cdot 2 \cdot 2 \cdot 2 = 32$
11. Adding: $15 + 14 = 29$
12. Dividing: $28 \div 7 = 4$
13. Dividing: $26 \div 2 = 13$
14. Subtracting: $125 - 81 = 44$
15. Dividing: $630 \div 63 = 10$
16. Dividing: $210 \div 10 = 21$
17. Multiplying: $2 \cdot 3 \cdot 5 \cdot 7 = 210$
18. Multiplying: $3 \cdot 7 \cdot 11 = 231$
19. Multiplying: $2 \cdot 3 \cdot 3 \cdot 5 \cdot 7 = 630$
20. Dividing: $24 \div 8 = 3$

1.1 Variables, Notation, and Symbols

1. The equivalent expression is $x + 5 = 14$.
3. The equivalent expression is $5y < 30$.
5. The equivalent expression is $3y \le y + 6$.
7. The equivalent expression is $\frac{x}{3} = x + 2$.
9. Expanding the expression: $3^2 = 3 \cdot 3 = 9$
11. Expanding the expression: $7^2 = 7 \cdot 7 = 49$
13. Expanding the expression: $2^3 = 2 \cdot 2 \cdot 2 = 8$
15. Expanding the expression: $4^3 = 4 \cdot 4 \cdot 4 = 64$
17. Expanding the expression: $2^4 = 2 \cdot 2 \cdot 2 \cdot 2 = 16$
19. Expanding the expression: $10^2 = 10 \cdot 10 = 100$
21. Expanding the expression: $11^2 = 11 \cdot 11 = 121$
23. Using the order of operations: $20 \div 2 \cdot 10 = 10 \cdot 10 = 100$
24. Using the order of operations: $40 \div 4 \cdot 5 = 10 \cdot 5 = 50$
25. Using the order of operations: $24 \div 8 \cdot 3 = 3 \cdot 3 = 9$
27. Using the order of operations: $36 \div 6 \cdot 3 = 6 \cdot 3 = 18$
29. Using the order of operations: $48 \div 12 \cdot 2 = 4 \cdot 2 = 8$
31. Using the order of operations: $16 - 8 + 4 = 8 + 4 = 12$
33. Using the order of operations: $16 - 4 + 6 = 12 + 6 = 18$
35. Using the order of operations: $36 - 6 + 12 = 30 + 12 = 42$
37. Using the order of operations: $48 - 12 + 17 = 36 + 17 = 53$
39. Using the order of operations: $2 \cdot 3 + 5 = 6 + 5 = 11$
41. Using the order of operations: $2(3+5) = 2(8) = 16$
43. Using the order of operations: $5 + 2 \cdot 6 = 5 + 12 = 17$
45. Using the order of operations: $(5+2) \cdot 6 = 7 \cdot 6 = 42$
47. Using the order of operations: $5 \cdot 4 + 5 \cdot 2 = 20 + 10 = 30$
49. Using the order of operations: $5(4+2) = 5(6) = 30$
51. Using the order of operations: $8 + 2(5+3) = 8 + 2(8) = 8 + 16 = 24$
53. Using the order of operations: $(8+2)(5+3) = (10)(8) = 80$

© 2012 Cengage Learning. All Rights Reserved. May not be scanned, copied or duplicated, or posted to a publicly accessible website, in whole or in part.

55. Using the order of operations: $20+2(8-5)+1=20+2(3)+1=20+6+1=27$

57. Using the order of operations: $5+2(3\cdot4-1)+8=5+2(12-1)+8=5+2(11)+8=5+22+8=35$

59. Using the order of operations: $8+10\div2=8+5=13$

61. Using the order of operations: $4+8\div4-2=4+2-2=4$

63. Using the order of operations: $3+12\div3+6\cdot5=3+4+30=37$

65. Using the order of operations: $3\cdot8+10\div2+4\cdot2=24+5+8=37$

67. Using the order of operations: $(5+3)(5-3)=(8)(2)=16$

69. Using the order of operations: $5^2-3^2=5\cdot5-3\cdot3=25-9=16$

71. Using the order of operations: $(4+5)^2=9^2=9\cdot9=81$

73. Using the order of operations: $4^2+5^2=4\cdot4+5\cdot5=16+25=41$

75. Using the order of operations: $3\cdot10^2+4\cdot10+5=300+40+5=345$

77. Using the order of operations: $2\cdot10^3+3\cdot10^2+4\cdot10+5=2000+300+40+5=2,345$

79. Using the order of operations: $10-2(4\cdot5-16)=10-2(20-16)=10-2(4)=10-8=2$

81. Using the order of operations: $4[7+3(2\cdot9-8)]=4[7+3(18-8)]=4[7+3(10)]=4(7+30)=4(37)=148$

83. Using the order of operations: $5(7-3)+8(6-4)=5(4)+8(2)=20+16=36$

85. Using the order of operations: $3(4\cdot5-12)+6(7\cdot6-40)=3(20-12)+6(42-40)=3(8)+6(2)=24+12=36$

87. Using the order of operations: $3^4+4^2\div2^3-5^2=81+16\div8-25=81+2-25=58$

89. Using the order of operations: $5^2+3^4\div9^2+6^2=25+81\div81+36=25+1+36=62$

91. The next number is 5.

93. The next number is 10.

95. The next number is $5^2=25$.

97. Since $2+2=4$ and $2+4=6$, the next number is $4+6=10$.

99. There are $5\cdot2=10$ cookies in the package.

101. The total number of calories is $210\cdot2=420$ calories.

103. Using the order of operations:

$$1(85,260)+2(63,935)+3(53,160)+1(31,885)=85,260+127,870+159,480+31,885=404,495$$

Their combined annual income is $404,495.

105. Completing the table:

Activity	Calories burned in 1 hour
Bicycling	374
Bowling	265
Handball	680
Jogging	680
Skiing	544

© 2012 Cengage Learning. All Rights Reserved. May not be scanned, copied or duplicated, or posted to a publicly accessible website, in whole or in part.

1.2 Real Numbers

1. Labeling the point:

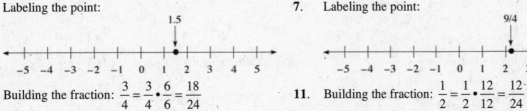

−5 −4 −3 −2 −1 0 1 2 3 4 5

3. Labeling the point:

−5 −4 −3 −2 −1 0 1 2 3 4 5

5. Labeling the point:

1.5

−5 −4 −3 −2 −1 0 1 2 3 4 5

7. Labeling the point:

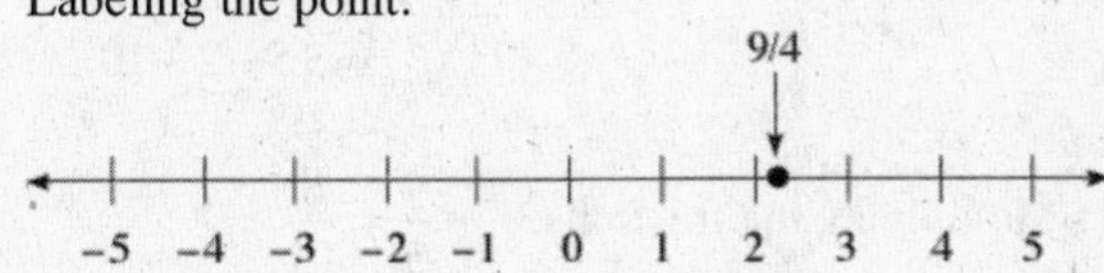

9. Building the fraction: $\frac{3}{4}=\frac{3}{4}\bullet\frac{6}{6}=\frac{18}{24}$

11. Building the fraction: $\frac{1}{2}=\frac{1}{2}\bullet\frac{12}{12}=\frac{12}{24}$

13. Building the fraction: $\frac{5}{8}=\frac{5}{8}\bullet\frac{3}{3}=\frac{15}{24}$

15. Building the fraction: $\frac{3}{5}=\frac{3}{5}\bullet\frac{12}{12}=\frac{36}{60}$

17. Building the fraction: $\frac{11}{30}=\frac{11}{30}\bullet\frac{2}{2}=\frac{22}{60}$

19. Building the fraction: $\frac{1}{2}=\frac{1}{2}\bullet\frac{2}{2}=\frac{2}{4}$, so the missing numerator is 2.

21. Building the fraction: $\frac{5}{9}=\frac{5}{9}\bullet\frac{5}{5}=\frac{25}{45}$, so the missing numerator is 25.

23. Building the fraction: $\frac{3}{4}=\frac{3}{4}\bullet\frac{2}{2}=\frac{6}{8}$, so the missing numerator is 6.

25. The opposite of 10 is –10, the reciprocal is $\frac{1}{10}$, and the absolute value is $|10|=10$.

27. The opposite of $\frac{3}{4}$ is $-\frac{3}{4}$, the reciprocal is $\frac{4}{3}$, and the absolute value is $\left|\frac{3}{4}\right|=\frac{3}{4}$.

29. The opposite of $\frac{11}{2}$ is $-\frac{11}{2}$, the reciprocal is $\frac{2}{11}$, and the absolute value is $\left|\frac{11}{2}\right|=\frac{11}{2}$.

31. The opposite of –3 is 3, the reciprocal is $-\frac{1}{3}$, and the absolute value is $|-3|=3$.

33. The opposite of $-\frac{2}{5}$ is $\frac{2}{5}$, the reciprocal is $-\frac{5}{2}$, and the absolute value is $\left|-\frac{2}{5}\right|=\frac{2}{5}$.

35. The opposite of x is $-x$, the reciprocal is $\frac{1}{x}$, and the absolute value is $|x|$.

37. The correct symbol is $<$: $-5<-3$

39. The correct symbol is $>$: $-3>-7$

41. Since $|-4|=4$ and $-|-4|=-4$, the correct symbol is $>$: $|-4|>-|-4|$

43. Since $-|-7|=-7$, the correct symbol is $>$: $7>-|-7|$

45. The correct symbol is $<$: $-\frac{3}{4}<-\frac{1}{4}$

47. The correct symbol is $<$: $-\frac{3}{2}<-\frac{3}{4}$

49. Simplifying the expression: $|8-2|=|6|=6$

51. Simplifying the expression: $|5\bullet 2^3-2\bullet 3^2|=|5\bullet 8-2\bullet 9|=|40-18|=|22|=22$

53. Simplifying the expression: $|7-2|-|4-2|=|5|-|2|=5-2=3$

55. Simplifying the expression: $10-|7-2(5-3)|=10-|7-2(2)|=10-|7-4|=10-|3|=10-3=7$

© 2012 Cengage Learning. All Rights Reserved. May not be scanned, copied or duplicated, or posted to a publicly accessible website, in whole or in part.

57. Simplifying the expression:

$$\begin{aligned}15-|8-2(3\cdot 4-9)|-10&=15-|8-2(12-9)|-10\\&=15-|8-2(3)|-10\\&=15-|8-6|-10\\&=15-|2|-10\\&=15-2-10\\&=3\end{aligned}$$

59. Multiplying the fractions: $\frac{2}{3}\cdot\frac{4}{5}=\frac{8}{15}$

61. Multiplying the fractions: $\frac{1}{2}(3)=\frac{1}{2}\cdot\frac{3}{1}=\frac{3}{2}$

63. Multiplying the fractions: $\frac{1}{4}(5)=\frac{1}{4}\cdot\frac{5}{1}=\frac{5}{4}$

65. Multiplying the fractions: $\frac{4}{3}\cdot\frac{3}{4}=\frac{12}{12}=1$

67. Multiplying the fractions: $6\left(\frac{1}{6}\right)=\frac{6}{1}\cdot\frac{1}{6}=1$

69. Multiplying the fractions: $3\cdot\frac{1}{3}=\frac{3}{1}\cdot\frac{1}{3}=\frac{3}{3}=1$

71. Since the pattern is to find reciprocals of odd numbers, the next number is $\frac{1}{9}$.

73. Since the pattern is to find reciprocals of squares, the next number is $\frac{1}{5^2}=\frac{1}{25}$.

75.
- **a.** Substituting $x=5$: $2x-6=2(5)-6=10-6=4$
- **b.** Substituting $x=10$: $2x-6=2(10)-6=20-6=14$
- **c.** Substituting $x=15$: $2x-6=2(15)-6=30-6=24$
- **d.** Substituting $x=20$: $2x-6=2(20)-6=40-6=34$

77.
- **a.** Substituting $x=10$: $x+2=10+2=12$
- **b.** Substituting $x=10$: $2x=2(10)=20$
- **c.** Substituting $x=10$: $x^2=(10)^2=100$
- **d.** Substituting $x=10$: $2^x=2^{10}=1{,}024$

79.
- **a.** Substituting $x=4$: $x^2+1=(4)^2+1=16+1=17$
- **b.** Substituting $x=4$: $(x+1)^2=(4+1)^2=(5)^2=25$
- **c.** Substituting $x=4$: $x^2+2x+1=(4)^2+2(4)+1=16+8+1=25$

81. The perimeter is $4(1\text{ in.})=4\text{ in.}$, and the area is $(1\text{ in.})^2=1\text{ in.}^2$.

83. The perimeter is $2(1.5\text{ in.})+2(0.75\text{ in.})=3.0\text{ in.}+1.5\text{ in.}=4.5\text{ in.}$, and the area is $(1.5\text{ in.})(0.75\text{ in.})=1.125\text{ in.}^2$.

85. The perimeter is $2.75\text{ cm}+4\text{ cm}+3.5\text{ cm}=10.25\text{ cm}$, and the area is $\frac{1}{2}(4\text{ cm})(2.5\text{ cm})=5.0\text{ cm}^2$.

87. A loss of 8 yards corresponds to –8 on a number line. The total yards gained corresponds to –2 yards.

89. The temperature can be represented as –64°. The new (warmer) temperature corresponds to –54°.

91. Finding the height: $\frac{23}{12}\cdot 600=\frac{23}{12}\cdot\frac{600}{1}=1{,}150$ feet

93. The wind chill temperature is –15°.

95. The wind chill temperature is –25°.

© 2012 Cengage Learning. All Rights Reserved. May not be scanned, copied or duplicated, or posted to a publicly accessible website, in whole or in part.

1.3 Addition and Subtraction of Real Numbers

1. Adding the numbers: $6+(-3)=3$

3. Adding the numbers: $18+(-32)=-14$

5. Adding the numbers: $-6+3=-3$

7. Adding the numbers: $-30+5=-25$

9. Adding the numbers: $-6+(-6)=-12$

11. Adding the numbers: $-10+(-15)=-25$

13. Adding the decimals: $3.9+7.1=11.0=11$

15. Adding the decimals: $8.1+2.7=10.8$

17. Adding all positive and negative combinations of 3 and 5:

$$3+5=8 \qquad 3+(-5)=-2$$
$$-3+5=2 \qquad (-3)+(-5)=-8$$

19. Adding all positive and negative combinations of 15 and 20:

$$15+20=35 \qquad 15+(-20)=-5$$
$$-15+20=5 \qquad (-15)+(-20)=-35$$

21. The pattern is to add 5, so the next two terms are $18+5=23$ and $23+5=28$.

23. The pattern is to add 5, so the next two terms are $25+5=30$ and $30+5=35$.

25. The pattern is to add –6, so the next two terms are $-6+(-6)=-12$ and $-12+(-6)=-18$.

27. Yes, since each successive odd number is 2 added to the previous one.

29. Subtracting the numbers: $5-8=5+(-8)=-3$

31. Subtracting the numbers: $5-5=5+(-5)=0$

33. Subtracting the numbers: $-8-2=-8+(-2)=-10$

35. Subtracting the numbers: $-4-12=-4+(-12)=-16$

37. Subtracting the numbers: $15-(-20)=15+20=35$

39. Subtracting the numbers: $-4-(-4)=-4+4=0$

41. Subtracting decimals: $-3.4-7.9=-3.4+(-7.9)=-11.3$

43. Subtracting decimals: $3.3-6.9=3.3+(-6.9)=-3.6$

45. Performing the additions: $5+(-6)+(-7)=5+(-13)=-8$

47. Performing the additions: $5+[6+(-2)]+(-3)=5+4+(-3)=9+(-3)=6$

49. Performing the additions: $[6+(-2)]+[3+(-1)]=4+2=6$

51. Performing the additions: $-3+(-2)+[5+(-4)]=-3+(-2)+1=-5+1=-4$

53. Using order of operations: $4-8-6=4+(-8)+(-6)=4+(-14)=-10$

55. Using order of operations: $-5-7-9=-5+(-7)+(-9)=-21$

57. Using order of operations: $-13+6-5=-13+6+(-5)=-18+6=-12$

59. Using order of operations: $15-(-3)-20=15+3+(-20)=18+(-20)=-2$

61. Using order of operations: $10-(4-6)-8=10-(-2)-8=10+2+(-8)=12+(-8)=4$

63. Using order of operations: $4-(3-7)-8=4-(-4)-8=4+4+(-8)=8+(-8)=0$

65. Using order of operations: $-(4-8)-(2-5)=-(-4)-(-3)=4+3=7$

67. Using order of operations: $-(3-7)-(1-2)=-(-4)-(-1)=4+1=5$

69. Using order of operations: $6-[(4-1)-9]=6-(3-9)=6-(-6)=6+6=12$

71. Using order of operations:

$$\begin{aligned}30-[-(10-5)-15]-25&=30-[-5-15]-25\\&=30-[-5+(-15)]-25\\&=30-(-20)-25\\&=30+20+(-25)\\&=50+(-25)\\&=25\end{aligned}$$

73. Using order of operations: $6\bullet 10-5\bullet 20=60-100=60+(-100)=-40$

75. Using order of operations: $5\bullet 9-3\bullet 8-4\bullet 5=45-24-20=45+(-24)+(-20)=45+(-44)=1$

77. Using order of operations: $3\bullet 7^2-2\bullet 8^2=3\bullet 49-2\bullet 64=147-128=147+(-128)=19$

79. Using order of operations:

$$3\bullet 6^2-2\bullet 3^2-8\bullet 6^2=3\bullet 36-2\bullet 9-8\bullet 36=108-18-288=108+(-18)+(-288)=108+(-306)=-198$$

81. The angles add to 90°, so $x=90°-45°=45°$.

83. The angles add to 180°, so $x=180°-30°=150°$.

85. Totaling the quantities: $235.1+234.2+108.4=577.7$ million users

© 2012 Cengage Learning. All Rights Reserved. May not be scanned, copied or duplicated, or posted to a publicly accessible website, in whole or in part.

87. The expression is $73° + 10° - 8° = 83° - 8° = 75°$.
89. The new balance is $-\$25 + \$75 - \$18 = \$32$.
91. Subtracting: $3,154 - 2,205 = 949$ strikeouts
93. The expression is $-\$35 + \$15 - \$20 = -\$35 + (-\$20) + \$15 = -\$55 + \$15 = -\$40$.
95. The addition problem is $8 + (-5) = 3$.
97. The sequence of values is \$4500, \$3950, \$3400, \$2850, and \$2300. This is an arithmetic sequence, since –\$550 is added to each value to obtain the new value.
99. The difference is $1000 \text{ feet} - 231 \text{ feet} = 769 \text{ feet}$.
101. He is 439 feet from the starting line.
103. 2 seconds have gone by.

1.4 Multiplication of Real Numbers

1. Finding the product: $7(-6) = -42$
3. Finding the product: $-8(2) = -16$
5. Finding the product: $-3(-1) = 3$
7. Finding the product: $-11(-11) = 121$
9. Using order of operations: $-3(2)(-1) = 6$
11. Using order of operations: $-3(-4)(-5) = -60$
13. Using order of operations: $-2(-4)(-3)(-1) = 24$
15. Using order of operations: $(-7)^2 = (-7)(-7) = 49$
17. Using order of operations: $(-3)^3 = (-3)(-3)(-3) = -27$
19. Using order of operations: $-2(2-5) = -2(-3) = 6$
21. Using order of operations: $-5(8-10) = -5(-2) = 10$
23. Using order of operations: $(4-7)(6-9) = (-3)(-3) = 9$
25. Using order of operations: $(-3-2)(-5-4) = (-5)(-9) = 45$
27. Using order of operations: $-3(-6) + 4(-1) = 18 + (-4) = 14$
29. Using order of operations: $2(3) - 3(-4) + 4(-5) = 6 + 12 + (-20) = 18 + (-20) = -2$
31. Using order of operations: $4(-3)^2 + 5(-6)^2 = 4(9) + 5(36) = 36 + 180 = 216$
33. Using order of operations: $7(-2)^3 - 2(-3)^3 = 7(-8) - 2(-27) = -56 + 54 = -2$
35. Using order of operations: $6 - 4(8-2) = 6 - 4(6) = 6 - 24 = 6 + (-24) = -18$
37. Using order of operations: $9 - 4(3-8) = 9 - 4(-5) = 9 + 20 = 29$
39. Using order of operations: $-4(3-8) - 6(2-5) = -4(-5) - 6(-3) = 20 + 18 = 38$
41. Using order of operations: $7 - 2[-6 - 4(-3)] = 7 - 2(-6 + 12) = 7 - 2(6) = 7 - 12 = 7 + (-12) = -5$
43. Using order of operations:
$$7 - 3[2(-4-4) - 3(-1-1)] = 7 - 3[2(-8) - 3(-2)] = 7 - 3(-16+6) = 7 - 3(-10) = 7 + 30 = 37$$
45. a. Simplifying: $5(-4)(-3) = (-20)(-3) = 60$
    b. Simplifying: $5(-4) - 3 = -20 - 3 = -23$
    c. Simplifying: $5 - 4(-3) = 5 - (-12) = 5 + 12 = 17$
    d. Simplifying: $5 - 4 - 3 = 1 - 3 = -2$
47. Multiplying the fractions: $-\frac{2}{3} \cdot \frac{5}{7} = -\frac{2 \cdot 5}{3 \cdot 7} = -\frac{10}{21}$
49. Multiplying the fractions: $-8\left(\frac{1}{2}\right) = -\frac{8}{1} \cdot \frac{1}{2} = -\frac{8}{2} = -4$
51. Multiplying the fractions: $\left(-\frac{3}{4}\right)^2 = \left(-\frac{3}{4}\right)\left(-\frac{3}{4}\right) = \frac{9}{16}$

© 2012 Cengage Learning. All Rights Reserved. May not be scanned, copied or duplicated, or posted to a publicly accessible website, in whole or in part.

53. a. Simplifying: $\frac{5}{8}(24)+\frac{3}{7}(28)=15+12=27$

b. Simplifying: $\frac{5}{8}(24)-\frac{3}{7}(28)=15-12=3$

c. Simplifying: $\frac{5}{8}(-24)+\frac{3}{7}(-28)=-15-12=-27$

d. Simplifying: $-\frac{5}{8}(24)-\frac{3}{7}(28)=-15-12=-27$

55. Simplifying: $\left(\frac{1}{2}\bullet 6\right)^2=(3)^2=(3)(3)=9$

57. Simplifying: $\left(\frac{1}{2}\bullet 5\right)^2=\left(\frac{5}{2}\right)^2=\left(\frac{5}{2}\right)\left(\frac{5}{2}\right)=\frac{25}{4}$

59. Simplifying: $\left(\frac{1}{2}(-4)\right)^2=(-2)^2=(-2)(-2)=4$

61. Simplifying: $\left(\frac{1}{2}(-3)\right)^2=\left(-\frac{3}{2}\right)^2=\left(-\frac{3}{2}\right)\left(-\frac{3}{2}\right)=\frac{9}{4}$

63. Multiplying the expressions: $-2(0)=0$

65. Multiplying the expressions: $-7x(0)=0$

67. Multiplying the expressions: $(3x)(-5)(0)=0$

69. The pattern is to multiply by 2, so the next number is $4\bullet 2=8$.

71. The pattern is to multiply by –2, so the next number is $40\bullet(-2)=-80$.

73. The pattern is to multiply by $\frac{1}{2}$, so the next number is $\frac{1}{4}\bullet\frac{1}{2}=\frac{1}{8}$.

75. The pattern is to multiply by –2, so the next number is $12\bullet(-2)=-24$.

77. Simplifying the expression: $2(5-5)+4=2(0)+4=0+4=4$

79. Simplifying the expression: $6(3+1)-5(0)=6(4)-0=24-0=24$

81. Simplifying the expression: $\left(\frac{1}{2}\bullet 18\right)^2=(9)^2=9\bullet 9=81$

83. Simplifying the expression: $\left(\frac{1}{2}\bullet 3\right)^2=\left(\frac{3}{2}\right)^2=\left(\frac{3}{2}\right)\left(\frac{3}{2}\right)=\frac{9}{4}$

84. Simplifying the expression: $\left(\frac{1}{2}\bullet 5\right)^2=\left(\frac{5}{2}\right)^2=\left(\frac{5}{2}\right)\left(\frac{5}{2}\right)=\frac{25}{4}$

85. Her charge for April is: 10($0.25) = $2.50

87. The amount lost is: 20($3) = $60

89. The temperature is: $25°-4(6°)=25°-24°=1°$ F

91. The total dosage is: 4(25 mg) = 100 mg

1.5 Division of Real Numbers

1. Finding the quotient: $\frac{8}{-4}=-2$

3. Finding the quotient: $\frac{-48}{16}=-3$

5. Finding the quotient: $\frac{-7}{21}=-\frac{1}{3}$

7. Finding the quotient: $\frac{-39}{-13}=3$

9. Finding the quotient: $\frac{-6}{-42}=\frac{1}{7}$

11. Finding the quotient: $\frac{0}{-32}=0$

13. Performing the operations: $-3+12=9$

15. Performing the operations: $-3-12=-3+(-12)=-15$

17. Performing the operations: $-3(12)=-36$

19. Performing the operations: $-3\div 12=\frac{-3}{12}=-\frac{1}{4}$

21. Dividing and reducing: $\frac{4}{5}\div\frac{3}{4}=\frac{4}{5}\bullet\frac{4}{3}=\frac{16}{15}$

23. Dividing and reducing: $-\frac{5}{6}\div\left(-\frac{5}{8}\right)=-\frac{5}{6}\bullet\left(-\frac{8}{5}\right)=\frac{40}{30}=\frac{4}{3}$

© 2012 Cengage Learning. All Rights Reserved. May not be scanned, copied or duplicated, or posted to a publicly accessible website, in whole or in part.

25. Dividing and reducing: $\frac{10}{13}\div\left(-\frac{5}{4}\right)=\frac{10}{13}\bullet\left(-\frac{4}{5}\right)=-\frac{40}{65}=-\frac{8}{13}$

27. Dividing and reducing: $-\frac{5}{6}\div\frac{5}{6}=-\frac{5}{6}\bullet\frac{6}{5}=-\frac{30}{30}=-1$

29. Dividing and reducing: $-\frac{3}{4}\div\left(-\frac{3}{4}\right)=-\frac{3}{4}\bullet\left(-\frac{4}{3}\right)=\frac{12}{12}=1$

31. Using order of operations: $\frac{3(-2)}{-10}=\frac{-6}{-10}=\frac{3}{5}$

33. Using order of operations: $\frac{-5(-5)}{-15}=\frac{25}{-15}=-\frac{5}{3}$

35. Using order of operations: $\frac{-8(-7)}{-28}=\frac{56}{-28}=-2$

37. Using order of operations: $\frac{27}{4-13}=\frac{27}{-9}=-3$

39. Using order of operations: $\frac{20-6}{5-5}=\frac{14}{0}$, which is undefined

41. Using order of operations: $\frac{-3+9}{2\bullet 5-10}=\frac{6}{10-10}=\frac{6}{0}$, which is undefined

43. Using order of operations: $\frac{15(-5)-25}{2(-10)}=\frac{-75-25}{-20}=\frac{-100}{-20}=5$

45. Using order of operations: $\frac{27-2(-4)}{-3(5)}=\frac{27+8}{-15}=\frac{35}{-15}=-\frac{7}{3}$

47. Using order of operations: $\frac{12-6(-2)}{12(-2)}=\frac{12+12}{-24}=\frac{24}{-24}=-1$

49. Using order of operations: $\frac{5^2-2^2}{-5+2}=\frac{25-4}{-3}=\frac{21}{-3}=-7$

51. Using order of operations: $\frac{8^2-2^2}{8^2+2^2}=\frac{64-4}{64+4}=\frac{60}{68}=\frac{15}{17}$

53. Using order of operations: $\frac{(5+3)^2}{-5^2-3^2}=\frac{8^2}{-25-9}=\frac{64}{-34}=-\frac{32}{17}$

55. Using order of operations: $\frac{(8-4)^2}{8^2-4^2}=\frac{4^2}{64-16}=\frac{16}{48}=\frac{1}{3}$

57. Using order of operations: $\frac{-4\bullet 3^2-5\bullet 2^2}{-8(7)}=\frac{-4\bullet 9-5\bullet 4}{-56}=\frac{-36-20}{-56}=\frac{-56}{-56}=1$

59. Using order of operations: $\frac{3\bullet 10^2+4\bullet 10+5}{345}=\frac{300+40+5}{345}=\frac{345}{345}=1$

61. Using order of operations: $\frac{7-[(2-3)-4]}{-1-2-3}=\frac{7-(-1-4)}{-6}=\frac{7-(-5)}{-6}=\frac{7+5}{-6}=\frac{12}{-6}=-2$

63. Using order of operations: $\frac{6(-4)-2(5-8)}{-6-3-5}=\frac{-24-2(-3)}{-14}=\frac{-24+6}{-14}=\frac{-18}{-14}=\frac{9}{7}$

65. Using order of operations: $\frac{3(-5-3)+4(7-9)}{5(-2)+3(-4)}=\frac{3(-8)+4(-2)}{-10+(-12)}=\frac{-24+(-8)}{-22}=\frac{-32}{-22}=\frac{16}{11}$

67. Using order of operations: $\frac{|3-9|}{3-9}=\frac{|-6|}{-6}=\frac{6}{-6}=-1$

69. Using order of operations: $\frac{2+0.15(10)}{10}=\frac{2+1.5}{10}=\frac{3.5}{10}=\frac{35}{100}=\frac{7}{20}=0.35$

71. Using order of operations: $\frac{1-3}{3-1}=\frac{-2}{2}=-1$

73. **a.** Simplifying: $\frac{5-2}{3-1}=\frac{3}{2}$

b. Simplifying: $\frac{2-5}{1-3}=\frac{-3}{-2}=\frac{3}{2}$

© 2012 Cengage Learning. All Rights Reserved. May not be scanned, copied or duplicated, or posted to a publicly accessible website, in whole or in part.

75. **a.** Simplifying: $\frac{-4-1}{5-(-2)}=\frac{-4+(-1)}{5+2}=\frac{-5}{7}=-\frac{5}{7}$ **b.** Simplifying: $\frac{1-(-4)}{-2-5}=\frac{1+4}{-2+(-5)}=\frac{5}{-7}=-\frac{5}{7}$

77. **a.** Simplifying: $\frac{3+2.236}{2}=\frac{5.236}{2}=2.618$ **b.** Simplifying: $\frac{3-2.236}{2}=\frac{0.764}{2}=0.382$

c. Simplifying: $\frac{3+2.236}{2}+\frac{3-2.236}{2}=\frac{5.236}{2}+\frac{0.764}{2}=2.618+0.382=3$

79. **a.** Simplifying: $20\div 4\bullet 5=5\bullet 5=25$ **b.** Simplifying: $-20\div 4\bullet 5=-5\bullet 5=-25$

c. Simplifying: $20\div(-4)\bullet 5=-5\bullet 5=-25$ **d.** Simplifying: $20\div 4(-5)=5(-5)=-25$

e. Simplifying: $-20\div 4(-5)=-5(-5)=25$

81. **a.** Simplifying: $8\div\frac{4}{5}=8\bullet\frac{5}{4}=10$ **b.** Simplifying: $8\div\frac{4}{5}-10=8\bullet\frac{5}{4}-10=10-10=0$

c. Simplifying: $8\div\frac{4}{5}(-10)=8\bullet\frac{5}{4}(-10)=10(-10)=-100$

d. Simplifying: $8\div\left(-\frac{4}{5}\right)-10=8\bullet\left(-\frac{5}{4}\right)-10=-10-10=-20$

83. The quotient is $\frac{-12}{-4}=3$. **85.** The number is –10, since $\frac{-10}{-5}=2$.

87. The number is –3, since $\frac{27}{-3}=-9$. **89.** The expression is: $\frac{-20}{4}-3=-5-3=-8$

91. Finding the average length: $\frac{7,643\text{ yards}}{18\text{ holes}}\approx 425$ yards per hole

93. Each person would lose: $\frac{13600-15000}{4}=\frac{-1400}{4}=-350=\$350$ loss

95. The change per hour is: $\frac{61°-75°}{4}=\frac{-14°}{4}=-3.5°$ per hour

97. Finding the number of capsules: $\frac{75\text{ mg}}{25\text{ mg/capsule}}=3$ capsules

99. Finding the strength of each tablet: $\frac{1.2\text{ mg}}{4\text{ tablets}}=0.3$ mg/tablet

101. **a.** Since they predict \$50 revenue for every 25 people, their projected revenue is:

$$\frac{\$50}{25\text{ people}}\bullet 10,000\text{ people}=\$20,000$$

b. Since they predict \$50 revenue for every 25 people, their projected revenue is:

$$\frac{\$50}{25\text{ people}}\bullet 25,000\text{ people}=\$50,000$$

c. Since they predict \$50 revenue for every 25 people, their projected revenue is:

$$\frac{\$50}{25\text{ people}}\bullet 5,000\text{ people}=\$10,000$$

Since this projected revenue is more than the \$5,000 cost for the list, it is a wise purchase.

© 2012 Cengage Learning. All Rights Reserved. May not be scanned, copied or duplicated, or posted to a publicly accessible website, in whole or in part.

1.6 Properties of Real Numbers

1. Simplifying the expression: $4+(2+x)=(4+2)+x=6+x$
3. Simplifying the expression: $(x+2)+7=x+(2+7)=x+9$
5. Simplifying the expression: $3(5x)=(3\cdot 5)x=15x$
7. Simplifying the expression: $9(6y)=(9\cdot 6)y=54y$
9. Simplifying the expression: $\frac{1}{2}(3a)=\left(\frac{1}{2}\cdot 3\right)a=\frac{3}{2}a$
11. Simplifying the expression: $\frac{1}{3}(3x)=\left(\frac{1}{3}\cdot 3\right)x=1x=x$
13. Simplifying the expression: $\frac{1}{2}(2y)=\left(\frac{1}{2}\cdot 2\right)y=1y=y$
15. Simplifying the expression: $\frac{3}{4}\left(\frac{4}{3}x\right)=\left(\frac{3}{4}\cdot\frac{4}{3}\right)x=1x=x$
17. Simplifying the expression: $\frac{6}{5}\left(\frac{5}{6}a\right)=\left(\frac{6}{5}\cdot\frac{5}{6}\right)a=1a=a$
19. Applying the distributive property: $8(x+2)=8\cdot x+8\cdot 2=8x+16$
21. Applying the distributive property: $8(x-2)=8\cdot x-8\cdot 2=8x-16$
23. Applying the distributive property: $4(y+1)=4\cdot y+4\cdot 1=4y+4$
25. Applying the distributive property: $3(6x+5)=3\cdot 6x+3\cdot 5=18x+15$
27. Applying the distributive property: $2(3a+7)=2\cdot 3a+2\cdot 7=6a+14$
29. Applying the distributive property: $9(6y-8)=9\cdot 6y-9\cdot 8=54y-72$
31. Applying the distributive property: $\frac{1}{2}(3x-6)=\frac{1}{2}\cdot 3x-\frac{1}{2}\cdot 6=\frac{3}{2}x-3$
33. Applying the distributive property: $\frac{1}{3}(3x+6)=\frac{1}{3}\cdot 3x+\frac{1}{3}\cdot 6=x+2$
35. Applying the distributive property: $3(x+y)=3x+3y$
37. Applying the distributive property: $8(a-b)=8a-8b$
39. Applying the distributive property: $6(2x+3y)=6\cdot 2x+6\cdot 3y=12x+18y$
41. Applying the distributive property: $4(3a-2b)=4\cdot 3a-4\cdot 2b=12a-8b$
43. Applying the distributive property: $\frac{1}{2}(6x+4y)=\frac{1}{2}\cdot 6x+\frac{1}{2}\cdot 4y=3x+2y$
45. Applying the distributive property: $4(a+4)+9=4a+16+9=4a+25$
47. Applying the distributive property: $2(3x+5)+2=6x+10+2=6x+12$
49. Applying the distributive property: $7(2x+4)+10=14x+28+10=14x+38$
51. Applying the distributive property: $0.09(x+2{,}000)=0.09x+180$
53. Applying the distributive property: $0.05(3x+1{,}500)=0.15x+75$
55. Applying the distributive property: $6\left(\frac{1}{2}x-\frac{1}{3}y\right)=6\cdot\frac{1}{2}x-6\cdot\frac{1}{3}y=3x-2y$
57. Applying the distributive property: $12\left(\frac{1}{3}x+\frac{1}{4}y\right)=12\cdot\frac{1}{3}x+12\cdot\frac{1}{4}y=4x+3y$
59. Applying the distributive property: $\frac{1}{2}(4x+2)=\frac{1}{2}\cdot 4x+\frac{1}{2}\cdot 2=2x+1$
61. Applying the distributive property: $\frac{3}{4}(8x-4)=\frac{3}{4}\cdot 8x-\frac{3}{4}\cdot 4=6x-3$
63. Applying the distributive property: $\frac{5}{6}(6x+12)=\frac{5}{6}\cdot 6x+\frac{5}{6}\cdot 12=5x+10$

© 2012 Cengage Learning. All Rights Reserved. May not be scanned, copied or duplicated, or posted to a publicly accessible website, in whole or in part.

65. Applying the distributive property: $10\left(\frac{3}{5}x+\frac{1}{2}\right)=10\bullet\frac{3}{5}x+10\bullet\frac{1}{2}=6x+5$

67. Applying the distributive property: $15\left(\frac{1}{3}x+\frac{2}{5}\right)=15\bullet\frac{1}{3}x+15\bullet\frac{2}{5}=5x+6$

69. Applying the distributive property: $12\left(\frac{1}{2}m-\frac{5}{12}\right)=12\bullet\frac{1}{2}m-12\bullet\frac{5}{12}=6m-5$

71. Applying the distributive property: $21\left(\frac{1}{3}+\frac{1}{7}x\right)=21\bullet\frac{1}{3}+21\bullet\frac{1}{7}x=7+3x$

73. Applying the distributive property: $12\left(\frac{1}{2}x-\frac{1}{3}y\right)=12\bullet\frac{1}{2}x-12\bullet\frac{1}{3}y=6x-4y$

75. Applying the distributive property: $0.15(x+600)=0.15x+90$

77. Applying the distributive property: $0.12(x+500)=0.12x+60$

79. Applying the distributive property: $a\left(1+\frac{1}{a}\right)=a\bullet1+a\bullet\frac{1}{a}=a+1$

81. Applying the distributive property: $a\left(\frac{1}{a}-1\right)=a\bullet\frac{1}{a}-a\bullet1=1-a$

83. commutative property (of addition)
85. multiplicative inverse property
87. commutative property (of addition)
89. distributive property
91. commutative and associative properties (of addition)
93. commutative and associative properties (of addition)
95. commutative property (of addition)
97. commutative and associative properties (of multiplication)
99. commutative property (of multiplication)
101. additive inverse property
103. The expression should read $3(x+2)=3x+6$.
105. The expression should read $9(a+b)=9a+9b$.
107. The expression should read $3(0)=0$.
109. The expression should read $3+(-3)=0$.
111. The expression should read $10(1)=10$.
113. No. The man cannot reverse the order of putting on his socks and putting on his shoes.
115. No. The skydiver must jump out of the plane before pulling the rip cord.
117. Calculating the cost: $3(6,200-100)=18,600-300=\$18,300$
119. Division is not a commutative operation. For example, $8\div4=2$ while $4\div8=\frac{1}{2}$.
121. Computing the yearly take-home pay:

$$12(2400-480)=12(1920)=\$23,040 \qquad 12\bullet2400-12\bullet480=28,800-5,760=\$23,040$$

1.7 Subsets of Real Numbers

1. The whole numbers are: 0, 1
3. The rational numbers are: $-3, -2.5, 0, 1, \frac{3}{2}$
5. The real numbers are: $-3, -2.5, 0, 1, \frac{3}{2}, \sqrt{15}$
7. The integers are: $-10, -8, -2, 9$
9. The irrational numbers are: π
11. true
13. false
15. false
17. true
19. This number is composite: $48=6\bullet8=(2\bullet3)\bullet(2\bullet2\bullet2)=2^4\bullet3$
21. This number is prime.
23. This number is composite: $1023=3\bullet341=3\bullet11\bullet31$
25. Factoring the number: $144=12\bullet12=(3\bullet4)\bullet(3\bullet4)=(3\bullet2\bullet2)\bullet(3\bullet2\bullet2)=2^4\bullet3^2$
27. Factoring the number: $38=2\bullet19$

© 2012 Cengage Learning. All Rights Reserved. May not be scanned, copied or duplicated, or posted to a publicly accessible website, in whole or in part.

29. Factoring the number: $105 = 5 \cdot 21 = 5 \cdot (3 \cdot 7) = 3 \cdot 5 \cdot 7$

31. Factoring the number: $180 = 10 \cdot 18 = (2 \cdot 5) \cdot (3 \cdot 6) = (2 \cdot 5) \cdot (3 \cdot 2 \cdot 3) = 2^2 \cdot 3^2 \cdot 5$

33. Factoring the number: $385 = 5 \cdot 77 = 5 \cdot (7 \cdot 11) = 5 \cdot 7 \cdot 11$

35. Factoring the number: $121 = 11 \cdot 11 = 11^2$

37. Factoring the number: $420 = 10 \cdot 42 = (2 \cdot 5) \cdot (7 \cdot 6) = (2 \cdot 5) \cdot (7 \cdot 2 \cdot 3) = 2^2 \cdot 3 \cdot 5 \cdot 7$

39. Factoring the number: $620 = 10 \cdot 62 = (2 \cdot 5) \cdot (2 \cdot 31) = 2^2 \cdot 5 \cdot 31$

41. Reducing the fraction: $\frac{105}{165} = \frac{3 \cdot 5 \cdot 7}{3 \cdot 5 \cdot 11} = \frac{7}{11}$

43. Reducing the fraction: $\frac{525}{735} = \frac{3 \cdot 5 \cdot 5 \cdot 7}{3 \cdot 5 \cdot 7 \cdot 7} = \frac{5}{7}$

45. Reducing the fraction: $\frac{385}{455} = \frac{5 \cdot 7 \cdot 11}{5 \cdot 7 \cdot 13} = \frac{11}{13}$

47. Reducing the fraction: $\frac{322}{345} = \frac{2 \cdot 7 \cdot 23}{3 \cdot 5 \cdot 23} = \frac{2 \cdot 7}{3 \cdot 5} = \frac{14}{15}$

49. Reducing the fraction: $\frac{205}{369} = \frac{5 \cdot 41}{3 \cdot 3 \cdot 41} = \frac{5}{3 \cdot 3} = \frac{5}{9}$

51. Reducing the fraction: $\frac{215}{344} = \frac{5 \cdot 43}{2 \cdot 2 \cdot 2 \cdot 43} = \frac{5}{2 \cdot 2 \cdot 2} = \frac{5}{8}$

53. **a.** Adding: $50 + (-80) = -30$

b. Subtracting: $50 - (-80) = 50 + 80 = 130$

c. Multiplying: $50(-80) = -4{,}000$

d. Dividing: $\frac{50}{-80} = -\frac{5}{8}$

55. Simplifying; $\frac{6.28}{9(3.14)} = \frac{6.28}{28.26} = \frac{628}{2826} = \frac{2 \cdot 314}{9 \cdot 314} = \frac{2}{9}$

57. Simplifying; $\frac{9.42}{2(3.14)} = \frac{9.42}{6.28} = \frac{942}{628} = \frac{3 \cdot 314}{2 \cdot 314} = \frac{3}{2}$

59. Simplifying; $\frac{32}{0.5} = \frac{320}{5} = 64$

61. Simplifying; $\frac{5{,}599}{11} = 509$

63. **a.** Substituting $x = 10$: $\frac{2 + 0.15x}{x} = \frac{2 + 0.15(10)}{10} = \frac{2 + 1.5}{10} = \frac{3.5}{10} = 0.35$

b. Substituting $x = 15$: $\frac{2 + 0.15x}{x} = \frac{2 + 0.15(15)}{15} = \frac{2 + 2.25}{15} = \frac{4.25}{15} \approx 0.28$

c. Substituting $x = 20$: $\frac{2 + 0.15x}{x} = \frac{2 + 0.15(20)}{20} = \frac{2 + 3}{20} = \frac{5}{20} = 0.25$

65. Factoring into prime numbers: $6^3 = (2 \cdot 3)^3 = 2^3 \cdot 3^3$

67. Factoring into prime numbers: $9^4 \cdot 16^2 = (3 \cdot 3)^4 \cdot (2 \cdot 2 \cdot 2 \cdot 2)^2 = 3^4 \cdot 3^4 \cdot 2^2 \cdot 2^2 \cdot 2^2 \cdot 2^2 = 2^8 \cdot 3^8$

69. Simplifying and factoring: $3 \cdot 8 + 3 \cdot 7 + 3 \cdot 5 = 24 + 21 + 15 = 60 = 6 \cdot 10 = (2 \cdot 3) \cdot (2 \cdot 5) = 2^2 \cdot 3 \cdot 5$

71. Factoring into primes: $273 = 3 \cdot 91 = 3 \cdot 7 \cdot 13$

73. They are not a subset of the irrational numbers.

75. 8, 21, and 34 are Fibonacci numbers that are composite numbers.

© 2012 Cengage Learning. All Rights Reserved. May not be scanned, copied or duplicated, or posted to a publicly accessible website, in whole or in part.

1.8 Addition and Subtraction of Fractions with Variables

1. Combining the fractions: $\frac{3}{6}+\frac{1}{6}=\frac{4}{6}=\frac{2}{3}$

3. Combining the fractions: $\frac{3}{8}-\frac{5}{8}=-\frac{2}{8}=-\frac{1}{4}$

5. Combining the fractions: $-\frac{1}{4}+\frac{3}{4}=\frac{2}{4}=\frac{1}{2}$

7. Combining the fractions: $\frac{x}{3}-\frac{1}{3}=\frac{x-1}{3}$

9. Combining the fractions: $\frac{1}{4}+\frac{2}{4}+\frac{3}{4}=\frac{6}{4}=\frac{3}{2}$

11. Combining the fractions: $\frac{x+7}{2}-\frac{1}{2}=\frac{x+7-1}{2}=\frac{x+6}{2}$

13. Combining the fractions: $\frac{1}{10}-\frac{3}{10}-\frac{4}{10}=-\frac{6}{10}=-\frac{3}{5}$

15. Combining the fractions: $\frac{1}{a}+\frac{4}{a}+\frac{5}{a}=\frac{10}{a}$

17. Combining the fractions: $\frac{1}{8}+\frac{3}{4}=\frac{1}{8}+\frac{3\cdot 2}{4\cdot 2}=\frac{1}{8}+\frac{6}{8}=\frac{7}{8}$

19. Combining the fractions: $\frac{3}{10}-\frac{1}{5}=\frac{3}{10}-\frac{1\cdot 2}{5\cdot 2}=\frac{3}{10}-\frac{2}{10}=\frac{1}{10}$

21. Combining the fractions: $\frac{4}{9}+\frac{1}{3}=\frac{4}{9}+\frac{1\cdot 3}{3\cdot 3}=\frac{4}{9}+\frac{3}{9}=\frac{7}{9}$

23. Combining the fractions: $2+\frac{1}{3}=\frac{2\cdot 3}{1\cdot 3}+\frac{1}{3}=\frac{6}{3}+\frac{1}{3}=\frac{7}{3}$

25. Combining the fractions: $-\frac{3}{4}+1=-\frac{3}{4}+\frac{1\cdot 4}{1\cdot 4}=-\frac{3}{4}+\frac{4}{4}=\frac{1}{4}$

27. Combining the fractions: $\frac{1}{2}+\frac{2}{3}=\frac{1\cdot 3}{2\cdot 3}+\frac{2\cdot 2}{3\cdot 2}=\frac{3}{6}+\frac{4}{6}=\frac{7}{6}$

29. Combining the fractions: $\frac{5}{12}-\left(-\frac{3}{8}\right)=\frac{5}{12}+\frac{3}{8}=\frac{5\cdot 2}{12\cdot 2}+\frac{3\cdot 3}{8\cdot 3}=\frac{10}{24}+\frac{9}{24}=\frac{19}{24}$

31. Combining the fractions: $-\frac{1}{20}+\frac{8}{30}=-\frac{1\cdot 3}{20\cdot 3}+\frac{8\cdot 2}{30\cdot 2}=-\frac{3}{60}+\frac{16}{60}=\frac{13}{60}$

33. First factor the denominators to find the LCM:

$30=2\cdot 3\cdot 5$

$42=2\cdot 3\cdot 7$

$\text{LCM}=2\cdot 3\cdot 5\cdot 7=210$

Combining the fractions: $\frac{17}{30}+\frac{11}{42}=\frac{17\cdot 7}{30\cdot 7}+\frac{11\cdot 5}{42\cdot 5}=\frac{119}{210}+\frac{55}{210}=\frac{174}{210}=\frac{2\cdot 3\cdot 29}{2\cdot 3\cdot 5\cdot 7}=\frac{29}{5\cdot 7}=\frac{29}{35}$

35. First factor the denominators to find the LCM:

$84=2\cdot 2\cdot 3\cdot 7$

$90=2\cdot 3\cdot 3\cdot 5$

$\text{LCM}=2\cdot 2\cdot 3\cdot 3\cdot 5\cdot 7=1260$

Combining the fractions: $\frac{25}{84}+\frac{41}{90}=\frac{25\cdot 15}{84\cdot 15}+\frac{41\cdot 14}{90\cdot 14}=\frac{375}{1,260}+\frac{574}{1,260}=\frac{949}{1,260}$

37. First factor the denominators to find the LCM:

$126=2\cdot 3\cdot 3\cdot 7$

$180=2\cdot 2\cdot 3\cdot 3\cdot 5$

$\text{LCM}=2\cdot 2\cdot 3\cdot 3\cdot 5\cdot 7=1260$

Combining the fractions:

$$\frac{13}{126}-\frac{13}{180}=\frac{13\cdot 10}{126\cdot 10}-\frac{13\cdot 7}{180\cdot 7}=\frac{130}{1260}-\frac{91}{1260}=\frac{39}{1260}=\frac{3\cdot 13}{2\cdot 2\cdot 3\cdot 3\cdot 5\cdot 7}=\frac{13}{2\cdot 2\cdot 3\cdot 5\cdot 7}=\frac{13}{420}$$

39. Combining the fractions: $\frac{3}{4}+\frac{1}{8}+\frac{5}{6}=\frac{3\cdot 6}{4\cdot 6}+\frac{1\cdot 3}{8\cdot 3}+\frac{5\cdot 4}{6\cdot 4}=\frac{18}{24}+\frac{3}{24}+\frac{20}{24}=\frac{41}{24}$

41. Combining the fractions: $\frac{1}{2}+\frac{1}{3}+\frac{1}{4}+\frac{1}{6}=\frac{1\cdot 6}{2\cdot 6}+\frac{1\cdot 4}{3\cdot 4}+\frac{1\cdot 3}{4\cdot 3}+\frac{1\cdot 2}{6\cdot 2}=\frac{6}{12}+\frac{4}{12}+\frac{3}{12}+\frac{2}{12}=\frac{15}{12}=\frac{5}{4}$

© 2012 Cengage Learning. All Rights Reserved. May not be scanned, copied or duplicated, or posted to a publicly accessible website, in whole or in part.

43. Combining the fractions: $1-\frac{5}{2}=1\cdot\frac{2}{2}-\frac{5}{2}=\frac{2}{2}-\frac{5}{2}=-\frac{3}{2}$

45. Combining the fractions: $1+\frac{1}{2}=1\cdot\frac{2}{2}+\frac{1}{2}=\frac{2}{2}+\frac{1}{2}=\frac{3}{2}$

47. The pattern is to add $-\frac{1}{3}$, so the fourth term is: $-\frac{1}{3}+\left(-\frac{1}{3}\right)=-\frac{2}{3}$

49. The pattern is to add $\frac{2}{3}$, so the fourth term is: $\frac{5}{3}+\frac{2}{3}=\frac{7}{3}$

51. The pattern is to multiply by $\frac{1}{5}$, so the fourth term is: $\frac{1}{25}\bullet\frac{1}{5}=\frac{1}{125}$

53. Using order of operations: $9-3\left(\frac{5}{3}\right)=9-5=4$

55. Using order of operations: $-\frac{1}{2}+2\left(-\frac{3}{4}\right)=-\frac{1}{2}-\frac{3}{2}=-\frac{4}{2}=-2$

57. Using order of operations: $\frac{3}{5}(-10)+\frac{4}{7}(-21)=-6-12=-18$

59. Using order of operations: $16\left(-\frac{1}{2}\right)^2-125\left(-\frac{2}{5}\right)^2=16\left(\frac{1}{4}\right)-125\left(\frac{4}{25}\right)=4-20=-16$

61. Using order of operations: $-\frac{4}{3}\div 2\cdot 3=-\frac{4}{3}\cdot\frac{1}{2}\cdot 3=-\frac{2}{3}\cdot 3=-2$

63. Using order of operations: $-\frac{4}{3}\div 2(-3)=-\frac{4}{3}\cdot\frac{1}{2}\cdot(-3)=-\frac{2}{3}\cdot(-3)=2$

65. Using order of operations: $-6\div\frac{1}{2}\cdot 12=-6\cdot 2\cdot 12=-144$

67. Using order of operations: $-15\div\frac{5}{3}\cdot 18=-15\cdot\frac{3}{5}\cdot 18=-9\cdot 18=-162$

69. Combining the fractions: $\frac{x}{4}+\frac{1}{5}=\frac{x}{4}\cdot\frac{5}{5}+\frac{1}{5}\cdot\frac{4}{4}=\frac{5x}{20}+\frac{4}{20}=\frac{5x+4}{20}$

71. Combining the fractions: $\frac{1}{3}+\frac{a}{12}=\frac{1}{3}\cdot\frac{4}{4}+\frac{a}{12}=\frac{4}{12}+\frac{a}{12}=\frac{a+4}{12}$

73. Combining the fractions: $\frac{x}{2}+\frac{1}{3}+\frac{x}{4}=\frac{x}{2}\cdot\frac{6}{6}+\frac{1}{3}\cdot\frac{4}{4}+\frac{x}{4}\cdot\frac{3}{3}=\frac{6x}{12}+\frac{4}{12}+\frac{3x}{12}=\frac{6x+4+3x}{12}=\frac{9x+4}{12}$

75. Combining the fractions: $\frac{2}{x}+\frac{3}{5}=\frac{2}{x}\cdot\frac{5}{5}+\frac{3}{5}\cdot\frac{x}{x}=\frac{10}{5x}+\frac{3x}{5x}=\frac{3x+10}{5x}$

77. Combining the fractions: $\frac{3}{7}+\frac{4}{y}=\frac{3}{7}\cdot\frac{y}{y}+\frac{4}{y}\cdot\frac{7}{7}=\frac{3y}{7y}+\frac{28}{7y}=\frac{3y+28}{7y}$

79. Combining the fractions: $\frac{3}{a}+\frac{3}{4}+\frac{1}{5}=\frac{3}{a}\cdot\frac{20}{20}+\frac{3}{4}\cdot\frac{5a}{5a}+\frac{1}{5}\cdot\frac{4a}{4a}=\frac{60}{20a}+\frac{15a}{20a}+\frac{4a}{20a}=\frac{60+15a+4a}{20a}=\frac{19a+60}{20a}$

81. Combining the fractions: $\frac{1}{2}x+\frac{1}{6}x=\left(\frac{1}{2}+\frac{1}{6}\right)x=\left(\frac{1}{2}\cdot\frac{3}{3}+\frac{1}{6}\right)x=\left(\frac{3}{6}+\frac{1}{6}\right)x=\frac{4}{6}x=\frac{2}{3}x$

83. Combining the fractions: $\frac{1}{2}x-\frac{3}{4}x=\left(\frac{1}{2}-\frac{3}{4}\right)x=\left(\frac{1}{2}\cdot\frac{2}{2}-\frac{3}{4}\right)x=\left(\frac{2}{4}-\frac{3}{4}\right)x=-\frac{1}{4}x$

85. Combining the fractions: $\frac{1}{3}x+\frac{3}{5}x=\left(\frac{1}{3}+\frac{3}{5}\right)x=\left(\frac{1}{3}\cdot\frac{5}{5}+\frac{3}{5}\cdot\frac{3}{3}\right)x=\left(\frac{5}{15}+\frac{9}{15}\right)x=\frac{14}{15}x$

© 2012 Cengage Learning. All Rights Reserved. May not be scanned, copied or duplicated, or posted to a publicly accessible website, in whole or in part.

87. Combining the fractions: $\frac{3x}{4}+\frac{x}{6}=\frac{3x}{4}\bullet\frac{3}{3}+\frac{x}{6}\bullet\frac{2}{2}=\frac{9x}{12}+\frac{2x}{12}=\frac{11x}{12}=\frac{11}{12}x$

89. Combining the fractions: $\frac{2x}{5}+\frac{5x}{8}=\frac{2x}{5}\bullet\frac{8}{8}+\frac{5x}{8}\bullet\frac{5}{5}=\frac{16x}{40}+\frac{25x}{40}=\frac{41x}{40}=\frac{41}{40}x$

91. Combining the fractions: $1-\frac{1}{x}=\frac{1}{1}\bullet\frac{x}{x}-\frac{1}{x}=\frac{x}{x}-\frac{1}{x}=\frac{x-1}{x}$

93. **a.** Adding the fractions: $\frac{3}{4}+\left(-\frac{1}{2}\right)=\frac{3}{4}-\frac{1}{2}\bullet\frac{2}{2}=\frac{3}{4}-\frac{2}{4}=\frac{1}{4}$

b. Subtracting the fractions: $\frac{3}{4}-\left(-\frac{1}{2}\right)=\frac{3}{4}+\frac{1}{2}\bullet\frac{2}{2}=\frac{3}{4}+\frac{2}{4}=\frac{5}{4}$

c. Multiplying the fractions: $\frac{3}{4}\left(-\frac{1}{2}\right)=-\frac{3}{8}$

d. Dividing the fractions: $\frac{3}{4}\div\left(-\frac{1}{2}\right)=\frac{3}{4}\bullet\left(-\frac{2}{1}\right)=-\frac{6}{4}=-\frac{3}{2}$

95. Simplifying: $\left(1-\frac{1}{2}\right)\left(1-\frac{1}{3}\right)=\left(\frac{2}{2}-\frac{1}{2}\right)\left(\frac{3}{3}-\frac{1}{3}\right)=\left(\frac{1}{2}\right)\left(\frac{2}{3}\right)=\frac{2}{6}=\frac{1}{3}$

97. Simplifying: $\left(1+\frac{1}{2}\right)\left(1-\frac{1}{2}\right)=\left(\frac{2}{2}+\frac{1}{2}\right)\left(\frac{2}{2}-\frac{1}{2}\right)=\left(\frac{3}{2}\right)\left(\frac{1}{2}\right)=\frac{3}{4}$

99. **a.** Substituting $x=2$: $1+\frac{1}{x}=1+\frac{1}{2}=\frac{2}{2}+\frac{1}{2}=\frac{3}{2}$

b. Substituting $x=3$: $1+\frac{1}{x}=1+\frac{1}{3}=\frac{3}{3}+\frac{1}{3}=\frac{4}{3}$

c. Substituting $x=4$: $1+\frac{1}{x}=1+\frac{1}{4}=\frac{4}{4}+\frac{1}{4}=\frac{5}{4}$

101. **a.** Substituting $x=1$: $2x+\frac{6}{x}=2(1)+\frac{6}{1}=2+6=8$

b. Substituting $x=2$: $2x+\frac{6}{x}=2(2)+\frac{6}{2}=4+3=7$

c. Substituting $x=3$: $2x+\frac{6}{x}=2(3)+\frac{6}{3}=6+2=8$

Chapter 1 Review

1. The expression is: $-7+(-10)=-17$

3. The expression is: $(-3+12)+5=9+5=14$

5. The expression is: $9-(-3)=9+3=12$

7. The expression is: $(-3)(-7)-6=21-6=15$

9. The expression is: $2[-8(3x)]=2(-24x)=-48x$

11. Simplifying: $|-1.8|=1.8$

13. The opposite is –6, and the reciprocal is $\frac{1}{6}$.

15. Multiplying the fractions: $\frac{1}{2}(-10)=\frac{1}{2}\bullet\left(-\frac{10}{1}\right)=-\frac{10}{2}=-5$

17. Adding: $-9+12=3$

19. Adding: $(-2)+(-8)+[-9+(-6)]=-2+(-8)+(-15)=-25$

21. Subtracting: $6-9=6+(-9)=-3$

23. Subtracting: $-12-(-8)=-12+8=-4$

25. Multiplying: $(-5)(6)=-30$

27. Multiplying: $-2(3)(4)=-24$

29. Finding the quotient: $\frac{12}{-3}=-4$

31. Simplifying using order of operations: $4\bullet5+3=20+3=23$

© 2012 Cengage Learning. All Rights Reserved. May not be scanned, copied or duplicated, or posted to a publicly accessible website, in whole or in part.

33. Simplifying using order of operations: $2^3 - 4 \bullet 3^2 + 5^2 = 8 - 4 \bullet 9 + 25 = 8 - 36 + 25 = 33 - 36 = -3$
35. Simplifying using order of operations: $20 + 8 \div 4 + 2 \bullet 5 = 20 + 2 + 10 = 32$
37. Simplifying using order of operations: $30 \div 3 \bullet 2 = 10 \bullet 2 = 20$
39. Simplifying using order of operations:
$$3(4-7)^2 - 5(3-8)^2 = 3(-3)^2 - 5(-5)^2 = 3 \bullet 9 - 5 \bullet 25 = 27 - 125 = 27 + (-125) = -98$$
41. Simplifying using order of operations: $\frac{4(-3)}{-6} = \frac{-12}{-6} = 2$
43. Simplifying using order of operations: $\frac{15-10}{6-6} = \frac{5}{0}$, which is undefined
45. associative property (of multiplication)
47. commutative property (of addition)
49. Simplifying the expression: $7 + (5 + x) = (7 + 5) + x = 12 + x$
51. Simplifying the expression: $\frac{1}{9}(9x) = \left(\frac{1}{9} \bullet 9\right)x = 1x = x$
53. Applying the distributive property: $7(2x+3) = 7 \bullet 2x + 7 \bullet 3 = 14x + 21$
55. Applying the distributive property: $\frac{1}{2}(5x-6) = \frac{1}{2} \bullet 5x - \frac{1}{2} \bullet 6 = \frac{5}{2}x - 3$
57. The rational numbers are: $-\frac{1}{3}, 0, 5, -4.5, \frac{2}{5}, -3$
59. The irrational numbers are: $\sqrt{7}, \pi$
61. Factoring the number: $90 = 9 \bullet 10 = (3 \bullet 3) \bullet (2 \bullet 5) = 2 \bullet 3^2 \bullet 5$
63. First factor the denominators to find the LCM:
$$35 = 5 \bullet 7$$
$$42 = 2 \bullet 3 \bullet 7$$
$$\text{LCM} = 2 \bullet 3 \bullet 5 \bullet 7 = 210$$
Combining the fractions: $\frac{18}{35} + \frac{13}{42} = \frac{18 \bullet 6}{35 \bullet 6} + \frac{13 \bullet 5}{42 \bullet 5} = \frac{108}{210} + \frac{65}{210} = \frac{173}{210}$

Chapter 1 Test

1. Translating into symbols: $x + 3 = 8$
2. Translating into symbols: $5y = 15$
3. Simplifying using order of operations: $5^2 + 3(9-7) + 3^2 = 5^2 + 3(2) + 3^2 = 25 + 6 + 9 = 40$
4. Simplifying using order of operations: $10 - 6 \div 3 + 2^3 = 10 - 6 \div 3 + 8 = 10 - 2 + 8 = 18 - 2 = 16$
5. The opposite of -4 is 4, the reciprocal is $-\frac{1}{4}$, and the absolute value is $|-4| = 4$.
6. The opposite of $\frac{3}{4}$ is $-\frac{3}{4}$, the reciprocal is $\frac{4}{3}$, and the absolute value is $\left|\frac{3}{4}\right| = \frac{3}{4}$.
7. Adding: $3 + (-7) = -4$
8. Adding: $|-9 + (-6)| + |-3 + 5| = |-15| + |2| = 15 + 2 = 17$
9. Subtracting: $-4 - 8 = -4 + (-8) = -12$
10. Subtracting: $9 - (7-2) - 4 = 9 - 5 - 4 = 9 + (-5) + (-4) = 9 + (-9) = 0$
11. c (associative property of addition)
12. e (distributive property)
13. d (associative property of multiplication)
14. a (commutative property of addition)
15. Multiplying: $-3(7) = -21$
16. Multiplying: $-4(8)(-2) = 64$
17. Multiplying: $8\left(-\frac{1}{4}\right) = \frac{8}{1} \bullet \left(-\frac{1}{4}\right) = -\frac{8}{4} = -2$
18. Multiplying: $\left(-\frac{2}{3}\right)^3 = \left(-\frac{2}{3}\right) \bullet \left(-\frac{2}{3}\right) \bullet \left(-\frac{2}{3}\right) = -\frac{8}{27}$
19. Simplifying using order of operations: $-3(-4) - 8 = 12 - 8 = 4$
20. Simplifying using order of operations: $5(-6)^2 - 3(-2)^3 = 5(36) - 3(-8) = 180 + 24 = 204$

© 2012 Cengage Learning. All Rights Reserved. May not be scanned, copied or duplicated, or posted to a publicly accessible website, in whole or in part.

21. Simplifying using order of operations: $7-3(2-8)=7-3(-6)=7+18=25$

22 Simplifying using order of operations:

$$4-2[-3(-1+5)+4(-3)]=4-2[-3(4)+4(-3)]=4-2(-12-12)=4-2(-24)=4+48=52$$

23. Simplifying using order of operations: $\dfrac{4(-5)-2(7)}{-10-7}=\dfrac{-20-14}{-17}=\dfrac{-34}{-17}=2$

24. Simplifying using order of operations: $\dfrac{2(-3-1)+4(-5+2)}{-3(2)-4}=\dfrac{2(-4)+4(-3)}{-6-4}=\dfrac{-8-12}{-10}=\dfrac{-20}{-10}=2$

25. Simplifying: $3+(5+2x)=(3+5)+2x=8+2x$

26. Simplifying: $-2(-5x)=[-2(-5)]x=10x$

27. Multiplying: $2(3x+5)=2\cdot 3x+2\cdot 5=6x+10$

28. Multiplying: $-\frac{1}{2}(4x-2)=-\frac{1}{2}(4x)-\left(-\frac{1}{2}\right)(2)=-2x+1$

29. The integers are –8 and 1.

30. The rational numbers are –8, $\frac{3}{4}$, 1, and 1.5.

31. The irrational numbers are $\sqrt{2}$.

32. The real numbers are –8, $\frac{3}{4}$, 1, $\sqrt{2}$, and 1.5.

33. Factoring the number: $592=4\cdot 148=(2\cdot 2)\cdot(4\cdot 37)=(2\cdot 2)\cdot(2\cdot 2\cdot 37)=2^4\cdot 37$

34. Factoring the number: $1340=10\cdot 134=(2\cdot 5)\cdot(2\cdot 67)=2^2\cdot 5\cdot 67$

35. First factor the denominators to find the LCM:

$15=3\cdot 5$

$42=2\cdot 3\cdot 7$

$\text{LCM}=2\cdot 3\cdot 5\cdot 7=210$

Combining the fractions: $\dfrac{5}{15}+\dfrac{11}{42}=\dfrac{5\cdot 14}{15\cdot 14}+\dfrac{11\cdot 5}{42\cdot 5}=\dfrac{70}{210}+\dfrac{55}{210}=\dfrac{125}{210}=\dfrac{5\cdot 25}{5\cdot 42}=\dfrac{25}{42}$

36. Combining the fractions: $\dfrac{5}{x}+\dfrac{3}{x}=\dfrac{5+3}{x}=\dfrac{8}{x}$

37. The expression is: $8+(-3)=5$

38. The expression is: $-24-2=-24+(-2)=-26$

39. The expression is: $-5(-4)=20$

40. The expression is: $\dfrac{-24}{-2}=12$

41. The pattern is to add 5, so the next term is $7+5=12$.

42. The pattern is to multiply by $-\frac{1}{2}$, so the next term is $-1\left(-\frac{1}{2}\right)=\frac{1}{2}$.

43. From the table, 425 million speak Spanish.

44. The total is: $1075+275=1,350$ million (1.35 billion)

45. Subtracting: $514-496=18$ million

© 2012 Cengage Learning. All Rights Reserved. May not be scanned, copied or duplicated, or posted to a publicly accessible website, in whole or in part.

Chapter 2
Linear Equations and Inequalities

Getting Ready for Chapter 2

1. Simplifying: $-3+7=4$
2. Simplifying: $-10-4=-14$
3. Simplifying: $9-(-24)=9+24=33$
4. Simplifying: $-6(5)-5=-30-5=-35$
5. Simplifying: $-3(2y+1)=-6y-3$
6. Simplifying: $8.1+2.7=10.8$
7. Simplifying: $-\frac{3}{4}+\left(-\frac{1}{2}\right)=-\frac{3}{4}-\frac{1}{2}\bullet\frac{2}{2}=-\frac{3}{4}-\frac{2}{4}=-\frac{5}{4}$
8. Simplifying: $-\frac{7}{10}+\left(-\frac{1}{2}\right)=-\frac{7}{10}-\frac{1}{2}\bullet\frac{5}{5}=-\frac{7}{10}-\frac{5}{10}=-\frac{12}{10}=-\frac{6}{5}$
9. Simplifying: $\frac{1}{5}(5x)=\left(\frac{1}{5}\bullet 5\right)x=x$
10. Simplifying: $4\left(\frac{1}{4}x\right)=\left(4\bullet\frac{1}{4}\right)x=x$
11. Simplifying: $\frac{3}{2}\left(\frac{2}{3}x\right)=\left(\frac{3}{2}\bullet\frac{2}{3}\right)x=x$
12. Simplifying: $\frac{5}{2}\left(\frac{2}{5}x\right)=\left(\frac{5}{2}\bullet\frac{2}{5}\right)x=x$
13. Simplifying: $-1(3x+4)=-3x-4$
14. Simplifying: $-2.4+(-7.3)=-9.7$
15. Simplifying: $0.04(x+7{,}000)=0.04x+280$
16. Simplifying: $0.09(x+2{,}000)=0.09x+180$
17. Applying the distributive property: $3(x-5)+4=3x-15+4=3x-11$
18. Applying the distributive property: $5(x-3)+2=5x-15+2=5x-13$
19. Applying the distributive property: $5(2x-8)-3=10x-40-3=10x-43$
20. Applying the distributive property: $4(3x-5)-2=12x-20-2=12x-22$

2.1 Simplifying Expressions

1. Simplifying the expression: $3x-6x=(3-6)x=-3x$
3. Simplifying the expression: $-2a+a=(-2+1)a=-a$
5. Simplifying the expression: $7x+3x+2x=(7+3+2)x=12x$
7. Simplifying the expression: $3a-2a+5a=(3-2+5)a=6a$
9. Simplifying the expression: $4x-3+2x=4x+2x-3=6x-3$
11. Simplifying the expression: $3a+4a+5=7a+5$
13. Simplifying the expression: $2x-3+3x-2=2x+3x-3-2=5x-5$
15. Simplifying the expression: $3a-1+a+3=3a+a-1+3=4a+2$
17. Simplifying the expression: $-4x+8-5x-10=-4x-5x+8-10=-9x-2$
19. Simplifying the expression: $7a+3+2a+3a=7a+2a+3a+3=12a+3$

© 2012 Cengage Learning. All Rights Reserved. May not be scanned, copied or duplicated, or posted to a publicly accessible website, in whole or in part.

21. Simplifying the expression: $5(2x-1)+4=10x-5+4=10x-1$
23. Simplifying the expression: $7(3y+2)-8=21y+14-8=21y+6$
25. Simplifying the expression: $-3(2x-1)+5=-6x+3+5=-6x+8$
27. Simplifying the expression: $5-2(a+1)=5-2a-2=-2a-2+5=-2a+3$
29. Simplifying the expression: $6-4(x-5)=6-4x+20=-4x+20+6=-4x+26$
31. Simplifying the expression: $-9-4(2-y)+1=-9-8+4y+1=4y+1-9-8=4y-16$
33. Simplifying the expression: $-6+2(2-3x)+1=-6+4-6x+1=-6x-6+4+1=-6x-1$
35. Simplifying the expression: $(4x-7)-(2x+5)=4x-7-2x-5=4x-2x-7-5=2x-12$
37. Simplifying the expression: $8(2a+4)-(6a-1)=16a+32-6a+1=16a-6a+32+1=10a+33$
39. Simplifying the expression: $3(x-2)+(x-3)=3x-6+x-3=3x+x-6-3=4x-9$
41. Simplifying the expression: $4(2y-8)-(y+7)=8y-32-y-7=8y-y-32-7=7y-39$
43. Simplifying the expression: $-9(2x+1)-(x+5)=-18x-9-x-5=-18x-x-9-5=-19x-14$
45. Evaluating when $x=2$: $3x-1=3(2)-1=6-1=5$
47. Evaluating when $x=2$: $-2x-5=-2(2)-5=-4-5=-9$
49. Evaluating when $x=2$: $x^2-8x+16=(2)^2-8(2)+16=4-16+16=4$
51. Evaluating when $x=2$: $(x-4)^2=(2-4)^2=(-2)^2=4$
53. Evaluating when $x=-5$: $7x-4-x-3=7(-5)-4-(-5)-3=-35-4+5-3=-42+5=-37$
    Now simplifying the expression: $7x-4-x-3=7x-x-4-3=6x-7$
    Evaluating when $x=-5$: $6x-7=6(-5)-7=-30-7=-37$
    Note that the two values are the same.
55. Evaluating when $x=-5$: $5(2x+1)+4=5[2(-5)+1]+4=5(-10+1)+4=5(-9)+4=-45+4=-41$
    Now simplifying the expression: $5(2x+1)+4=10x+5+4=10x+9$
    Evaluating when $x=-5$: $10x+9=10(-5)+9=-50+9=-41$
    Note that the two values are the same.
57. Evaluating when $x=-3$ and $y=5$: $x^2-2xy+y^2=(-3)^2-2(-3)(5)+(5)^2=9+30+25=64$
59. Evaluating when $x=-3$ and $y=5$: $(x-y)^2=(-3-5)^2=(-8)^2=64$
61. Evaluating when $x=-3$ and $y=5$: $x^2+6xy+9y^2=(-3)^2+6(-3)(5)+9(5)^2=9-90+225=144$
63. Evaluating when $x=-3$ and $y=5$: $(x+3y)^2=[-3+3(5)]^2=(-3+15)^2=(12)^2=144$
65. Evaluating when $x=\frac{1}{2}$: $12x-3=12\left(\frac{1}{2}\right)-3=6-3=3$
67. Evaluating when $x=\frac{1}{4}$: $12x-3=12\left(\frac{1}{4}\right)-3=3-3=0$
69. Evaluating when $x=\frac{3}{2}$: $12x-3=12\left(\frac{3}{2}\right)-3=18-3=15$
71. Evaluating when $x=\frac{3}{4}$: $12x-3=12\left(\frac{3}{4}\right)-3=9-3=6$
73. **a.** Substituting the values for n:

n	1	2	3	4
$3n$	3	6	9	12

b. Substituting the values for n:

n	1	2	3	4
n^3	1	8	27	64

© 2012 Cengage Learning. All Rights Reserved. May not be scanned, copied or duplicated, or posted to a publicly accessible website, in whole or in part.

75. Substituting $n = 1, 2, 3, 4$:
$n=1: \quad 3(1)-2=3-2=1$
$n=2: \quad 3(2)-2=6-2=4$
$n=3: \quad 3(3)-2=9-2=7$
$n=4: \quad 3(4)-2=12-2=10$
The sequence is 1, 4, 7, 10, ..., which is an arithmetic sequence.

77. Substituting $n = 1, 2, 3, 4$:
$n=1: \quad (1)^2-2(1)+1=1-2+1=0$
$n=2: \quad (2)^2-2(2)+1=4-4+1=1$
$n=3: \quad (3)^2-2(3)+1=9-6+1=4$
$n=4: \quad (4)^2-2(4)+1=16-8+1=9$
The sequence is 0, 1, 4, 9, ..., which is a sequence of squares.

79. Simplifying: $7-3(2y+1)=7-6y-3=-6y+4$

81. Simplifying: $0.08x+0.09x=0.17x$

83. Simplifying: $(x+y)+(x-y)=x+y+x-y=2x$

85. Simplifying: $3x+2(x-2)=3x+2x-4=5x-4$

87. Simplifying: $4(x+1)+3(x-3)=4x+4+3x-9=7x-5$

89. Simplifying: $x+(x+3)(-3)=x-3x-9=-2x-9$

91. Simplifying: $3(4x-2)-(5x-8)=12x-6-5x+8=7x+2$

93. Simplifying: $-(3x+1)-(4x-7)=-3x-1-4x+7=-7x+6$

95. Simplifying: $(x+3y)+3(2x-y)=x+3y+6x-3y=7x$

97. Simplifying: $3(2x+3y)-2(3x+5y)=6x+9y-6x-10y=-y$

99. Simplifying: $-6\left(\frac{1}{2}x-\frac{1}{3}y\right)+12\left(\frac{1}{4}x+\frac{2}{3}y\right)=-3x+2y+3x+8y=10y$

101. Simplifying: $0.08x+0.09(x+2{,}000)=0.08x+0.09x+180=0.17x+180$

103. Simplifying: $0.10x+0.12(x+500)=0.10x+0.12x+60=0.22x+60$

105. For the y terms to cancel, we must have $4y-ay=0$, so $a=4$.
Simplifying: $(5x+4y)+4(2x-y)=5x+4y+8x-4y=13x$

107. Evaluating the expression: $b^2-4ac=(-5)^2-4(1)(-6)=25-(-24)=25+24=49$

109. Evaluating the expression: $b^2-4ac=(4)^2-4(2)(-3)=16-(-24)=16+24=40$

111. **a.** Substituting $x = 8{,}000$: $-0.0035(8000)+70=42°\text{F}$
b. Substituting $x = 12{,}000$: $-0.0035(12000)+70=28°\text{F}$
c. Substituting $x = 24{,}000$: $-0.0035(24000)+70=-14°\text{F}$

113. **a.** Substituting $t = 10$: $35+0.25(10)=\$37.50$
b. Substituting $t = 20$: $35+0.25(20)=\$40.00$
c. Substituting $t = 30$: $35+0.25(30)=\$42.50$

115. Simplifying the expression: $G-0.21G-0.08G=0.71G$. Substituting $G = \$1{,}250$: $0.71(\$1{,}250)=\$887.50$

117. Computing the speed: $s=\frac{1}{2}(248)-56=124-56=68$ mph

119. Simplifying: $17-5=12$

121. Simplifying: $2-5=-3$

123. Simplifying: $-2.4+(-7.3)=-9.7$

© 2012 Cengage Learning. All Rights Reserved. May not be scanned, copied or duplicated, or posted to a publicly accessible website, in whole or in part.

125. Simplifying: $-\frac{1}{2}+\left(-\frac{3}{4}\right)=-\frac{1}{2}\bullet\frac{2}{2}-\frac{3}{4}=-\frac{2}{4}-\frac{3}{4}=-\frac{5}{4}$

127. Simplifying: $4(2\bullet 9-3)-7\bullet 9=4(18-3)-63=4(15)-63=60-63=-3$

129. Simplifying: $4(2a-3)-7a=8a-12-7a=a-12$

131. Subtracting: $-3-\frac{1}{2}=\frac{-3}{1}-\frac{1}{2}=\frac{-3\bullet 2}{1\bullet 2}-\frac{1}{2}=\frac{-6}{2}-\frac{1}{2}=-\frac{7}{2}$

133. Adding: $\frac{4}{5}+\frac{1}{10}+\frac{3}{8}=\frac{4\bullet 8}{5\bullet 8}+\frac{1\bullet 4}{10\bullet 4}+\frac{3\bullet 5}{8\bullet 5}=\frac{32}{40}+\frac{4}{40}+\frac{15}{40}=\frac{51}{40}$

135. Evaluating when $x=5$: $2(5)-3=10-3=7$

137. Simplifying: $\frac{1}{8}-\frac{1}{6}=\frac{1}{8}\bullet\frac{3}{3}-\frac{1}{6}\bullet\frac{4}{4}=\frac{3}{24}-\frac{4}{24}=-\frac{1}{24}$

139. Simplifying: $\frac{5}{9}-\frac{4}{3}=\frac{5}{9}-\frac{4}{3}\bullet\frac{3}{3}=\frac{5}{9}-\frac{12}{9}=-\frac{7}{9}$

141. Simplifying: $-\frac{7}{30}+\frac{5}{28}=-\frac{7}{30}\bullet\frac{14}{14}+\frac{5}{28}\bullet\frac{15}{15}=-\frac{98}{420}+\frac{75}{420}=-\frac{23}{420}$

2.2 Addition Property of Equality

1. Solving the equation:

$$\begin{aligned}x-3&=8\\x-3+3&=8+3\\x&=11\end{aligned}$$

3. Solving the equation:

$$\begin{aligned}x+2&=6\\x+2+(-2)&=6+(-2)\\x&=4\end{aligned}$$

5. Solving the equation:

$$\begin{aligned}a+\frac{1}{2}&=-\frac{1}{4}\\a+\frac{1}{2}+\left(-\frac{1}{2}\right)&=-\frac{1}{4}+\left(-\frac{1}{2}\right)\\a&=-\frac{1}{4}+\left(-\frac{2}{4}\right)\\a&=-\frac{3}{4}\end{aligned}$$

7. Solving the equation:

$$\begin{aligned}x+2.3&=-3.5\\x+2.3+(-2.3)&=-3.5+(-2.3)\\x&=-5.8\end{aligned}$$

9. Solving the equation:

$$\begin{aligned}y+11&=-6\\y+11+(-11)&=-6+(-11)\\y&=-17\end{aligned}$$

11. Solving the equation:

$$\begin{aligned}x-\frac{5}{8}&=-\frac{3}{4}\\x-\frac{5}{8}+\frac{5}{8}&=-\frac{3}{4}+\frac{5}{8}\\x&=-\frac{6}{8}+\frac{5}{8}\\x&=-\frac{1}{8}\end{aligned}$$

13. Solving the equation:

$$\begin{aligned}m-6&=2m\\m-6-m&=2m-m\\m&=-6\end{aligned}$$

15. Solving the equation:

$$\begin{aligned}6.9+x&=3.3\\-6.9+6.9+x&=-6.9+3.3\\x&=-3.6\end{aligned}$$

© 2012 Cengage Learning. All Rights Reserved. May not be scanned, copied or duplicated, or posted to a publicly accessible website, in whole or in part.

17. Solving the equation:

$$\begin{aligned} 5a &= 4a-7 \\ 5a-4a &= 4a-4a-7 \\ a &= -7 \end{aligned}$$

19. Solving the equation:

$$\begin{aligned} -\frac{5}{9} &= x-\frac{2}{5} \\ -\frac{5}{9}+\frac{2}{5} &= x-\frac{2}{5}+\frac{2}{5} \\ -\frac{25}{45}+\frac{18}{45} &= x \\ x &= -\frac{7}{45} \end{aligned}$$

21. Solving the equation:

$$\begin{aligned} 4x+2-3x &= 4+1 \\ x+2 &= 5 \\ x+2+(-2) &= 5+(-2) \\ x &= 3 \end{aligned}$$

23. Solving the equation:

$$\begin{aligned} 8a-\frac{1}{2}-7a &= \frac{3}{4}+\frac{1}{8} \\ a-\frac{1}{2} &= \frac{6}{8}+\frac{1}{8} \\ a-\frac{1}{2} &= \frac{7}{8} \\ a-\frac{1}{2}+\frac{1}{2} &= \frac{7}{8}+\frac{1}{2} \\ a &= \frac{7}{8}+\frac{4}{8} \\ a &= \frac{11}{8} \end{aligned}$$

25. Solving the equation:

$$\begin{aligned} -3-4x+5x &= 18 \\ -3+x &= 18 \\ 3-3+x &= 3+18 \\ x &= 21 \end{aligned}$$

27. Solving the equation:

$$\begin{aligned} -11x+2+10x+2x &= 9 \\ x+2 &= 9 \\ x+2+(-2) &= 9+(-2) \\ x &= 7 \end{aligned}$$

29. Solving the equation:

$$\begin{aligned} -2.5+4.8 &= 8x-1.2-7x \\ 2.3 &= x-1.2 \\ 2.3+1.2 &= x-1.2+1.2 \\ x &= 3.5 \end{aligned}$$

31. Solving the equation:

$$\begin{aligned} 2y-10+3y-4y &= 18-6 \\ y-10 &= 12 \\ y-10+10 &= 12+10 \\ y &= 22 \end{aligned}$$

33. Solving the equation:

$$\begin{aligned} 2(x+3)-x &= 4 \\ 2x+6-x &= 4 \\ x+6 &= 4 \\ x+6+(-6) &= 4+(-6) \\ x &= -2 \end{aligned}$$

35. Solving the equation:

$$\begin{aligned} -3(x-4)+4x &= 3-7 \\ -3x+12+4x &= -4 \\ x+12 &= -4 \\ x+12+(-12) &= -4+(-12) \\ x &= -16 \end{aligned}$$

37. Solving the equation:

$$\begin{aligned} 5(2a+1)-9a &= 8-6 \\ 10a+5-9a &= 2 \\ a+5 &= 2 \\ a+5+(-5) &= 2+(-5) \\ a &= -3 \end{aligned}$$

39. Solving the equation:

$$\begin{aligned} -(x+3)+2x-1 &= 6 \\ -x-3+2x-1 &= 6 \\ x-4 &= 6 \\ x-4+4 &= 6+4 \\ x &= 10 \end{aligned}$$

41. Solving the equation:

$$\begin{aligned} 4y-3(y-6)+2 &= 8 \\ 4y-3y+18+2 &= 8 \\ y+20 &= 8 \\ y+20+(-20) &= 8+(-20) \\ y &= -12 \end{aligned}$$

43. Solving the equation:

$$\begin{aligned} -3(2m-9)+7(m-4) &= 12-9 \\ -6m+27+7m-28 &= 3 \\ m-1 &= 3 \\ m-1+1 &= 3+1 \\ m &= 4 \end{aligned}$$

45. Solving the equation:

$$\begin{aligned} 4x &= 3x+2 \\ 4x+(-3x) &= 3x+(-3x)+2 \\ x &= 2 \end{aligned}$$

47. Solving the equation:

$$\begin{aligned} 8a &= 7a-5 \\ 8a+(-7a) &= 7a+(-7a)-5 \\ a &= -5 \end{aligned}$$

© 2012 Cengage Learning. All Rights Reserved. May not be scanned, copied or duplicated, or posted to a publicly accessible website, in whole or in part.

49. Solving the equation:

$$\begin{aligned} 2x &= 3x+1 \\ (-2x)+2x &= (-2x)+3x+1 \\ 0 &= x+1 \\ 0+(-1) &= x+1+(-1) \\ x &= -1 \end{aligned}$$

51. Solving the equation:

$$\begin{aligned} 2y+1 &= 3y+4 \\ 2y+(-2y)+1 &= 3y+(-2y)+4 \\ 1 &= y+4 \\ 1+(-4) &= y+4+(-4) \\ y &= -3 \end{aligned}$$

53. Solving the equation:

$$\begin{aligned} 2m-3 &= m+5 \\ 2m+(-m)-3 &= m+(-m)+5 \\ m-3 &= 5 \\ m-3+3 &= 5+3 \\ m &= 8 \end{aligned}$$

55. Solving the equation:

$$\begin{aligned} 4x-7 &= 5x+1 \\ 4x+(-4x)-7 &= 5x+(-4x)+1 \\ -7 &= x+1 \\ -7+(-1) &= x+1+(-1) \\ x &= -8 \end{aligned}$$

57. Solving the equation:

$$\begin{aligned} 4x+\frac{4}{3} &= 5x-\frac{2}{3} \\ 4x+(-4x)+\frac{4}{3} &= 5x+(-4x)-\frac{2}{3} \\ \frac{4}{3} &= x-\frac{2}{3} \\ \frac{4}{3}+\frac{2}{3} &= x-\frac{2}{3}+\frac{2}{3} \\ x &= \frac{6}{3} = 2 \end{aligned}$$

59. Solving the equation:

$$\begin{aligned} 8a-7.1 &= 7a+3.9 \\ 8a+(-7a)-7.1 &= 7a+(-7a)+3.9 \\ a-7.1 &= 3.9 \\ a-7.1+7.1 &= 3.9+7.1 \\ a &= 11 \end{aligned}$$

61. Solving the equation:

$$\begin{aligned} 12x-5.8 &= 11x+4.2 \\ 12x+(-11x)-5.8 &= 11x+(-11x)+4.2 \\ x-5.8 &= 4.2 \\ x-5.8+5.8 &= 4.2+5.8 \\ x &= 10 \end{aligned}$$

63. **a.** Solving for *R*:

$$\begin{aligned} T+R+A &= 100 \\ 88+R+6 &= 100 \\ 94+R &= 100 \\ R &= 6\% \end{aligned}$$

b. Solving for *R*:

$$\begin{aligned} T+R+A &= 100 \\ 0+R+95 &= 100 \\ 95+R &= 100 \\ R &= 5\% \end{aligned}$$

c. Solving for *A*:

$$\begin{aligned} T+R+A &= 100 \\ 0+98+A &= 100 \\ 98+A &= 100 \\ A &= 2\% \end{aligned}$$

d. Solving for *R*:

$$\begin{aligned} T+R+A &= 100 \\ 0+R+25 &= 100 \\ 25+R &= 100 \\ R &= 75\% \end{aligned}$$

65. Solving for *x*:

$$\begin{aligned} x+55+55 &= 180 \\ x+110 &= 180 \\ x &= 70° \end{aligned}$$

67. **a.** Using *B*, *N*, and *R* as our variables, the combined market share was: $B+N+R = 33.3+23.6+9.0 = 65.9\%$

b. The combined market share was: $B+N+R = 24.7+30.0+16.8 = 71.5\%$

c. The total market share for all other movie companies was: $100\% - 65.9\% = 34.1\%$

d. The total market share for all other movie companies was: $100\% - 71.5\% = 28.5\%$

69. Simplifying: $\frac{3}{2}\left(\frac{2}{3}y\right) = y$

71. Simplifying: $\frac{1}{5}(5x) = x$

73. Simplifying: $\frac{1}{5}(30) = 6$

75. Simplifying: $\frac{3}{2}(4) = \frac{12}{2} = 6$

© 2012 Cengage Learning. All Rights Reserved. May not be scanned, copied or duplicated, or posted to a publicly accessible website, in whole or in part.

77. Simplifying: $12\left(-\frac{3}{4}\right)=-\frac{36}{4}=-9$

79. Simplifying: $\frac{3}{2}\left(-\frac{5}{4}\right)=-\frac{15}{8}$

81. Simplifying: $-13+(-5)=-18$

83. Simplifying: $-\frac{3}{4}+\left(-\frac{1}{2}\right)=-\frac{3}{4}-\frac{1}{2}\bullet\frac{2}{2}=-\frac{3}{4}-\frac{2}{4}=-\frac{5}{4}$

85. Simplifying: $7x+(-4x)=3x$

87. Applying the associative property: $3(6x)=(3\bullet 6)x=18x$

89. Applying the associative property: $\frac{1}{5}(5x)=\left(\frac{1}{5}\bullet 5\right)x=1x=x$

91. Applying the associative property: $8\left(\frac{1}{8}y\right)=\left(8\bullet\frac{1}{8}\right)y=1y=y$

93. Applying the associative property: $-2\left(-\frac{1}{2}x\right)=\left[-2\bullet\left(-\frac{1}{2}\right)\right]x=1x=x$

95. Applying the associative property: $-\frac{4}{3}\left(-\frac{3}{4}a\right)=\left[-\frac{4}{3}\bullet\left(-\frac{3}{4}\right)\right]a=1a=a$

2.3 Multiplication Property of Equality

1. Solving the equation:
$$5x=10$$
$$\frac{1}{5}(5x)=\frac{1}{5}(10)$$
$$x=2$$

3. Solving the equation:
$$7a=28$$
$$\frac{1}{7}(7a)=\frac{1}{7}(28)$$
$$a=4$$

5. Solving the equation:
$$-8x=4$$
$$-\frac{1}{8}(-8x)=-\frac{1}{8}(4)$$
$$x=-\frac{1}{2}$$

7. Solving the equation:
$$8m=-16$$
$$\frac{1}{8}(8m)=\frac{1}{8}(-16)$$
$$m=-2$$

9. Solving the equation:
$$-3x=-9$$
$$-\frac{1}{3}(-3x)=-\frac{1}{3}(-9)$$
$$x=3$$

11. Solving the equation:
$$-7y=-28$$
$$-\frac{1}{7}(-7y)=-\frac{1}{7}(-28)$$
$$y=4$$

13. Solving the equation:
$$2x=0$$
$$\frac{1}{2}(2x)=\frac{1}{2}(0)$$
$$x=0$$

15. Solving the equation:
$$-5x=0$$
$$-\frac{1}{5}(-5x)=-\frac{1}{5}(0)$$
$$x=0$$

17. Solving the equation:
$$\frac{x}{3}=2$$
$$3\left(\frac{x}{3}\right)=3(2)$$
$$x=6$$

19. Solving the equation:
$$-\frac{m}{5}=10$$
$$-5\left(-\frac{m}{5}\right)=-5(10)$$
$$m=-50$$

© 2012 Cengage Learning. All Rights Reserved. May not be scanned, copied or duplicated, or posted to a publicly accessible website, in whole or in part.

21. Solving the equation:

$$-\frac{x}{2}=-\frac{3}{4}$$
$$-2\left(-\frac{x}{2}\right)=-2\left(-\frac{3}{4}\right)$$
$$x=\frac{3}{2}$$

23. Solving the equation:

$$\frac{2}{3}a=8$$
$$\frac{3}{2}\left(\frac{2}{3}a\right)=\frac{3}{2}(8)$$
$$a=12$$

25. Solving the equation:

$$-\frac{3}{5}x=\frac{9}{5}$$
$$-\frac{5}{3}\left(-\frac{3}{5}x\right)=-\frac{5}{3}\left(\frac{9}{5}\right)$$
$$x=-3$$

27. Solving the equation:

$$-\frac{5}{8}y=-20$$
$$-\frac{8}{5}\left(-\frac{5}{8}y\right)=-\frac{8}{5}(-20)$$
$$y=32$$

29. Simplifying and then solving the equation:

$$-4x-2x+3x=24$$
$$-3x=24$$
$$-\frac{1}{3}(-3x)=-\frac{1}{3}(24)$$
$$x=-8$$

31. Simplifying and then solving the equation:

$$4x+8x-2x=15-10$$
$$10x=5$$
$$\frac{1}{10}(10x)=\frac{1}{10}(5)$$
$$x=\frac{1}{2}$$

33. Simplifying and then solving the equation:

$$-3-5=3x+5x-10x$$
$$-8=-2x$$
$$-\frac{1}{2}(-8)=-\frac{1}{2}(-2x)$$
$$x=4$$

35. Using Method 2 to eliminate fractions:

$$18-13=\frac{1}{2}a+\frac{3}{4}a-\frac{5}{8}a$$
$$8(5)=8\left(\frac{1}{2}a+\frac{3}{4}a-\frac{5}{8}a\right)$$
$$40=4a+6a-5a$$
$$40=5a$$
$$\frac{1}{5}(40)=\frac{1}{5}(5a)$$
$$a=8$$

37. Solving by multiplying both sides of the equation by –1:

$$-x=4$$
$$-1(-x)=-1(4)$$
$$x=-4$$

39. Solving by multiplying both sides of the equation by –1:

$$-x=-4$$
$$-1(-x)=-1(-4)$$
$$x=4$$

41. Solving by multiplying both sides of the equation by –1:

$$15=-a$$
$$-1(15)=-1(-a)$$
$$a=-15$$

43. Solving by multiplying both sides of the equation by –1:

$$-y=\frac{1}{2}$$
$$-1(-y)=-1\left(\frac{1}{2}\right)$$
$$y=-\frac{1}{2}$$

© 2012 Cengage Learning. All Rights Reserved. May not be scanned, copied or duplicated, or posted to a publicly accessible website, in whole or in part.

45. Solving the equation:

$$\begin{aligned} 3x-2&=7\\ 3x-2+2&=7+2\\ 3x&=9\\ \frac{1}{3}(3x)&=\frac{1}{3}(9)\\ x&=3 \end{aligned}$$

47. Solving the equation:

$$\begin{aligned} 2a+1&=3\\ 2a+1+(-1)&=3+(-1)\\ 2a&=2\\ \frac{1}{2}(2a)&=\frac{1}{2}(2)\\ a&=1 \end{aligned}$$

49. Using Method 2 to eliminate fractions:

$$\begin{aligned} \frac{1}{8}+\frac{1}{2}x&=\frac{1}{4}\\ 8\left(\frac{1}{8}+\frac{1}{2}x\right)&=8\left(\frac{1}{4}\right)\\ 1+4x&=2\\ (-1)+1+4x&=(-1)+2\\ 4x&=1\\ \frac{1}{4}(4x)&=\frac{1}{4}(1)\\ x&=\frac{1}{4} \end{aligned}$$

51. Solving the equation:

$$\begin{aligned} 6x&=2x-12\\ 6x+(-2x)&=2x+(-2x)-12\\ 4x&=-12\\ \frac{1}{4}(4x)&=\frac{1}{4}(-12)\\ x&=-3 \end{aligned}$$

53. Solving the equation:

$$\begin{aligned} 2y&=-4y+18\\ 2y+4y&=-4y+4y+18\\ 6y&=18\\ \frac{1}{6}(6y)&=\frac{1}{6}(18)\\ y&=3 \end{aligned}$$

55. Solving the equation:

$$\begin{aligned} -7x&=-3x-8\\ -7x+3x&=-3x+3x-8\\ -4x&=-8\\ -\frac{1}{4}(-4x)&=-\frac{1}{4}(-8)\\ x&=2 \end{aligned}$$

57. Solving the equation:

$$\begin{aligned} 2x-5&=8x+4\\ 2x+(-8x)-5&=8x+(-8x)+4\\ -6x-5&=4\\ -6x-5+5&=4+5\\ -6x&=9\\ -\frac{1}{6}(-6x)&=-\frac{1}{6}(9)\\ x&=-\frac{3}{2} \end{aligned}$$

59. Using Method 2 to eliminate fractions:

$$\begin{aligned} x+\frac{1}{2}&=\frac{1}{4}x-\frac{5}{8}\\ 8\left(x+\frac{1}{2}\right)&=8\left(\frac{1}{4}x-\frac{5}{8}\right)\\ 8x+4&=2x-5\\ 8x+(-2x)+4&=2x+(-2x)-5\\ 6x+4&=-5\\ 6x+4+(-4)&=-5+(-4)\\ 6x&=-9\\ \frac{1}{6}(6x)&=\frac{1}{6}(-9)\\ x&=-\frac{3}{2} \end{aligned}$$

61. Solving the equation:

$$\begin{aligned} m+2&=6m-3\\ m+(-6m)+2&=6m+(-6m)-3\\ -5m+2&=-3\\ -5m+2+(-2)&=-3+(-2)\\ -5m&=-5\\ -\frac{1}{5}(-5m)&=-\frac{1}{5}(-5)\\ m&=1 \end{aligned}$$

63. Using Method 2 to eliminate fractions:

$$\begin{aligned} \frac{1}{2}m-\frac{1}{4}&=\frac{1}{12}m+\frac{1}{6}\\ 12\left(\frac{1}{2}m-\frac{1}{4}\right)&=12\left(\frac{1}{12}m+\frac{1}{6}\right)\\ 6m-3&=m+2\\ 6m+(-m)-3&=m+(-m)+2\\ 5m-3&=2\\ 5m-3+3&=2+3\\ 5m&=5\\ \frac{1}{5}(5m)&=\frac{1}{5}(5)\\ m&=1 \end{aligned}$$

© 2012 Cengage Learning. All Rights Reserved. May not be scanned, copied or duplicated, or posted to a publicly accessible website, in whole or in part.

65. Solving the equation:

$$\begin{aligned}6y-4&=9y+2\\6y+(-9y)-4&=9y+(-9y)+2\\-3y-4&=2\\-3y-4+4&=2+4\\-3y&=6\\-\frac{1}{3}(-3y)&=-\frac{1}{3}(6)\\y&=-2\end{aligned}$$

67. Using Method 2 to eliminate fractions:

$$\begin{aligned}\frac{3}{2}y+\frac{1}{3}&=y-\frac{2}{3}\\6\left(\frac{3}{2}y+\frac{1}{3}\right)&=6\left(y-\frac{2}{3}\right)\\9y+2&=6y-4\\9y+(-6y)+2&=6y+(-6y)-4\\3y+2&=-4\\3y+2+(-2)&=-4+(-2)\\3y&=-6\\\frac{1}{3}(3y)&=\frac{1}{3}(-6)\\y&=-2\end{aligned}$$

69. Solving the equation:

$$\begin{aligned}5x+6&=2\\5x+6+(-6)&=2+(-6)\\5x&=-4\\x&=-\frac{4}{5}\end{aligned}$$

71. Solving the equation:

$$\begin{aligned}\frac{x}{2}&=\frac{6}{12}\\2\bullet\frac{x}{2}&=2\bullet\frac{6}{12}\\x&=\frac{12}{12}=1\end{aligned}$$

73. Solving the equation:

$$\begin{aligned}\frac{3}{x}&=\frac{6}{7}\\7x\bullet\frac{3}{x}&=7x\bullet\frac{6}{7}\\21&=6x\\x&=\frac{21}{6}=\frac{7}{2}\end{aligned}$$

75. Solving the equation:

$$\begin{aligned}\frac{a}{3}&=\frac{5}{12}\\3\bullet\frac{a}{3}&=3\bullet\frac{5}{12}\\a&=\frac{15}{12}=\frac{5}{4}\end{aligned}$$

77. Solving the equation:

$$\begin{aligned}\frac{10}{20}&=\frac{20}{x}\\20x\bullet\frac{10}{20}&=20x\bullet\frac{20}{x}\\10x&=400\\x&=\frac{400}{10}=40\end{aligned}$$

79. Solving the equation:

$$\begin{aligned}\frac{2}{x}&=\frac{6}{7}\\7x\bullet\frac{2}{x}&=7x\bullet\frac{6}{7}\\14&=6x\\x&=\frac{14}{6}=\frac{7}{3}\end{aligned}$$

81. Solving the equation:

$$\begin{aligned}7.5x&=1500\\\frac{7.5x}{7.5}&=\frac{1500}{7.5}\\x&=200\end{aligned}$$

The break-even point is 200 tickets.

83. Solving the equation:

$$\begin{aligned}G-0.21G-0.08G&=987.5\\0.71G&=987.5\\G&\approx 1,391\end{aligned}$$

Your gross income is approximately $1,391.

85. Solving the equation:

$$\begin{aligned}2x&=1450\\\frac{1}{2}(2x)&=\frac{1}{2}(1450)\\x&=725\end{aligned}$$

The Hoover Dam is 725 feet tall.

87. Solving the equation:

$$\begin{aligned}2x&=4\\\frac{1}{2}(2x)&=\frac{1}{2}(4)\\x&=2\end{aligned}$$

© 2012 Cengage Learning. All Rights Reserved. May not be scanned, copied or duplicated, or posted to a publicly accessible website, in whole or in part.

89. Solving the equation:
$$30 = 5x$$
$$5x = 30$$
$$\frac{1}{5}(5x) = \frac{1}{5}(30)$$
$$x = 6$$

91. Solving the equation:
$$0.17x = 510$$
$$x = \frac{510}{0.17} = 3{,}000$$

93. Simplifying: $3(x-5)+4 = 3x-15+4 = 3x-11$

95. Simplifying: $0.09(x+2{,}000) = 0.09x+180$

97. Simplifying: $5-2(3y+1) = 5-6y-2 = 3-6y = -6y+3$

99. Simplifying: $3(2x-5)-(2x-4) = 6x-15-2x+4 = 4x-11$

101. Simplifying: $10x+(-5x) = 5x$

103. Simplifying: $0.08x+0.09x = 0.17x$

105. Applying the distributive property: $2(3x-5) = 2 \bullet 3x - 2 \bullet 5 = 6x-10$

107. Applying the distributive property: $\frac{1}{2}(3x+6) = \frac{1}{2} \bullet 3x + \frac{1}{2} \bullet 6 = \frac{3}{2}x+3$

109. Applying the distributive property: $\frac{1}{3}(-3x+6) = \frac{1}{3} \bullet (-3x) + \frac{1}{3} \bullet 6 = -x+2$

111. Using the distributive property and combining like terms: $5(2x-8)-3 = 10x-40-3 = 10x-43$

113. Using the distributive property and combining like terms:
$$-2(3x+5)+3(x-1) = -6x-10+3x-3 = -6x+3x-10-3 = -3x-13$$

115. Using the distributive property and combining like terms: $7-3(2y+1) = 7-6y-3 = -6y+7-3 = -6y+4$

2.4 Solving Linear Equations

1. Solving the equation:
$$2(x+3) = 12$$
$$2x+6 = 12$$
$$2x+6+(-6) = 12+(-6)$$
$$2x = 6$$
$$\frac{1}{2}(2x) = \frac{1}{2}(6)$$
$$x = 3$$

3. Solving the equation:
$$6(x-1) = -18$$
$$6x-6 = -18$$
$$6x-6+6 = -18+6$$
$$6x = -12$$
$$\frac{1}{6}(6x) = \frac{1}{6}(-12)$$
$$x = -2$$

5. Solving the equation:
$$2(4a+1) = -6$$
$$8a+2 = -6$$
$$8a+2+(-2) = -6+(-2)$$
$$8a = -8$$
$$\frac{1}{8}(8a) = \frac{1}{8}(-8)$$
$$a = -1$$

7. Solving the equation:
$$14 = 2(5x-3)$$
$$14 = 10x-6$$
$$14+6 = 10x-6+6$$
$$20 = 10x$$
$$\frac{1}{10}(20) = \frac{1}{10}(10x)$$
$$x = 2$$

9. Solving the equation:
$$-2(3y+5) = 14$$
$$-6y-10 = 14$$
$$-6y-10+10 = 14+10$$
$$-6y = 24$$
$$-\frac{1}{6}(-6y) = -\frac{1}{6}(24)$$
$$y = -4$$

11. Solving the equation:
$$-5(2a+4) = 0$$
$$-10a-20 = 0$$
$$-10a-20+20 = 0+20$$
$$-10a = 20$$
$$-\frac{1}{10}(-10a) = -\frac{1}{10}(20)$$
$$a = -2$$

© 2012 Cengage Learning. All Rights Reserved. May not be scanned, copied or duplicated, or posted to a publicly accessible website, in whole or in part.

13. Solving the equation:

$$\begin{aligned} 1 &= \tfrac{1}{2}(4x+2) \\ 1 &= 2x+1 \\ 1+(-1) &= 2x+1+(-1) \\ 0 &= 2x \\ \frac{1}{2}(0) &= \frac{1}{2}(2x) \\ x &= 0 \end{aligned}$$

15. Solving the equation:

$$\begin{aligned} 3(t-4)+5 &= -4 \\ 3t-12+5 &= -4 \\ 3t-7 &= -4 \\ 3t-7+7 &= -4+7 \\ 3t &= 3 \\ \frac{1}{3}(3t) &= \frac{1}{3}(3) \\ t &= 1 \end{aligned}$$

17. Solving the equation:

$$\begin{aligned} 4(2y+1)-7 &= 1 \\ 8y+4-7 &= 1 \\ 8y-3 &= 1 \\ 8y-3+3 &= 1+3 \\ 8y &= 4 \\ \frac{1}{8}(8y) &= \frac{1}{8}(4) \\ y &= \frac{1}{2} \end{aligned}$$

19. Solving the equation:

$$\begin{aligned} \frac{1}{2}(x-3) &= \frac{1}{4}(x+1) \\ \frac{1}{2}x-\frac{3}{2} &= \frac{1}{4}x+\frac{1}{4} \\ 4\left(\frac{1}{2}x-\frac{3}{2}\right) &= 4\left(\frac{1}{4}x+\frac{1}{4}\right) \\ 2x-6 &= x+1 \\ 2x+(-x)-6 &= x+(-x)+1 \\ x-6 &= 1 \\ x-6+6 &= 1+6 \\ x &= 7 \end{aligned}$$

21. Solving the equation:

$$\begin{aligned} -0.7(2x-7) &= 0.3(11-4x) \\ -1.4x+4.9 &= 3.3-1.2x \\ -1.4x+1.2x+4.9 &= 3.3-1.2x+1.2x \\ -0.2x+4.9 &= 3.3 \\ -0.2x+4.9+(-4.9) &= 3.3+(-4.9) \\ -0.2x &= -1.6 \\ \frac{-0.2x}{-0.2} &= \frac{-1.6}{-0.2} \\ x &= 8 \end{aligned}$$

23. Solving the equation:

$$\begin{aligned} -2(3y+1) &= 3(1-6y)-9 \\ -6y-2 &= 3-18y-9 \\ -6y-2 &= -18y-6 \\ -6y+18y-2 &= -18y+18y-6 \\ 12y-2 &= -6 \\ 12y-2+2 &= -6+2 \\ 12y &= -4 \\ \frac{1}{12}(12y) &= \frac{1}{12}(-4) \\ y &= -\frac{1}{3} \end{aligned}$$

25. Solving the equation:

$$\begin{aligned} \frac{3}{4}(8x-4)+3 &= \frac{2}{5}(5x+10)-1 \\ 6x-3+3 &= 2x+4-1 \\ 6x &= 2x+3 \\ 6x+(-2x) &= 2x+(-2x)+3 \\ 4x &= 3 \\ \frac{1}{4}(4x) &= \frac{1}{4}(3) \\ x &= \frac{3}{4} \end{aligned}$$

27. Solving the equation:

$$\begin{aligned} 0.06x+0.08(100-x) &= 6.5 \\ 0.06x+8-0.08x &= 6.5 \\ -0.02x+8 &= 6.5 \\ -0.02x+8+(-8) &= 6.5+(-8) \\ -0.02x &= -1.5 \\ \frac{-0.02x}{-0.02} &= \frac{-1.5}{-0.02} \\ x &= 75 \end{aligned}$$

© 2012 Cengage Learning. All Rights Reserved. May not be scanned, copied or duplicated, or posted to a publicly accessible website, in whole or in part.

29. Solving the equation:

$$\begin{aligned}6-5(2a-3)&=1\\6-10a+15&=1\\-10a+21&=1\\-10a+21+(-21)&=1+(-21)\\-10a&=-20\\-\frac{1}{10}(-10a)&=-\frac{1}{10}(-20)\\a&=2\end{aligned}$$

31. Solving the equation:

$$\begin{aligned}0.2x-0.5&=0.5-0.2(2x-13)\\0.2x-0.5&=0.5-0.4x+2.6\\0.2x-0.5&=-0.4x+3.1\\0.2x+0.4x-0.5&=-0.4x+0.4x+3.1\\0.6x-0.5&=3.1\\0.6x-0.5+0.5&=3.1+0.5\\0.6x&=3.6\\\frac{0.6x}{0.6}&=\frac{3.6}{0.6}\\x&=6\end{aligned}$$

33. Solving the equation:

$$\begin{aligned}2(t-3)+3(t-2)&=28\\2t-6+3t-6&=28\\5t-12&=28\\5t-12+12&=28+12\\5t&=40\\\frac{1}{5}(5t)&=\frac{1}{5}(40)\\t&=8\end{aligned}$$

35. Solving the equation:

$$\begin{aligned}5(x-2)-(3x+4)&=3(6x-8)+10\\5x-10-3x-4&=18x-24+10\\2x-14&=18x-14\\2x+(-18x)-14&=18x+(-18x)-14\\-16x-14&=-14\\-16x-14+14&=-14+14\\-16x&=0\\-\frac{1}{16}(-16x)&=-\frac{1}{16}(0)\\x&=0\end{aligned}$$

37. Solving the equation:

$$\begin{aligned}2(5x-3)-(2x-4)&=5-(6x+1)\\10x-6-2x+4&=5-6x-1\\8x-2&=-6x+4\\8x+6x-2&=-6x+6x+4\\14x-2&=4\\14x-2+2&=4+2\\14x&=6\\\frac{1}{14}(14x)&=\frac{1}{14}(6)\\x&=\frac{3}{7}\end{aligned}$$

39. Solving the equation:

$$\begin{aligned}-(3x+1)-(4x-7)&=4-(3x+2)\\-3x-1-4x+7&=4-3x-2\\-7x+6&=-3x+2\\-7x+3x+6&=-3x+3x+2\\-4x+6&=2\\-4x+6+(-6)&=2+(-6)\\-4x&=-4\\-\frac{1}{4}(-4x)&=-\frac{1}{4}(-4)\\x&=1\end{aligned}$$

41. Solving the equation:

$$\begin{aligned}x+(2x-1)&=2\\3x-1&=2\\3x-1+1&=2+1\\3x&=3\\\frac{1}{3}(3x)&=\frac{1}{3}(3)\\x&=1\end{aligned}$$

43. Solving the equation:

$$\begin{aligned}x-(3x+5)&=-3\\x-3x-5&=-3\\-2x-5&=-3\\-2x-5+5&=-3+5\\-2x&=2\\-\frac{1}{2}(-2x)&=-\frac{1}{2}(2)\\x&=-1\end{aligned}$$

45. Solving the equation:

$$\begin{aligned}15&=3(x-1)\\15&=3x-3\\15+3&=3x-3+3\\18&=3x\\\frac{1}{3}(18)&=\frac{1}{3}(3x)\\x&=6\end{aligned}$$

47. Solving the equation:

$$\begin{aligned}4x-(-4x+1)&=5\\4x+4x-1&=5\\8x-1&=5\\8x-1+1&=5+1\\8x&=6\\\frac{1}{8}(8x)&=\frac{1}{8}(6)\\x&=\frac{3}{4}\end{aligned}$$

© 2012 Cengage Learning. All Rights Reserved. May not be scanned, copied or duplicated, or posted to a publicly accessible website, in whole or in part.

49. Solving the equation:

$$\begin{aligned} 5x-8(2x-5)&=7\\ 5x-16x+40&=7\\ -11x+40&=7\\ -11x+40-40&=7-40\\ -11x&=-33\\ -\frac{1}{11}(-11x)&=-\frac{1}{11}(-33)\\ x&=3 \end{aligned}$$

51. Solving the equation:

$$\begin{aligned} 7(2y-1)-6y&=-1\\ 14y-7-6y&=-1\\ 8y-7&=-1\\ 8y-7+7&=-1+7\\ 8y&=6\\ \frac{1}{8}(8y)&=\frac{1}{8}(6)\\ y&=\frac{3}{4} \end{aligned}$$

53. Solving the equation:

$$\begin{aligned} 0.2x+0.5(12-x)&=3.6\\ 0.2x+6-0.5x&=3.6\\ -0.3x+6&=3.6\\ -0.3x+6-6&=3.6-6\\ -0.3x&=-2.4\\ x&=\frac{-2.4}{-0.3}=8 \end{aligned}$$

55. Solving the equation:

$$\begin{aligned} 0.5x+0.2(18-x)&=5.4\\ 0.5x+3.6-0.2x&=5.4\\ 0.3x+3.6&=5.4\\ 0.3x+3.6-3.6&=5.4-3.6\\ 0.3x&=1.8\\ x&=\frac{1.8}{0.3}=6 \end{aligned}$$

57. Solving the equation:

$$\begin{aligned} x+(x+3)(-3)&=x-3\\ x-3x-9&=x-3\\ -2x-9&=x-3\\ -2x-x-9&=x-x-3\\ -3x-9&=-3\\ -3x-9+9&=-3+9\\ -3x&=6\\ -\frac{1}{3}(-3x)&=-\frac{1}{3}(6)\\ x&=-2 \end{aligned}$$

59. Solving the equation:

$$\begin{aligned} 5(x+2)+3(x-1)&=-9\\ 5x+10+3x-3&=-9\\ 8x+7&=-9\\ 8x+7-7&=-9-7\\ 8x&=-16\\ \frac{1}{8}(8x)&=\frac{1}{8}(-16)\\ x&=-2 \end{aligned}$$

61. Solving the equation:

$$\begin{aligned} 3(x-3)+2(2x)&=5\\ 3x-9+4x&=5\\ 7x-9&=5\\ 7x-9+9&=5+9\\ 7x&=14\\ \frac{1}{7}(7x)&=\frac{1}{7}(14)\\ x&=2 \end{aligned}$$

63. Solving the equation:

$$\begin{aligned} 5(y+2)&=4(y+1)\\ 5y+10&=4y+4\\ 5y-4y+10&=4y-4y+4\\ y+10&=4\\ y+10-10&=4-10\\ y&=-6 \end{aligned}$$

65. Solving the equation:

$$\begin{aligned} 3x+2(x-2)&=6\\ 3x+2x-4&=6\\ 5x-4&=6\\ 5x-4+4&=6+4\\ 5x&=10\\ \frac{1}{5}(5x)&=\frac{1}{5}(10)\\ x&=2 \end{aligned}$$

67. Solving the equation:

$$\begin{aligned} 50(x-5)&=30(x+5)\\ 50x-250&=30x+150\\ 50x-30x-250&=30x-30x+150\\ 20x-250&=150\\ 20x-250+250&=150+250\\ 20x&=400\\ \frac{1}{20}(20x)&=\frac{1}{20}(400)\\ x&=20 \end{aligned}$$

© 2012 Cengage Learning. All Rights Reserved. May not be scanned, copied or duplicated, or posted to a publicly accessible website, in whole or in part.

69. Solving the equation:

$$\begin{aligned} 0.08x+0.09(x+2,000)&=860\\ 0.08x+0.09x+180&=860\\ 0.17x+180&=860\\ 0.17x+180-180&=860-180\\ 0.17x&=680\\ x&=\frac{680}{0.17}=4,000 \end{aligned}$$

71. Solving the equation:

$$\begin{aligned} 0.10x+0.12(x+500)&=214\\ 0.10x+0.12x+60&=214\\ 0.22x+60&=214\\ 0.22x+60-60&=214-60\\ 0.22x&=154\\ x&=\frac{154}{0.22}=700 \end{aligned}$$

73. Solving the equation:

$$\begin{aligned} 5x+10(x+8)&=245\\ 5x+10x+80&=245\\ 15x+80&=245\\ 15x+80-80&=245-80\\ 15x&=165\\ x&=\frac{165}{15}=11 \end{aligned}$$

75. Solving the equation:

$$\begin{aligned} 5x+10(x+3)+25(x+5)&=435\\ 5x+10x+30+25x+125&=435\\ 40x+155&=435\\ 40x+155-155&=435-155\\ 40x&=280\\ x&=\frac{280}{40}=7 \end{aligned}$$

77. **a.** Solving the equation:

$$\begin{aligned} 4x-5&=0\\ 4x-5+5&=0+5\\ 4x&=5\\ \frac{1}{4}(4x)&=\frac{1}{4}(5)\\ x&=\frac{5}{4}=1.25 \end{aligned}$$

b. Solving the equation:

$$\begin{aligned} 4x-5&=25\\ 4x-5+5&=25+5\\ 4x&=30\\ \frac{1}{4}(4x)&=\frac{1}{4}(30)\\ x&=\frac{15}{2}=7.5 \end{aligned}$$

c. Adding: $(4x-5)+(2x+25)=6x+20$

d. Solving the equation:

$$\begin{aligned} 4x-5&=2x+25\\ 4x-2x-5&=2x-2x+25\\ 2x-5&=25\\ 2x-5+5&=25+5\\ 2x&=30\\ \frac{1}{2}(2x)&=\frac{1}{2}(30)\\ x&=15 \end{aligned}$$

e. Multiplying: $4(x-5)=4x-20$

f. Solving the equation:

$$\begin{aligned} 4(x-5)&=2x+25\\ 4x-20&=2x+25\\ 4x-2x-20&=2x-2x+25\\ 2x-20&=25\\ 2x-20+20&=25+20\\ 2x&=45\\ \frac{1}{2}(2x)&=\frac{1}{2}(45)\\ x&=\frac{45}{2}=22.5 \end{aligned}$$

79. Solving the equation:

$$\begin{aligned} 48&=7x-13980\\ 48+13980&=7x-13980+13980\\ 14028&=7x\\ x&=\frac{14028}{7}=2004 \end{aligned}$$

The year is 2004.

81. Solving the equation:

$$\begin{aligned} 40&=2x+12\\ 2x+12&=40\\ 2x&=28\\ x&=14 \end{aligned}$$

© 2012 Cengage Learning. All Rights Reserved. May not be scanned, copied or duplicated, or posted to a publicly accessible website, in whole or in part.

83. Solving the equation:
$$\begin{aligned} 12+2y&=6 \\ 2y&=-6 \\ y&=-3 \end{aligned}$$

85. Solving the equation:
$$\begin{aligned} 24x&=6 \\ x&=\frac{6}{24}=\frac{1}{4} \end{aligned}$$

87. Solving the equation:
$$\begin{aligned} 70&=x\bullet 210 \\ x&=\frac{70}{210}=\frac{1}{3} \end{aligned}$$

89. Simplifying: $\frac{1}{2}(-3x+6)=\frac{1}{2}(-3x)+\frac{1}{2}(6)=-\frac{3}{2}x+3$

91. Multiplying the fractions: $\frac{1}{2}(3)=\frac{1}{2}\bullet\frac{3}{1}=\frac{3}{2}$

93. Multiplying the fractions: $\frac{2}{3}(6)=\frac{2}{3}\bullet\frac{6}{1}=\frac{12}{3}=4$

95. Multiplying the fractions: $\frac{5}{9}\bullet\frac{9}{5}=\frac{45}{45}=1$

97. Completing the table:

x	$3(x+2)$	$3x+2$	$3x+6$
0	6	2	6
1	9	5	9
2	12	8	12
3	15	11	15

99. Completing the table:

a	$(2a+1)^2$	$4a^2+4a+1$
1	9	9
2	25	25
3	49	49

2.5 Formulas

1. Using the perimeter formula:
$$\begin{aligned} P&=2l+2w \\ 300&=2l+2(50) \\ 300&=2l+100 \\ 200&=2l \\ l&=100 \end{aligned}$$
The length is 100 feet.

3. **a.** Substituting $x=3$:
$$\begin{aligned} 2(0)+3y&=6 \\ 0+3y&=6 \\ 3y&=6 \\ y&=2 \end{aligned}$$

b. Substituting $y=1$:
$$\begin{aligned} 2x+3(1)&=6 \\ 2x+3&=6 \\ 2x&=3 \\ x&=\frac{3}{2} \end{aligned}$$

c. Substituting $x=3$:
$$\begin{aligned} 2(3)+3y&=6 \\ 6+3y&=6 \\ 6+(-6)+3y&=6+(-6) \\ 3y&=0 \\ y&=0 \end{aligned}$$

© 2012 Cengage Learning. All Rights Reserved. May not be scanned, copied or duplicated, or posted to a publicly accessible website, in whole or in part.

5. **a.** Substituting $x = 0$: $y = -\frac{1}{3}(0) + 2 = 0 + 2 = 2$

b. Substituting $y = 3$:

$$\begin{aligned} 3 &= -\frac{1}{3}x + 2 \\ 1 &= -\frac{1}{3}x \\ 3 &= -x \\ x &= -3 \end{aligned}$$

c. Substituting $x = 3$: $y = -\frac{1}{3}(3) + 2 = -1 + 2 = 1$

7. Substituting $x = 3$:

$$\begin{aligned} 3(3) + 3y &= 6 \\ 9 + 3y &= 6 \\ 9 + (-9) + 3y &= 6 + (-9) \\ 3y &= -3 \\ y &= -1 \end{aligned}$$

9. Substituting $x = 0$:

$$\begin{aligned} 3(0) + 3y &= 6 \\ 0 + 3y &= 6 \\ 3y &= 6 \\ y &= 2 \end{aligned}$$

11. Substituting $y = 2$:

$$\begin{aligned} 2x - 2(2) &= 20 \\ 2x - 4 &= 20 \\ 2x - 4 + 4 &= 20 + 4 \\ 2x &= 24 \\ x &= 12 \end{aligned}$$

13. Substituting $y = 0$:

$$\begin{aligned} 2x - 2(0) &= 20 \\ 2x - 0 &= 20 \\ 2x &= 20 \\ x &= 10 \end{aligned}$$

15. Substituting $x = -2$: $y = (-2+1)^2 - 3 = (-1)^2 - 3 = 1 - 3 = -2$

17. Substituting $x = 1$: $y = (1+1)^2 - 3 = (2)^2 - 3 = 4 - 3 = 1$

19. **a.** Substituting $x = 10$: $y = \frac{20}{10} = 2$

b. Substituting $x = 5$: $y = \frac{20}{5} = 4$

21. **a.** Substituting $y = 15$ and $x = 3$:

$$\begin{aligned} 15 &= K(3) \\ K &= \frac{15}{3} = 5 \end{aligned}$$

b. Substituting $y = 72$ and $x = 4$:

$$\begin{aligned} 72 &= K(4) \\ K &= \frac{72}{4} = 18 \end{aligned}$$

23. **a.** Substituting $x = 5$ and $y = 4$:

$$\begin{aligned} 4 &= \frac{K}{5} \\ K &= 20 \end{aligned}$$

b. Substituting $x = 5$ and $y = 15$:

$$\begin{aligned} 15 &= \frac{K}{5} \\ K &= 75 \end{aligned}$$

25. Solving for l:

$$\begin{aligned} lw &= A \\ \frac{lw}{w} &= \frac{A}{w} \\ l &= \frac{A}{w} \end{aligned}$$

27. Solving for h:

$$\begin{aligned} lwh &= V \\ \frac{lwh}{lw} &= \frac{V}{lw} \\ h &= \frac{V}{lw} \end{aligned}$$

29. Solving for a:

$$\begin{aligned} a + b + c &= P \\ a + b + c - b - c &= P - b - c \\ a &= P - b - c \end{aligned}$$

31. Solving for x:

$$\begin{aligned} x - 3y &= -1 \\ x - 3y + 3y &= -1 + 3y \\ x &= 3y - 1 \end{aligned}$$

© 2012 Cengage Learning. All Rights Reserved. May not be scanned, copied or duplicated, or posted to a publicly accessible website, in whole or in part.

33. Solving for y:

$$\begin{aligned} -3x+y&=6 \\ -3x+3x+y&=6+3x \\ y&=3x+6 \end{aligned}$$

35. Solving for y:

$$\begin{aligned} 2x+3y&=6 \\ -2x+2x+3y&=-2x+6 \\ 3y&=-2x+6 \\ \frac{1}{3}(3y)&=\frac{1}{3}(-2x+6) \\ y&=-\frac{2}{3}x+2 \end{aligned}$$

37. Solving for y:

$$\begin{aligned} y-3&=-2(x+4) \\ y-3&=-2x-8 \\ y&=-2x-5 \end{aligned}$$

39. Solving for y:

$$\begin{aligned} y-3&=-\frac{2}{3}(x+3) \\ y-3&=-\frac{2}{3}x-2 \\ y&=-\frac{2}{3}x+1 \end{aligned}$$

41. Solving for w:

$$\begin{aligned} 2l+2w&=P \\ 2l-2l+2w&=P-2l \\ 2w&=P-2l \\ \frac{2w}{2}&=\frac{P-2l}{2} \\ w&=\frac{P-2l}{2} \end{aligned}$$

43. Solving for v:

$$\begin{aligned} vt+16t^2&=h \\ vt+16t^2-16t^2&=h-16t^2 \\ vt&=h-16t^2 \\ \frac{vt}{t}&=\frac{h-16t^2}{t} \\ v&=\frac{h-16t^2}{t} \end{aligned}$$

45. Solving for h:

$$\begin{aligned} \pi r^2+2\pi rh&=A \\ \pi r^2-\pi r^2+2\pi rh&=A-\pi r^2 \\ 2\pi rh&=A-\pi r^2 \\ \frac{2\pi rh}{2\pi r}&=\frac{A-\pi r^2}{2\pi r} \\ h&=\frac{A-\pi r^2}{2\pi r} \end{aligned}$$

47. **a.** Solving for y:

$$\begin{aligned} y-3&=-2(x+4) \\ y-3&=-2x-8 \\ y&=-2x-5 \end{aligned}$$

b. Solving for y:

$$\begin{aligned} y-5&=4(x-3) \\ y-5&=4x-12 \\ y&=4x-7 \end{aligned}$$

49. **a.** Solving for y:

$$\begin{aligned} y-1&=\frac{3}{4}(x-1) \\ y-1&=\frac{3}{4}x-\frac{3}{4} \\ y&=\frac{3}{4}x+\frac{1}{4} \end{aligned}$$

b. Solving for y:

$$\begin{aligned} y+2&=\frac{3}{4}(x-4) \\ y+2&=\frac{3}{4}x-3 \\ y&=\frac{3}{4}x-5 \end{aligned}$$

51. **a.** Solving for y:

$$\begin{aligned} \frac{y-1}{x}&=\frac{3}{5} \\ 5y-5&=3x \\ 5y&=3x+5 \\ y&=\frac{3}{5}x+1 \end{aligned}$$

b. Solving for y:

$$\begin{aligned} \frac{y-2}{x}&=\frac{1}{2} \\ 2y-4&=x \\ 2y&=x+4 \\ y&=\frac{1}{2}x+2 \end{aligned}$$

© 2012 Cengage Learning. All Rights Reserved. May not be scanned, copied or duplicated, or posted to a publicly accessible website, in whole or in part.

53. Solving for y:

$$\frac{x}{7}-\frac{y}{3}=1$$
$$-\frac{x}{7}+\frac{x}{7}-\frac{y}{3}=-\frac{x}{7}+1$$
$$-\frac{y}{3}=-\frac{x}{7}+1$$
$$-3\left(-\frac{y}{3}\right)=-3\left(-\frac{x}{7}+1\right)$$
$$y=\frac{3}{7}x-3$$

55. Solving for y:

$$-\frac{1}{4}x+\frac{1}{8}y=1$$
$$-\frac{1}{4}x+\frac{1}{4}x+\frac{1}{8}y=1+\frac{1}{4}x$$
$$\frac{1}{8}y=\frac{1}{4}x+1$$
$$8\left(\frac{1}{8}y\right)=8\left(\frac{1}{4}x+1\right)$$
$$y=2x+8$$

57. **a.** Solving the equation:

$$4x+5=20$$
$$4x+5-5=20-5$$
$$4x=15$$
$$\frac{1}{4}(4x)=\frac{1}{4}(15)$$
$$x=\frac{15}{4}=3.75$$

b. Evaluating when $x = 3$:

$$4x+5=4(3)+5=12+5=17$$

c. Solving for y:

$$4x+5y=20$$
$$4x-4x+5y=20-4x$$
$$5y=-4x+20$$
$$\frac{1}{5}(5y)=\frac{1}{5}(-4x+20)$$
$$y=-\frac{4}{5}x+4$$

d. Solving for x:

$$4x+5y=20$$
$$4x+5y-5y=20-5y$$
$$4x=-5y+20$$
$$\frac{1}{4}(4x)=\frac{1}{4}(-5y+20)$$
$$x=-\frac{5}{4}y+5$$

59. The complement of 30° is 90° – 30° = 60°, and the supplement is 180° – 30° = 150°.

61. The complement of 45° is 90° – 45° = 45°, and the supplement is 180° – 45° = 135°.

63. Translating into an equation and solving:

$$x=0.25\bullet 40$$
$$x=10$$

The number 10 is 25% of 40.

65. Translating into an equation and solving:

$$x=0.12\bullet 2000$$
$$x=240$$

The number 240 is 12% of 2000.

67. Translating into an equation and solving:

$$x\bullet 28=7$$
$$28x=7$$
$$\frac{1}{28}(28x)=\frac{1}{28}(7)$$
$$x=0.25=25\%$$

The number 7 is 25% of 28.

69. Translating into an equation and solving:

$$x\bullet 40=14$$
$$40x=14$$
$$\frac{1}{40}(40x)=\frac{1}{40}(14)$$
$$x=0.35=35\%$$

The number 14 is 35% of 40.

71. Translating into an equation and solving:

$$0.50\bullet x=32$$
$$\frac{0.50x}{0.50}=\frac{32}{0.50}$$
$$x=64$$

The number 32 is 50% of 64.

73. Translating into an equation and solving:

$$0.12\bullet x=240$$
$$\frac{0.12x}{0.12}=\frac{240}{0.12}$$
$$x=2{,}000$$

The number 240 is 12% of 2,000.

75. Substituting $F=212$: $C=\frac{5}{9}(212-32)=\frac{5}{9}(180)=100°\text{C}$. This value agrees with the information in Table 1.

77. Substituting $F=68$: $C=\frac{5}{9}(68-32)=\frac{5}{9}(36)=20°\text{C}$. This value agrees with the information in Table 1.

© 2012 Cengage Learning. All Rights Reserved. May not be scanned, copied or duplicated, or posted to a publicly accessible website, in whole or in part.

79. Solving for C:

$$\frac{9}{5}C+32=F$$
$$\frac{9}{5}C+32-32=F-32$$
$$\frac{9}{5}C=F-32$$
$$\frac{5}{9}\left(\frac{9}{5}C\right)=\frac{5}{9}(F-32)$$
$$C=\frac{5}{9}(F-32)$$

81. We need to find what percent of 36.5 is 3.80 + 1.26 = 5.06:

$$x \bullet 36.5=5.06$$
$$x=\frac{5.06}{36.5}$$
$$x \approx 0.139=13.9\%$$

83. We need to find what percent of 150 is 90:

$$x \bullet 150=90$$
$$\frac{1}{150}(150x)=\frac{1}{150}(90)$$
$$x=0.60=60\%$$

So 60% of the calories in one serving of vanilla ice cream are fat calories.

85. We need to find what percent of 98 is 26:

$$x \bullet 98=26$$
$$\frac{1}{98}(98x)=\frac{1}{98}(26)$$
$$x \approx 0.265=26.5\%$$

So 26.5% of one serving of frozen yogurt are carbohydrates.

87. Solving for r:

$$2\pi r=C$$
$$2 \bullet \frac{22}{7}r=44$$
$$\frac{44}{7}r=44$$
$$\frac{7}{44}\left(\frac{44}{7}r\right)=\frac{7}{44}(44)$$
$$r=7 \text{ meters}$$

89. Solving for r:

$$2\pi r=C$$
$$2 \bullet 3.14r=9.42$$
$$6.28r=9.42$$
$$\frac{6.28r}{6.28}=\frac{9.42}{6.28}$$
$$r=1.5 \text{ inches}$$

91. Solving for h:

$$\pi r^2 h=V$$
$$\frac{22}{7}\left(\frac{7}{22}\right)^2 h=42$$
$$\frac{7}{22}h=42$$
$$\frac{22}{7}\left(\frac{7}{22}h\right)=\frac{22}{7}(42)$$
$$h=132 \text{ feet}$$

93. Solving for h:

$$\pi r^2 h=V$$
$$3.14(3)^2 h=6.28$$
$$28.26h=6.28$$
$$\frac{28.26h}{28.26}=\frac{6.28}{28.26}$$
$$h=\frac{2}{9} \text{ centimeters}$$

95. The sum of 4 and 1 is 5.

97. The difference of 6 and 2 is 4.

99. The difference of a number and 5 is –12.

101. The sum of a number and 3 is four times the difference of that number and 3.

103. An equivalent expression is: $2(6+3)=2(9)=18$

105. An equivalent expression is: $2(5)+3=10+3=13$

© 2012 Cengage Learning. All Rights Reserved. May not be scanned, copied or duplicated, or posted to a publicly accessible website, in whole or in part.

107. An equivalent expression is: $x+5=13$

109. An equivalent expression is: $5(x+7)=30$

111. **a.** Simplifying: $27-(-68)=27+68=95$

b. Simplifying: $27+(-68)=27-68=-41$

c. Simplifying: $-27-68=-95$

d. Simplifying: $-27+68=41$

113. **a.** Simplifying: $-32-(-41)=-32+41=9$

b. Simplifying: $-32+(-41)=-32-41=-73$

c. Simplifying: $-32+41=9$

d. Simplifying: $-32-41=-73$

2.6 Applications

1. Let x represent the number. The equation is:

$$x+5=13$$
$$x=8$$

The number is 8.

3. Let x represent the number. The equation is:

$$2x+4=14$$
$$2x=10$$
$$x=5$$

The number is 5.

5. Let x represent the number. The equation is:

$$5(x+7)=30$$
$$5x+35=30$$
$$5x=-5$$
$$x=-1$$

The number is -1.

7. Let x and $x+2$ represent the two numbers. The equation is:

$$x+x+2=8$$
$$2x+2=8$$
$$2x=6$$
$$x=3$$
$$x+2=5$$

The two numbers are 3 and 5.

9. Let x and $3x-4$ represent the two numbers. The equation is:

$$(x+3x-4)+5=25$$
$$4x+1=25$$
$$4x=24$$
$$x=6$$
$$3x-4=3(6)-4=14$$

The two numbers are 6 and 14.

11. Completing the table:

	Four Years Ago	Now
Shelly	$x+3-4=x-1$	$x+3$
Michele	$x-4$	x

The equation is:

$$x-1+x-4=67$$
$$2x-5=67$$
$$2x=72$$
$$x=36$$
$$x+3=39$$

Shelly is 39 and Michele is 36.

13. Completing the table:

	Three Years Ago	Now
Cody	$2x-3$	$2x$
Evan	$x-3$	x

The equation is:

$$2x-3+x-3=27$$
$$3x-6=27$$
$$3x=33$$
$$x=11$$
$$2x=22$$

Evan is 11 and Cody is 22.

© 2012 Cengage Learning. All Rights Reserved. May not be scanned, copied or duplicated, or posted to a publicly accessible website, in whole or in part.

15. Completing the table:

	Five Years Ago	Now
Fred	$x+4-5=x-1$	$x+4$
Barney	$x-5$	x

The equation is:

$$\begin{aligned} x-1+x-5&=48\\ 2x-6&=48\\ 2x&=54\\ x&=27\\ x+4&=31 \end{aligned}$$

Barney is 27 and Fred is 31.

17. Completing the table:

	Now	Three Years from Now
Jack	$2x$	$2x+3$
Lacy	x	$x+3$

The equation is:

$$\begin{aligned} 2x+3+x+3&=54\\ 3x+6&=54\\ 3x&=48\\ x&=16\\ 2x&=32 \end{aligned}$$

Lacy is 16 and Jack is 32.

19. Completing the table:

	Now	Two Years from Now
Pat	$x+20$	$x+20+2=x+22$
Patrick	x	$x+2$

The equation is:

$$\begin{aligned} x+22&=2(x+2)\\ x+22&=2x+4\\ 22&=x+4\\ x&=18\\ x+20&=38 \end{aligned}$$

Patrick is 18 and Pat is 38.

21. Using the formula $P=4s$, the equation is:

$$\begin{aligned} 4s&=36\\ s&=9 \end{aligned}$$

The length of each side is 9 inches.

23. Using the formula $P=4s$, the equation is:

$$\begin{aligned} 4s&=60\\ s&=15 \end{aligned}$$

The length of each side is 15 feet.

25. Let x, $3x$, and $x+7$ represent the sides of the triangle. The equation is:

$$\begin{aligned} x+3x+x+7&=62\\ 5x+7&=62\\ 5x&=55\\ x&=11\\ x+7&=18\\ 3x&=33 \end{aligned}$$

The sides are 11 feet, 18 feet, and 33 feet.

27. Let x, $2x$, and $2x-12$ represent the sides of the triangle. The equation is:

$$\begin{aligned} x+2x+2x-12&=53\\ 5x-12&=53\\ 5x&=65\\ x&=13\\ 2x&=26\\ 2x-12&=14 \end{aligned}$$

The sides are 13 feet, 14 feet, and 26 feet.

29. Let w represent the width and $w+5$ represent the length. The equation is:

$$\begin{aligned} 2w+2(w+5)&=34\\ 2w+2w+10&=34\\ 4w+10&=34\\ 4w&=24\\ w&=6\\ w+5&=11 \end{aligned}$$

The length is 11 inches and the width is 6 inches.

© 2012 Cengage Learning. All Rights Reserved. May not be scanned, copied or duplicated, or posted to a publicly accessible website, in whole or in part.

31. Let w represent the width and $2w + 7$ represent the length. The equation is:

$$2w+2(2w+7)=68$$
$$2w+4w+14=68$$
$$6w+14=68$$
$$6w=54$$
$$w=9$$
$$2w+7=2(9)+7=25$$

The length is 25 inches and the width is 9 inches.

33. Let w represent the width and $3w + 6$ represent the length. The equation is:

$$2w+2(3w+6)=36$$
$$2w+6w+12=36$$
$$8w+12=36$$
$$8w=24$$
$$w=3$$
$$3w+6=3(3)+6=15$$

The length is 15 feet and the width is 3 feet.

35. Completing the table:

	Dimes	Quarters
Number	x	$x+5$
Value (cents)	$10(x)$	$25(x+5)$

The equation is:

$$10(x)+25(x+5)=440$$
$$10x+25x+125=440$$
$$35x+125=440$$
$$35x=315$$
$$x=9$$
$$x+5=14$$

Marissa has 9 dimes and 14 quarters.

37. Completing the table:

	Nickels	Quarters
Number	$x+15$	x
Value (cents)	$5(x+15)$	$25(x)$

The equation is:

$$5(x+15)+25(x)=435$$
$$5x+75+25x=435$$
$$30x+75=435$$
$$30x=360$$
$$x=12$$
$$x+15=27$$

Tanner has 12 quarters and 27 nickels.

39. Completing the table:

	Nickels	Dimes
Number	x	$x+9$
Value (cents)	$5(x)$	$10(x+9)$

The equation is:

$$5(x)+10(x+9)=210$$
$$5x+10x+90=210$$
$$15x+90=210$$
$$15x=120$$
$$x=8$$
$$x+9=17$$

Sue has 8 nickels and 17 dimes.

41. Completing the table:

	Nickels	Dimes	Quarters
Number	x	$x+3$	$x+5$
Value (cents)	$5(x)$	$10(x+3)$	$25(x+5)$

The equation is:

$$5(x)+10(x+3)+25(x+5)=435$$
$$5x+10x+30+25x+125=435$$
$$40x+155=435$$
$$40x=280$$
$$x=7$$
$$x+3=10$$
$$x+5=12$$

Katie has 7 nickels, 10 dimes, and 12 quarters.

© 2012 Cengage Learning. All Rights Reserved. May not be scanned, copied or duplicated, or posted to a publicly accessible website, in whole or in part.

43. Completing the table:

	Nickels	Dimes	Quarters
Number	x	$x+6$	$2x$
Value (cents)	$5(x)$	$10(x+6)$	$25(2x)$

The equation is:

$$\begin{aligned} 5(x)+10(x+6)+25(2x) &= 255 \\ 5x+10x+60+50x &= 255 \\ 65x+60 &= 255 \\ 65x &= 195 \\ x &= 3 \\ x+6 &= 9 \\ 2x &= 6 \end{aligned}$$

Cory has 3 nickels, 9 dimes, and 6 quarters.

45. Simplifying: $x+2x+2x=5x$

47. Simplifying: $x+0.075x=1.075x$

49. Simplifying: $0.09(x+2{,}000)=0.09x+180$

51. Solving the equation:

$$\begin{aligned} 0.05x+0.06(x-1{,}500) &= 570 \\ 0.05x+0.06x-90 &= 570 \\ 0.11x &= 660 \\ x &= 6{,}000 \end{aligned}$$

53. Solving the equation:

$$\begin{aligned} x+2x+3x &= 180 \\ 6x &= 180 \\ x &= 30 \end{aligned}$$

55. The statement is: 4 is less than 10

57. The statement is: 9 is greater than or equal to –5

59. The correct statement is: $12 < 20$

61. The correct statement is: $-8 < -6$

63. Simplifying: $|8-3|-|5-2|=|5|-|3|=5-3=2$

65. Simplifying: $15-|9-3(7-5)|=15-|9-3(2)|=15-|9-6|=15-|3|=15-3=12$

67.
- **a.** The letter e has the largest area.
- **b.** The letter i has a larger area than f.
- **c.** Probably the letter x which is used in equations.
- **d.** The larger the area of space used to store the letter, the more often it is used.

2.7 More Applications

1. Let x and $x + 1$ represent the two numbers. The equation is:

$$\begin{aligned} x+x+1 &= 11 \\ 2x+1 &= 11 \\ 2x &= 10 \\ x &= 5 \\ x+1 &= 6 \end{aligned}$$

The numbers are 5 and 6.

3. Let x and $x + 1$ represent the two numbers. The equation is:

$$\begin{aligned} x+x+1 &= -9 \\ 2x+1 &= -9 \\ 2x &= -10 \\ x &= -5 \\ x+1 &= -4 \end{aligned}$$

The numbers are –5 and –4.

5. Let x and $x + 2$ represent the two numbers. The equation is:

$$\begin{aligned} x+x+2 &= 28 \\ 2x+2 &= 28 \\ 2x &= 26 \\ x &= 13 \\ x+2 &= 15 \end{aligned}$$

The numbers are 13 and 15.

© 2012 Cengage Learning. All Rights Reserved. May not be scanned, copied or duplicated, or posted to a publicly accessible website, in whole or in part.

7. Let x and $x + 2$ represent the two numbers. The equation is:

$$\begin{aligned} x+x+2 &= 106 \\ 2x+2 &= 106 \\ 2x &= 104 \\ x &= 52 \\ x+2 &= 54 \end{aligned}$$

The numbers are 52 and 54.

9. Let x and $x + 2$ represent the two numbers. The equation is:

$$\begin{aligned} x+x+2 &= -30 \\ 2x+2 &= -30 \\ 2x &= -322 \\ x &= -16 \\ x+2 &= -14 \end{aligned}$$

The numbers are –16 and –14.

11. Let x, $x + 2$, and $x + 4$ represent the three numbers. The equation is:

$$\begin{aligned} x+x+2+x+4 &= 57 \\ 3x+6 &= 57 \\ 3x &= 51 \\ x &= 17 \\ x+2 &= 19 \\ x+4 &= 21 \end{aligned}$$

The numbers are 17, 19, and 21.

13. Let x, $x + 2$, and $x + 4$ represent the three numbers. The equation is:

$$\begin{aligned} x+x+2+x+4 &= 132 \\ 3x+6 &= 132 \\ 3x &= 126 \\ x &= 42 \\ x+2 &= 44 \\ x+4 &= 46 \end{aligned}$$

The numbers are 42, 44, and 46.

15. Completing the table:

	Dollars Invested at 8%	Dollars Invested at 9%
Number of	x	$x+2000$
Interest on	$0.08(x)$	$0.09(x+2000)$

The equation is:

$$\begin{aligned} 0.08(x)+0.09(x+2000) &= 860 \\ 0.08x+0.09x+180 &= 860 \\ 0.17x+180 &= 860 \\ 0.17x &= 680 \\ x &= 4{,}000 \\ x+2{,}000 &= 6{,}000 \end{aligned}$$

You have \$4,000 invested at 8% and \$6,000 invested at 9%.

© 2012 Cengage Learning. All Rights Reserved. May not be scanned, copied or duplicated, or posted to a publicly accessible website, in whole or in part.

17. Completing the table:

	Dollars Invested at 10%	Dollars Invested at 12%
Number of	x	$x+500$
Interest on	$0.10(x)$	$0.12(x+500)$

The equation is:

$$\begin{aligned} 0.10(x)+0.12(x+500) &= 214 \\ 0.10x+0.12x+60 &= 214 \\ 0.22x+60 &= 214 \\ 0.22x &= 154 \\ x &= 700 \\ x+500 &= 1,200 \end{aligned}$$

Tyler has \$700 invested at 10% and \$1,200 invested at 12%.

19. Completing the table:

	Dollars Invested at 8%	Dollars Invested at 9%	Dollars Invested at 10%
Number of	x	$2x$	$3x$
Interest on	$0.08(x)$	$0.09(2x)$	$0.10(3x)$

The equation is:

$$\begin{aligned} 0.08(x)+0.09(2x)+0.10(3x) &= 280 \\ 0.08x+0.18x+0.30x &= 280 \\ 0.56x &= 280 \\ x &= 500 \\ 2x &= 1,000 \\ 3x &= 1,500 \end{aligned}$$

She has \$500 invested at 8%, \$1,000 invested at 9%, and \$1,500 invested at 10%.

21. Let x represent the measure of the two equal angles, so $x+x=2x$ represents the measure of the third angle. Since the sum of the three angles is 180°, the equation is:

$$\begin{aligned} x+x+2x &= 180^\circ \\ 4x &= 180^\circ \\ x &= 45^\circ \\ 2x &= 90^\circ \end{aligned}$$

The measures of the three angles are 45°, 45°, and 90°.

23. Let x represent the measure of the largest angle. Then $\frac{1}{5}x$ represents the measure of the smallest angle, and $2\left(\frac{1}{5}x\right)=\frac{2}{5}x$ represents the measure of the other angle. Since the sum of the three angles is 180°, the equation is:

$$\begin{aligned} x+\frac{1}{5}x+\frac{2}{5}x &= 180^\circ \\ \frac{5}{5}x+\frac{1}{5}x+\frac{2}{5}x &= 180^\circ \\ \frac{8}{5}x &= 180^\circ \\ x &= 112.5^\circ \\ \frac{1}{5}x &= 22.5^\circ \\ \frac{2}{5}x &= 45^\circ \end{aligned}$$

The measures of the three angles are 22.5°, 45°, and 112.5°.

© 2012 Cengage Learning. All Rights Reserved. May not be scanned, copied or duplicated, or posted to a publicly accessible website, in whole or in part.

25. Let x represent the measure of the other acute angle, and 90° is the measure of the right angle. Since the sum of the three angles is 180°, the equation is:

$$\begin{aligned} x+37°+90° &= 180° \\ x+127° &= 180° \\ x &= 53° \end{aligned}$$

The other two angles are 53° and 90°.

27. Let x represent the measure of the smallest angle, so $x + 20$ represents the measure of the second angle and $2x$ represents the measure of the third angle. Since the sum of the three angles is 180°, the equation is:

$$\begin{aligned} x+x+20+2x &= 180° \\ 4x+20 &= 180° \\ 4x &= 160° \\ x &= 40° \\ x+20 &= 60° \\ 2x &= 80° \end{aligned}$$

The measures of the three angles are 40°, 60°, and 80°.

29. Completing the table:

	Adult	Child
Number	x	$x+6$
Income	$6(x)$	$4(x+6)$

The equation is:

$$\begin{aligned} 6(x)+4(x+6) &= 184 \\ 6x+4x+24 &= 184 \\ 10x+24 &= 184 \\ 10x &= 160 \\ x &= 16 \\ x+6 &= 22 \end{aligned}$$

Miguel sold 16 adult and 22 children's tickets.

31. Let x represent the total minutes for the call. Then \$0.41 is charged for the first minute, and \$0.32 is charged for the additional $x - 1$ minutes. The equation is:

$$\begin{aligned} 0.41(1)+0.32(x-1) &= 5.21 \\ 0.41+0.32x-0.32 &= 5.21 \\ 0.32x+0.09 &= 5.21 \\ 0.32x &= 5.12 \\ x &= 16 \end{aligned}$$

The call was 16 minutes long.

33. Let x represent the hours Jo Ann worked that week. Then \$12/hour is paid for the first 35 hours and \$18/hour is paid for the additional $x - 35$ hours. The equation is:

$$\begin{aligned} 12(35)+18(x-35) &= 492 \\ 420+18x-630 &= 492 \\ 18x-210 &= 492 \\ 18x &= 702 \\ x &= 39 \end{aligned}$$

Jo Ann worked 39 hours that week.

35. Let x and $x + 2$ represent the two office numbers. The equation is:

$$\begin{aligned} x+x+2 &= 14,660 \\ 2x+2 &= 14,660 \\ 2x &= 14,658 \\ x &= 7329 \\ x+2 &= 7331 \end{aligned}$$

They are in offices 7329 and 7331.

© 2012 Cengage Learning. All Rights Reserved. May not be scanned, copied or duplicated, or posted to a publicly accessible website, in whole or in part.

37. Let x represent Kendra's age and $x + 2$ represent Marissa's age. The equation is:

$$\begin{aligned} x+2+2x&=26 \\ 3x+2&=26 \\ 3x&=24 \\ x&=8 \\ x+2&=10 \end{aligned}$$

Kendra is 8 years old and Marissa is 10 years old.

39. For Jeff, the total time traveled is $\frac{425 \text{ miles}}{55 \text{ miles/hour}} \approx 7.72 \text{ hours} \approx 463 \text{ minutes}$. Since he left at 11:00 AM, he will arrive at 6:43 PM. For Carla, the total time traveled is $\frac{425 \text{ miles}}{65 \text{ miles/hour}} \approx 6.54 \text{ hours} \approx 392 \text{ minutes}$. Since she left at 1:00 PM, she will arrive at 7:32 PM. Thus Jeff will arrive in Lake Tahoe first.

41. Since $\frac{1}{5} \text{ mile} = 0.2 \text{ mile}$, the taxi charge is \$1.25 for the first $\frac{1}{5}$ mile and \$0.25 per fifth mile for the remaining 7.3 miles. Since $7.3 \text{ miles} = \frac{7.3}{0.2} = 36.5 \text{ fifths}$, the total charge is: $\$1.25 + \$0.25(36.5) \approx \$10.38$

43. Let w represent the width and $w + 2$ represent the length. The equation is:

$$\begin{aligned} 2w+2(w+2)&=44 \\ 2w+2w+4&=44 \\ 4w+4&=44 \\ 4w&=40 \\ w&=10 \\ w+2&=12 \end{aligned}$$

The length is 12 meters and the width is 10 meters.

45. Let x, $x + 1$, and $x + 2$ represent the measures of the three angles. Since the sum of the three angles is 180°, the equation is:

$$\begin{aligned} x+x+1+x+2&=180^\circ \\ 3x+3&=180^\circ \\ 3x&=177^\circ \\ x&=59^\circ \\ x+1&=60^\circ \\ x+2&=61^\circ \end{aligned}$$

The measures of the three angles are 59°, 60°, and 61°.

47. If all 36 people are Elk's Lodge members (which would be the least amount), the cost of the lessons would be $\$3(36) = \$108$. Since half of the money is paid to Ike and Nancy, the least amount they could make is $\frac{1}{2}(\$108) = \$54$.

49. Yes. The total receipts were \$160, which is possible if there were 10 Elk's members and 26 nonmembers. Computing the total receipts: $10(\$3) + 26(\$5) = \$30 + \$130 = \$160$

51.

a. Solving the equation:

$$\begin{aligned} x-3&=6 \\ x&=9 \end{aligned}$$

b. Solving the equation:

$$\begin{aligned} x+3&=6 \\ x&=3 \end{aligned}$$

c. Solving the equation:

$$\begin{aligned} -x-3&=6 \\ -x&=9 \\ x&=-9 \end{aligned}$$

d. Solving the equation:

$$\begin{aligned} -x+3&=6 \\ -x&=3 \\ x&=-3 \end{aligned}$$

© 2012 Cengage Learning. All Rights Reserved. May not be scanned, copied or duplicated, or posted to a publicly accessible website, in whole or in part.

53. **a.** Solving the equation:

$$\frac{x}{4}=-2$$
$$x=-2(4)=-8$$

b. Solving the equation:

$$-\frac{x}{4}=-2$$
$$x=-2(-4)=8$$

c. Solving the equation:

$$\frac{x}{4}=2$$
$$x=2(4)=8$$

d. Solving the equation:

$$-\frac{x}{4}=2$$
$$x=2(-4)=-8$$

55. Solving the equation:

$$2.5x-3.48=4.9x+2.07$$
$$-2.4x-3.48=2.07$$
$$-2.4x=5.55$$
$$x=-2.3125$$

57. Solving the equation:

$$3(x-4)=-2$$
$$3x-12=-2$$
$$3x=10$$
$$x=\frac{10}{3}$$

59. Substituting: $36x-12=36\left(\frac{1}{4}\right)-12=9-12=-3$

61. Substituting: $36x-12=36\left(\frac{1}{9}\right)-12=4-12=-8$

63. Substituting: $36x-12=36\left(\frac{1}{3}\right)-12=12-12=0$

65. Substituting: $36x-12=36\left(\frac{5}{9}\right)-12=20-12=8$

67. Evaluating when $x=-4$: $3(x-4)=3(-4-4)=3(-8)=-24$

69. Evaluating when $x=-4$: $-5x+8=-5(-4)+8=20+8=28$

71. Evaluating when $x=-4$: $\frac{x-14}{36}=\frac{-4-14}{36}=-\frac{18}{36}=-\frac{1}{2}$

73. Evaluating when $x=-4$: $\frac{16}{x}+3x=\frac{16}{-4}+3(-4)=-4-12=-16$

75. Evaluating when $x=-4$: $7x-\frac{12}{x}=7(-4)-\frac{12}{-4}=-28+3=-25$

77. Evaluating when $x=-4$: $8\left(\frac{x}{2}+5\right)=8\left(\frac{-4}{2}+5\right)=8(-2+5)=8(3)=24$

2.8 Linear Inequalities

1. Solving the inequality:

$$x-5<7$$
$$x-5+5<7+5$$
$$x<12$$

Graphing the solution set:

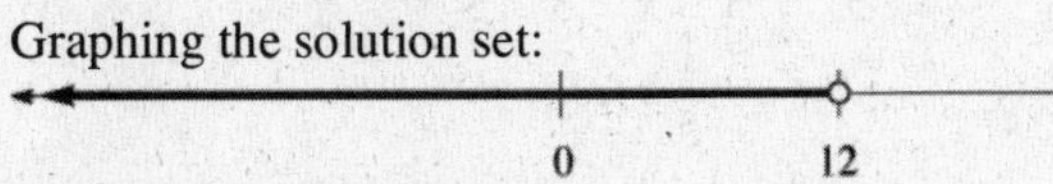

3. Solving the inequality:

$$a-4\le 8$$
$$a-4+4\le 8+4$$
$$a\le 12$$

Graphing the solution set:

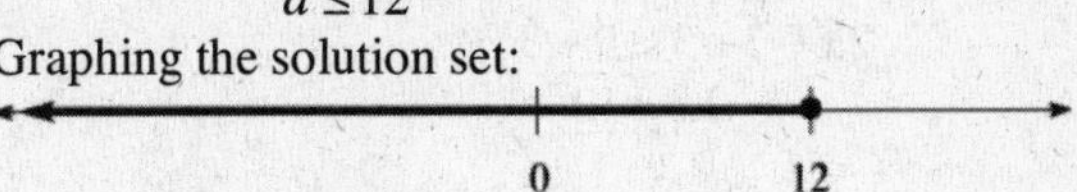

5. Solving the inequality:

$$x-4.3>8.7$$
$$x-4.3+4.3>8.7+4.3$$
$$x>13$$

Graphing the solution set:

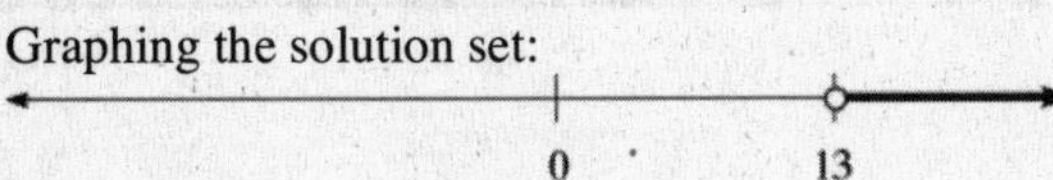

7. Solving the inequality:

$$y+6\ge 10$$
$$y+6+(-6)\ge 10+(-6)$$
$$y\ge 4$$

Graphing the solution set:

© 2012 Cengage Learning. All Rights Reserved. May not be scanned, copied or duplicated, or posted to a publicly accessible website, in whole or in part.

9. Solving the inequality:

$$2 < x - 7$$
$$2 + 7 < x - 7 + 7$$
$$9 < x$$
$$x > 9$$

Graphing the solution set:

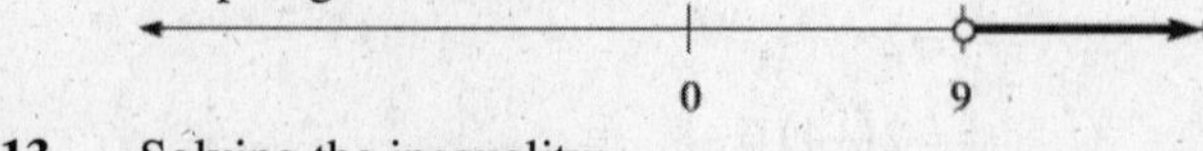

11. Solving the inequality:

$$3x < 6$$
$$\frac{1}{3}(3x) < \frac{1}{3}(6)$$
$$x < 2$$

Graphing the solution set:

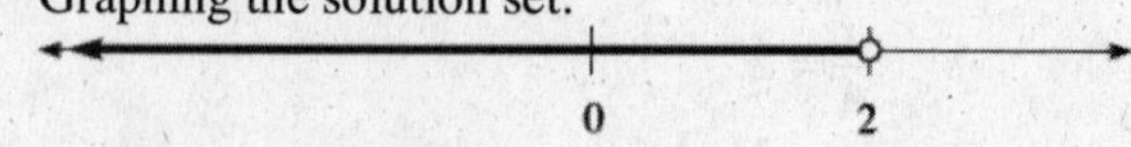

13. Solving the inequality:

$$5a \le 25$$
$$\frac{1}{5}(5a) \le \frac{1}{5}(25)$$
$$a \le 5$$

Graphing the solution set:

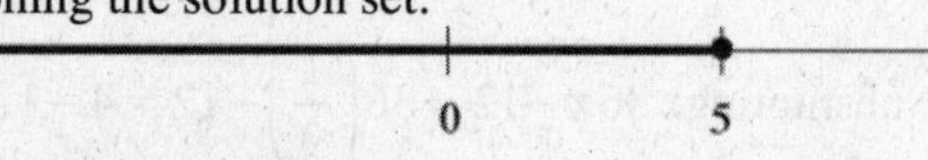

15. Solving the inequality:

$$\frac{x}{3} > 5$$
$$3\left(\frac{x}{3}\right) > 3(5)$$
$$x > 15$$

Graphing the solution set:

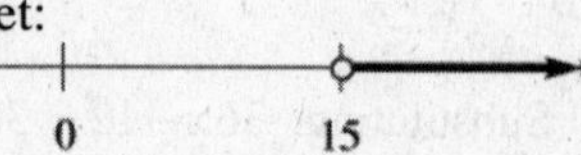

17. Solving the inequality:

$$-2x > 6$$
$$-\frac{1}{2}(-2x) < -\frac{1}{2}(6)$$
$$x < -3$$

Graphing the solution set:

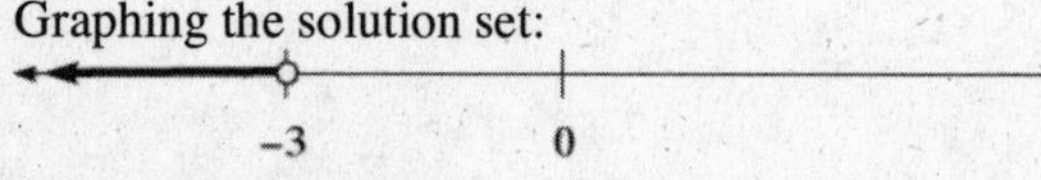

19. Solving the inequality:

$$-3x \ge -18$$
$$-\frac{1}{3}(-3x) \le -\frac{1}{3}(-18)$$
$$x \le 6$$

Graphing the solution set:

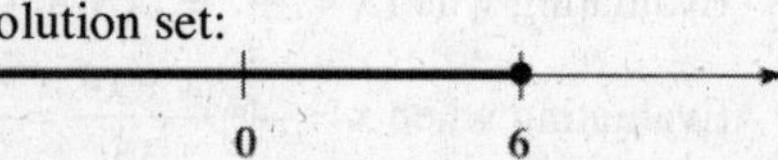

21. Solving the inequality:

$$-\frac{x}{5} \le 10$$
$$-5\left(-\frac{x}{5}\right) \ge -5(10)$$
$$x \ge -50$$

Graphing the solution set:

-50 0

23. Solving the inequality:

$$-\frac{2}{3}y > 4$$
$$-\frac{3}{2}\left(-\frac{2}{3}y\right) < -\frac{3}{2}(4)$$
$$y < -6$$

Graphing the solution set:

-6 0

25. Solving the inequality:

$$2x - 3 < 9$$
$$2x - 3 + 3 < 9 + 3$$
$$2x < 12$$
$$\frac{1}{2}(2x) < \frac{1}{2}(12)$$
$$x < 6$$

Graphing the solution set:

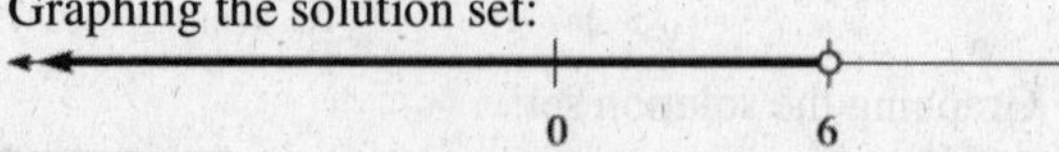

27. Solving the inequality:

$$-\frac{1}{5}y - \frac{1}{3} \le \frac{2}{3}$$
$$-\frac{1}{5}y - \frac{1}{3} + \frac{1}{3} \le \frac{2}{3} + \frac{1}{3}$$
$$-\frac{1}{5}y \le 1$$
$$-5\left(-\frac{1}{5}y\right) \ge -5(1)$$
$$y \ge -5$$

Graphing the solution set:

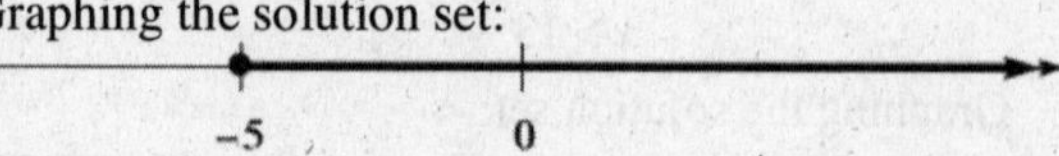

© 2012 Cengage Learning. All Rights Reserved. May not be scanned, copied or duplicated, or posted to a publicly accessible website, in whole or in part.

29. Solving the inequality:

$$\begin{aligned} -7.2x+1.8 &> -19.8 \\ -7.2x+1.8-1.8 &> -19.8-1.8 \\ -7.2x &> -21.6 \\ \frac{-7.2x}{-7.2} &< \frac{-21.6}{-7.2} \\ x &< 3 \end{aligned}$$

Graphing the solution set:

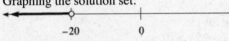

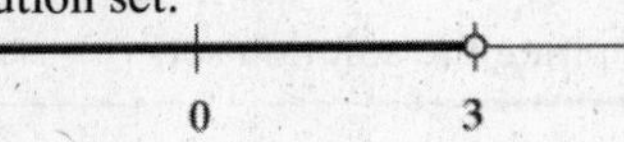

31. Solving the inequality:

$$\begin{aligned} \frac{2}{3}x-5 &\le 7 \\ \frac{2}{3}x-5+5 &\le 7+5 \\ \frac{2}{3}x &\le 12 \\ \frac{3}{2}\left(\frac{2}{3}x\right) &\le \frac{3}{2}(12) \\ x &\le 18 \end{aligned}$$

Graphing the solution set:

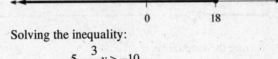

0 18

33. Solving the inequality:

$$\begin{aligned} -\frac{2}{5}a-3 &> 5 \\ -\frac{2}{5}a-3+3 &> 5+3 \\ -\frac{2}{5}a &> 8 \\ -\frac{5}{2}\left(-\frac{2}{5}a\right) &< -\frac{5}{2}(8) \\ a &< -20 \end{aligned}$$

Graphing the solution set:

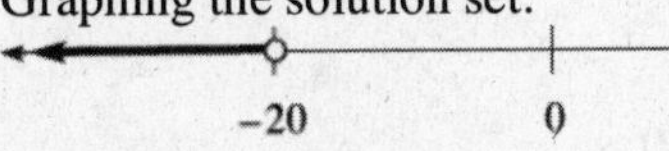

35. Solving the inequality:

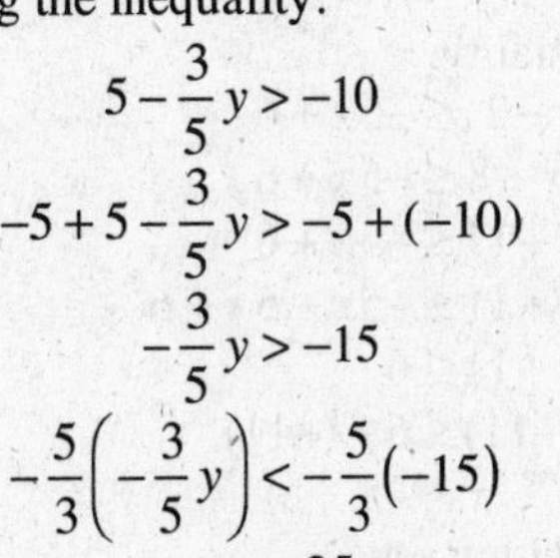

$$\begin{aligned} 5-\frac{3}{5}y &> -10 \\ -5+5-\frac{3}{5}y &> -5+(-10) \\ -\frac{3}{5}y &> -15 \\ -\frac{5}{3}\left(-\frac{3}{5}y\right) &< -\frac{5}{3}(-15) \\ y &< 25 \end{aligned}$$

Graphing the solution set:

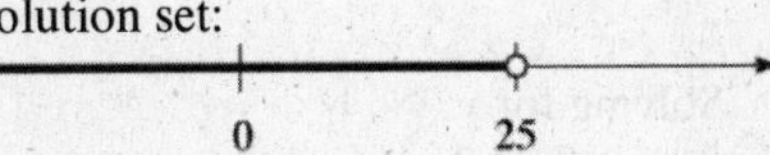

37. Solving the inequality:

$$\begin{aligned} 0.3(a+1) &\le 1.2 \\ 0.3a+0.3 &\le 1.2 \\ 0.3a+0.3+(-0.3) &\le 1.2+(-0.3) \\ 0.3a &\le 0.9 \\ \frac{0.3a}{0.3} &\le \frac{0.9}{0.3} \\ a &\le 3 \end{aligned}$$

Graphing the solution set:

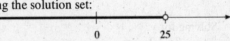

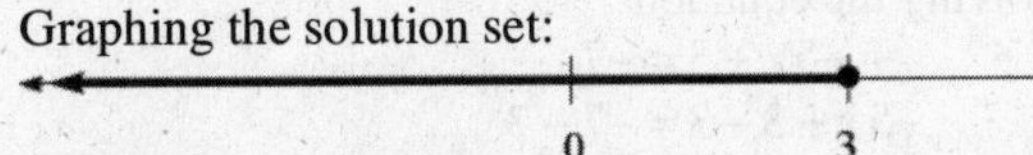

39. Solving the inequality:

$$\begin{aligned} 2(5-2x) &\le -20 \\ 10-4x &\le -20 \\ -10+10-4x &\le -10+(-20) \\ -4x &\le -30 \\ -\frac{1}{4}(-4x) &\ge -\frac{1}{4}(-30) \\ x &\ge \frac{15}{2} \end{aligned}$$

Graphing the solution set:

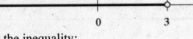

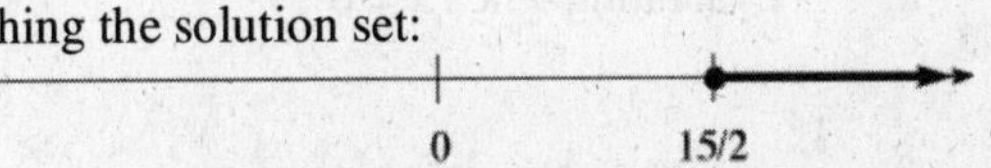

41. Solving the inequality:

$$\begin{aligned} 3x-5 &> 8x \\ -3x+3x-5 &> -3x+8x \\ -5 &> 5x \\ \frac{1}{5}(-5) &> \frac{1}{5}(5x) \\ -1 &> x \\ x &< -1 \end{aligned}$$

Graphing the solution set:

−1 0

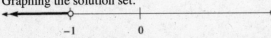

43. First multiply by 6 to clear the inequality of fractions:

$$\begin{aligned} \frac{1}{3}y-\frac{1}{2} &\le \frac{5}{6}y+\frac{1}{2} \\ 6\left(\frac{1}{3}y-\frac{1}{2}\right) &\le 6\left(\frac{5}{6}y+\frac{1}{2}\right) \\ 2y-3 &\le 5y+3 \\ -5y+2y-3 &\le -5y+5y+3 \\ -3y-3 &\le 3 \\ -3y-3+3 &\le 3+3 \\ -3y &\le 6 \\ -\frac{1}{3}(-3y) &\ge -\frac{1}{3}(6) \\ y &\ge -2 \end{aligned}$$

Graphing the solution set:

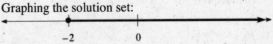

−2 0

© 2012 Cengage Learning. All Rights Reserved. May not be scanned, copied or duplicated, or posted to a publicly accessible website, in whole or in part.

45. Solving the inequality:

$$\begin{aligned}-2.8x+8.4&<-14x-2.8\\ -2.8x+14x+8.4&<-14x+14x-2.8\\ 11.2x+8.4&<-2.8\\ 11.2x+8.4-8.4&<-2.8-8.4\\ 11.2x&<-11.2\\ \frac{11.2x}{11.2}&<\frac{-11.2}{11.2}\\ x&<-1\end{aligned}$$

Graphing the solution set:

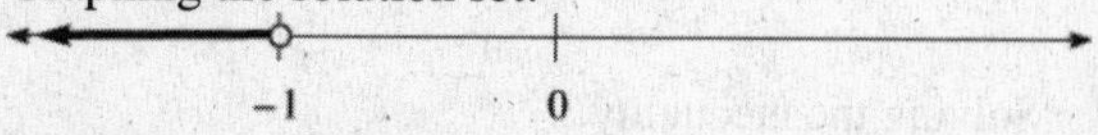

47. Solving the inequality:

$$\begin{aligned}3(m-2)-4&\geq 7m+14\\ 3m-6-4&\geq 7m+14\\ 3m-10&\geq 7m+14\\ -7m+3m-10&\geq -7m+7m+14\\ -4m-10&\geq 14\\ -4m-10+10&\geq 14+10\\ -4m&\geq 24\\ -\frac{1}{4}(-4m)&\leq -\frac{1}{4}(24)\\ m&\leq -6\end{aligned}$$

Graphing the solution set:

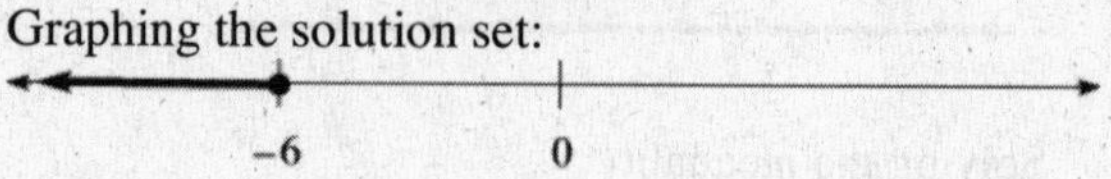

49. Solving the inequality:

$$\begin{aligned}3-4(x-2)&\leq -5x+6\\ 3-4x+8&\leq -5x+6\\ -4x+11&\leq -5x+6\\ -4x+5x+11&\leq -5x+5x+6\\ x+11&\leq 6\\ x+11+(-11)&\leq 6+(-11)\\ x&\leq -5\end{aligned}$$

Graphing the solution set:

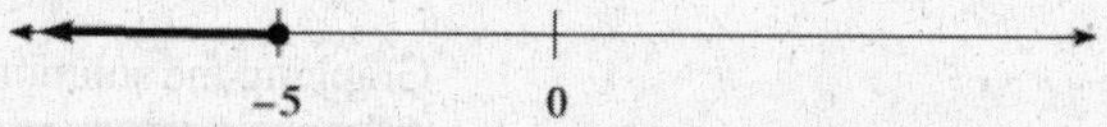

51. Solving for y:

$$\begin{aligned}3x+2y&<6\\ 2y&<-3x+6\\ y&<-\frac{3}{2}x+3\end{aligned}$$

53. Solving for y:

$$\begin{aligned}2x-5y&>10\\ -5y&>-2x+10\\ y&<\frac{2}{5}x-2\end{aligned}$$

55. Solving for y:

$$\begin{aligned}-3x+7y&\leq 21\\ 7y&\leq 3x+21\\ y&\leq \frac{3}{7}x+3\end{aligned}$$

57. Solving for y:

$$\begin{aligned}2x-4y&\geq -4\\ -4y&\geq -2x-4\\ y&\leq \frac{1}{2}x+1\end{aligned}$$

59. **a.** Evaluating when $x=0$:

$$-5x+3=-5(0)+3=0+3=3$$

b. Solving the equation:

$$\begin{aligned}-5x+3&=-7\\ -5x+3-3&=-7-3\\ -5x&=-10\\ -\frac{1}{5}(-5x)&=-\frac{1}{5}(-10)\\ x&=2\end{aligned}$$

c. Substituting $x=0$:

$$\begin{aligned}-5(0)+3&<-7\\ 3&<-7 \text{ (false)}\end{aligned}$$

No, $x=0$ is not a solution to the inequality.

d. Solving the inequality:

$$\begin{aligned}-5x+3&<-7\\ -5x+3-3&<-7-3\\ -5x&<-10\\ -\frac{1}{5}(-5x)&>-\frac{1}{5}(-10)\\ x&>2\end{aligned}$$

61. The inequality is $x<3$.

63. The inequality is $x\geq 3$.

65. The inequality is $x\leq 50$.

© 2012 Cengage Learning. All Rights Reserved. May not be scanned, copied or duplicated, or posted to a publicly accessible website, in whole or in part.

67. Let x and $x + 1$ represent the integers. Solving the inequality:
$$x+x+1\geq 583$$
$$2x+1\geq 583$$
$$2x\geq 582$$
$$x\geq 291$$
The two numbers are at least 291.

69. Let x represent the number. Solving the inequality:
$$2x+6<10$$
$$2x<4$$
$$x<2$$

71. Let x represent the number. Solving the inequality:
$$4x>x-8$$
$$3x>-8$$
$$x>-\frac{8}{3}$$

73. Let w represent the width, so $3w$ represents the length. Using the formula for perimeter:
$$2(w)+2(3w)\geq 48$$
$$2w+6w\geq 48$$
$$8w\geq 48$$
$$w\geq 6$$
The width is at least 6 meters.

75. Let x, $x + 2$, and $x + 4$ represent the sides of the triangle. The inequality is:
$$x+(x+2)+(x+4)>24$$
$$3x+6>24$$
$$3x>18$$
$$x>6$$
The shortest side is an even number greater than 6 inches (greater than or equal to 8 inches).

77. The inequality is $t\geq 100$.

79. Let n represent the number of tickets they sell. The inequality is:
$$7.5n<1500$$
$$n<200$$
They will lose money if they sell less than 200 tickets. They will make a profit if they sell more than 200 tickets.

81. Solving the inequality:
$$2x-1\geq 3$$
$$2x\geq 4$$
$$x\geq 2$$

83. Solving the inequality:
$$-2x>-8$$
$$x<4$$

85. Solving the inequality:
$$-3\leq 4x+1$$
$$4x\geq -4$$
$$x\geq -1$$

87. Simplifying: $\frac{1}{6}(12x+6)=\frac{1}{6}\bullet 12x+\frac{1}{6}\bullet 6=2x+1$

89. Simplifying: $\frac{2}{3}(-3x-6)=\frac{2}{3}\bullet(-3x)-\frac{2}{3}\bullet 6=-2x-4$

91. Simplifying: $3\left(\frac{5}{6}a+\frac{4}{9}\right)=3\bullet\frac{5}{6}a+3\bullet\frac{4}{9}=\frac{5}{2}a+\frac{4}{3}$

93. Simplifying: $-3\left(\frac{2}{3}a+\frac{5}{6}\right)=-3\bullet\frac{2}{3}a-3\bullet\frac{5}{6}=-2a-\frac{5}{2}$

95. Simplifying: $\frac{1}{2}x+\frac{1}{6}x=\frac{3}{3}\bullet\frac{1}{2}x+\frac{1}{6}x=\frac{3}{6}x+\frac{1}{6}x=\frac{4}{6}x=\frac{2}{3}x$

97. Simplifying: $\frac{2}{3}x-\frac{5}{6}x=\frac{2}{2}\bullet\frac{2}{3}x-\frac{5}{6}x=\frac{4}{6}x-\frac{5}{6}x=-\frac{1}{6}x$

99. Simplifying: $\frac{3}{4}x+\frac{1}{6}x=\frac{3}{3}\bullet\frac{3}{4}x+\frac{2}{2}\bullet\frac{1}{6}x=\frac{9}{12}x+\frac{2}{12}x=\frac{11}{12}x$

101. Simplifying: $\frac{2}{5}x+\frac{5}{8}x=\frac{8}{8}\bullet\frac{2}{5}x+\frac{5}{5}\bullet\frac{5}{8}x=\frac{16}{40}x+\frac{25}{40}x=\frac{41}{40}x$

© 2012 Cengage Learning. All Rights Reserved. May not be scanned, copied or duplicated, or posted to a publicly accessible website, in whole or in part.

2.9 Compound Inequalities

1. Graphing the solution set:

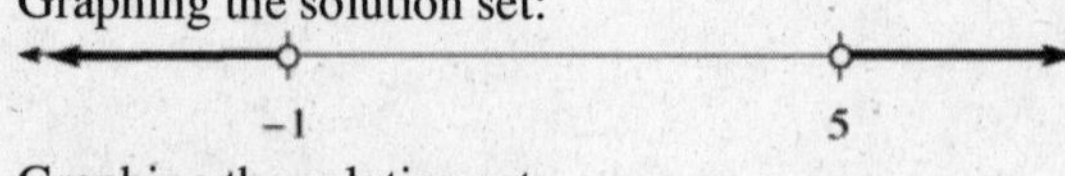

3. Graphing the solution set:

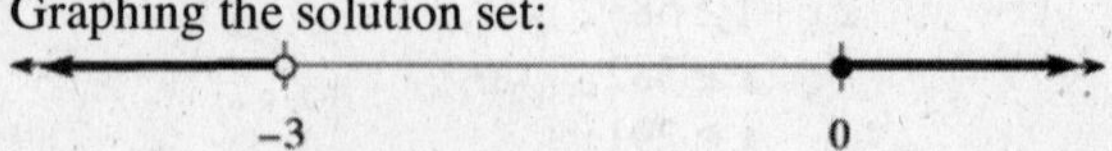

5. Graphing the solution set:

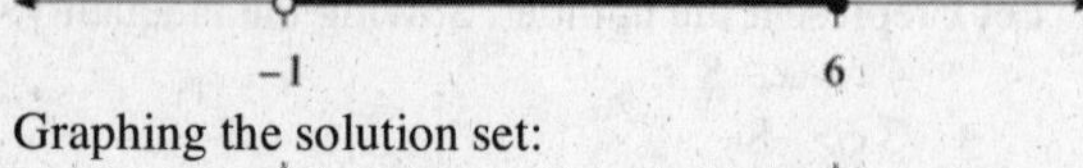

7. Graphing the solution set:

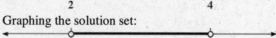

9. Graphing the solution set:

−2 4

11. Graphing the solution set:

−1 5

13. Graphing the solution set:

−1 3

15. Graphing the solution set:

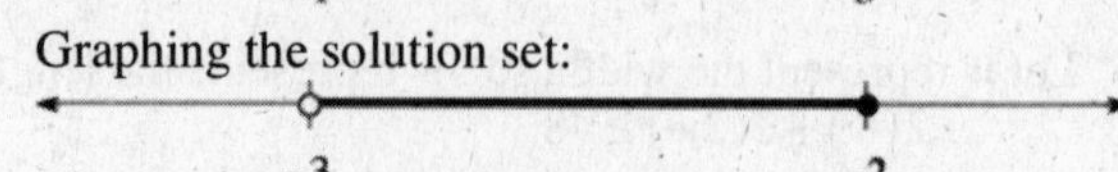

17. Solving the compound inequality:

$$3x-1<5 \quad \text{or} \quad 5x-5>10$$
$$3x<6 \qquad 5x>15$$
$$x<2 \qquad x>3$$

Graphing the solution set:

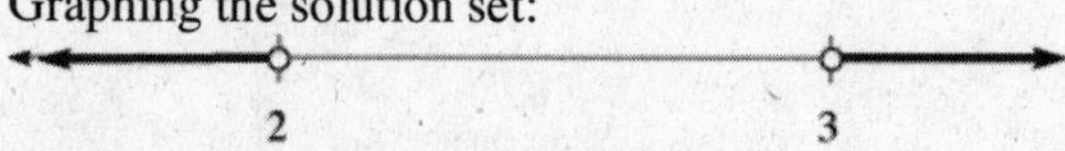

19. Solving the compound inequality:

$$x-2>-5 \quad \text{and} \quad x+7<13$$
$$x>-3 \qquad x<6$$

Graphing the solution set:

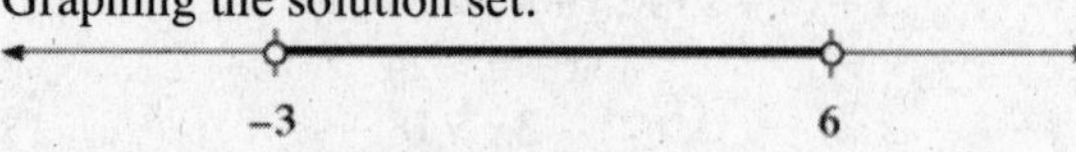

21. Solving the compound inequality:

$$11x<22 \quad \text{or} \quad 12x>36$$
$$x<2 \qquad x>3$$

Graphing the solution set:

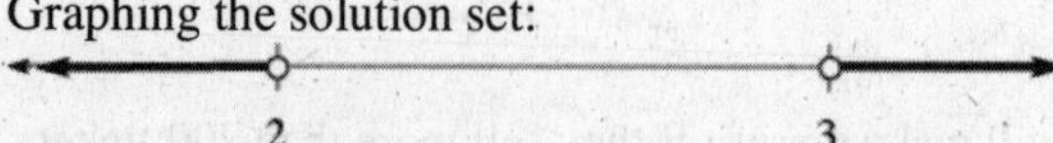

23. Solving the compound inequality:

$$3x-5<10 \quad \text{and} \quad 2x+1>-5$$
$$3x<15 \qquad 2x>-6$$
$$x<5 \qquad x>-3$$

Graphing the solution set:

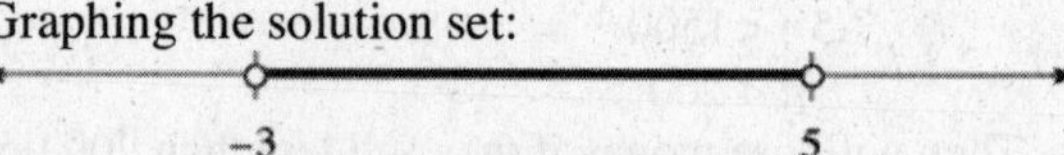

25. Solving the compound inequality:

$$2x-3<8 \quad \text{and} \quad 3x+1>-10$$
$$2x<11 \qquad 3x>-11$$
$$x<\frac{11}{2} \qquad x>-\frac{11}{3}$$

Graphing the solution set:

−11/3 11/2

27. Solving the compound inequality:

$$2x-1<3 \quad \text{and} \quad 3x-2>1$$
$$2x<4 \qquad 3x>3$$
$$x<2 \qquad x>1$$

Graphing the solution set:

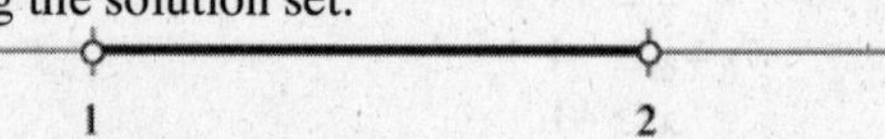

29. Solving the compound inequality:

$$-1 \le x-5 \le 2$$
$$4 \le x \le 7$$

Graphing the solution set:

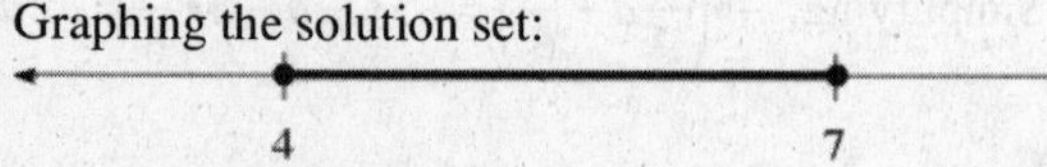

31. Solving the compound inequality:

$$-4 \le 2x \le 6$$
$$-2 \le x \le 3$$

Graphing the solution set:

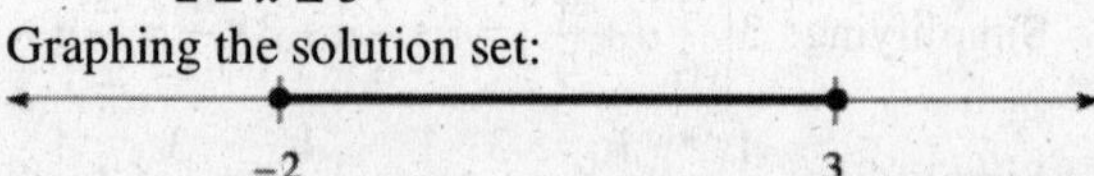

33. Solving the compound inequality:

$$-3<2x+1<5$$
$$-4<2x<4$$
$$-2<x<2$$

Graphing the solution set:

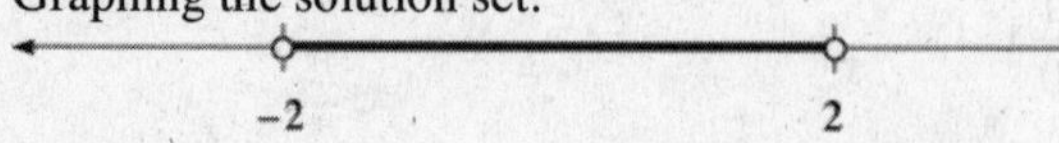

35. Solving the compound inequality:

$$0 \le 3x+2 \le 7$$
$$-2 \le 3x \le 5$$
$$-\frac{2}{3} \le x \le \frac{5}{3}$$

Graphing the solution set:

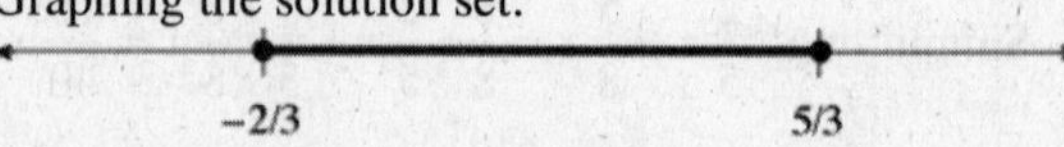

© 2012 Cengage Learning. All Rights Reserved. May not be scanned, copied or duplicated, or posted to a publicly accessible website, in whole or in part.

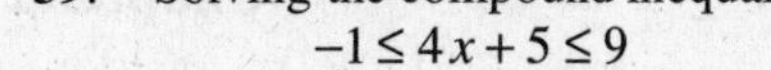

37. Solving the compound inequality:

$$-7 < 2x + 3 < 11$$
$$-10 < 2x < 8$$
$$-5 < x < 4$$

Graphing the solution set:

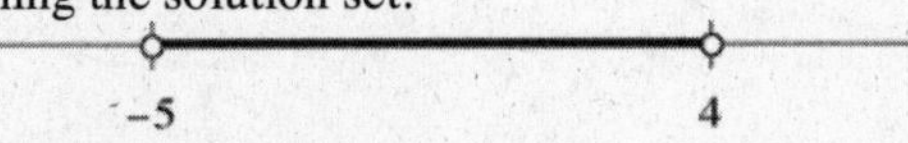

39. Solving the compound inequality:

$$-1 \le 4x + 5 \le 9$$
$$-6 \le 4x \le 4$$
$$-\frac{3}{2} \le x \le 1$$

Graphing the solution set:

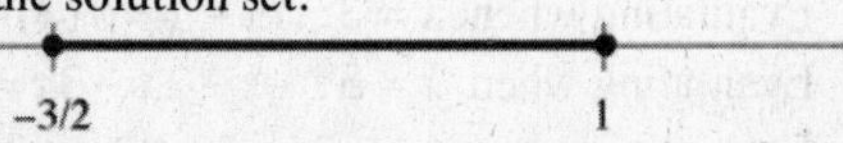

41. The inequality is $-2 < x < 3$.

43. The inequality is $-2 \le x \le 3$.

45. The inequality is $4 \le d \le 13$.

47. **a.** The three inequalities are $2x + x > 10$, $2x + 10 > x$, and $x + 10 > 2x$.

b. For $x + 10 > 2x$ we have $x < 10$. For $3x > 10$ we have $x > \frac{10}{3}$. The compound inequality is $\frac{10}{3} < x < 10$.

49. Graphing the inequality:

51. Let x represent the number. Solving the inequality:

$$5 < 2x - 3 < 7$$
$$8 < 2x < 10$$
$$4 < x < 5$$

The number is between 4 and 5.

53. Let w represent the width and $w + 4$ represent the length. Using the perimeter formula:

$$20 < 2w + 2(w+4) < 30$$
$$20 < 2w + 2w + 8 < 30$$
$$20 < 4w + 8 < 30$$
$$12 < 4w < 22$$
$$3 < w < \frac{11}{2}$$

The width is between 3 inches and $\frac{11}{2} = 5\frac{1}{2}$ inches.

55. Finding the number: $0.25(32) = 8$

57. Finding the number: $0.20(120) = 24$

59. Finding the percent:

$$p \cdot 36 = 9$$
$$p = \frac{9}{36} = 0.25 = 25\%$$

61. Finding the percent:

$$p \cdot 50 = 5$$
$$p = \frac{5}{50} = 0.10 = 10\%$$

63. Finding the number:

$$0.20n = 16$$
$$n = \frac{16}{0.20} = 80$$

65. Finding the number:

$$0.02n = 8$$
$$n = \frac{8}{0.02} = 400$$

67. Simplifying the expression: $-|-5| = -(5) = -5$

69. Simplifying the expression: $-3 - 4(-2) = -3 + 8 = 5$

71. Simplifying the expression: $5|3-8| - 6|2-5| = 5|-5| - 6|-3| = 5(5) - 6(3) = 25 - 18 = 7$

73. Simplifying the expression: $5 - 2[-3(5-7) - 8] = 5 - 2[-3(-2) - 8] = 5 - 2(6-8) = 5 - 2(-2) = 5 + 4 = 9$

75. The expression is: $-3 - (-9) = -3 + 9 = 6$

77. Applying the distributive property: $\frac{1}{2}(4x - 6) = \frac{1}{2} \cdot 4x - \frac{1}{2} \cdot 6 = 2x - 3$

79. The integers are: $-3, 0, 2$

© 2012 Cengage Learning. All Rights Reserved. May not be scanned, copied or duplicated, or posted to a publicly accessible website, in whole or in part.

Chapter 2 Review

1. Simplifying the expression: $5x-8x=(5-8)x=-3x$
3. Simplifying the expression: $-a+2+5a-9=-a+5a+2-9=4a-7$
5. Simplifying the expression: $6-2(3y+1)-4=6-6y-2-4=-6y+6-2-4=-6y$
7. Evaluating when $x=3$: $7x-2=7(3)-2=21-2=19$
9. Evaluating when $x=3$: $-x-2x-3x=-6x=-6(3)=-18$
11. Evaluating when $x=-2$: $-3x+2=-3(-2)+2=6+2=8$
13. Solving the equation:
$$\begin{aligned} x+2&=-6\\ x+2+(-2)&=-6+(-2)\\ x&=-8 \end{aligned}$$
15. Solving the equation:
$$\begin{aligned} 10-3y+4y&=12\\ 10+y&=12\\ -10+10+y&=-10+12\\ y&=2 \end{aligned}$$
17. Solving the equation:
$$\begin{aligned} 2x&=-10\\ \frac{1}{2}(2x)&=\frac{1}{2}(-10)\\ x&=-5 \end{aligned}$$
19. Solving the equation:
$$\begin{aligned} \frac{x}{3}&=4\\ 3\left(\frac{x}{3}\right)&=3(4)\\ x&=12 \end{aligned}$$
21. Solving the equation:
$$\begin{aligned} 3a-2&=5a\\ -3a+3a-2&=-3a+5a\\ -2&=2a\\ \frac{1}{2}(-2)&=\frac{1}{2}(2a)\\ a&=-1 \end{aligned}$$
23. Solving the equation:
$$\begin{aligned} 3x+2&=5x-8\\ 3x+(-5x)+2&=5x+(-5x)-8\\ -2x+2&=-8\\ -2x+2+(-2)&=-8+(-2)\\ -2x&=-10\\ -\frac{1}{2}(-2x)&=-\frac{1}{2}(-10)\\ x&=5 \end{aligned}$$
25. Solving the equation:
$$\begin{aligned} 0.7x-0.1&=0.5x-0.1\\ 0.7x+(-0.5x)-0.1&=0.5x+(-0.5x)-0.1\\ 0.2x-0.1&=-0.1\\ 0.2x-0.1+0.1&=-0.1+0.1\\ 0.2x&=0\\ \frac{0.2x}{0.2}&=\frac{0}{0.2}\\ x&=0 \end{aligned}$$
27. Simplifying and then solving the equation:
$$\begin{aligned} 2(x-5)&=10\\ 2x-10&=10\\ 2x-10+10&=10+10\\ 2x&=20\\ \frac{1}{2}(2x)&=\frac{1}{2}(20)\\ x&=10 \end{aligned}$$
29. Simplifying and then solving the equation:
$$\begin{aligned} \frac{1}{2}(3t-2)+\frac{1}{2}&=\frac{5}{2}\\ \frac{3}{2}t-1+\frac{1}{2}&=\frac{5}{2}\\ \frac{3}{2}t-\frac{1}{2}&=\frac{5}{2}\\ \frac{3}{2}t-\frac{1}{2}+\frac{1}{2}&=\frac{5}{2}+\frac{1}{2}\\ \frac{3}{2}t&=3\\ \frac{2}{3}\left(\frac{3}{2}t\right)&=\frac{2}{3}(3)\\ t&=2 \end{aligned}$$
31. Simplifying and then solving the equation:
$$\begin{aligned} 2(3x+7)&=4(5x-1)+18\\ 6x+14&=20x-4+18\\ 6x+14&=20x+14\\ 6x+(-20x)+14&=20x+(-20x)+14\\ -14x+14&=14\\ -14x+14+(-14)&=14+(-14)\\ -14x&=0\\ -\frac{1}{14}(-14x)&=-\frac{1}{14}(0)\\ x&=0 \end{aligned}$$

© 2012 Cengage Learning. All Rights Reserved. May not be scanned, copied or duplicated, or posted to a publicly accessible website, in whole or in part.

33. Substituting $x=5$:

$$\begin{aligned} 4(5)-5y&=20 \\ 20-5y&=20 \\ -5y&=0 \\ y&=0 \end{aligned}$$

35. Substituting $x=-5$:

$$\begin{aligned} 4(-5)-5y&=20 \\ -20-5y&=20 \\ -5y&=40 \\ y&=-8 \end{aligned}$$

37. Solving for y:

$$\begin{aligned} 2x-5y&=10 \\ -5y&=-2x+10 \\ y&=\frac{2}{5}x-2 \end{aligned}$$

39. Solving for h:

$$\begin{aligned} \pi r^2 h&=V \\ \frac{\pi r^2 h}{\pi r^2}&=\frac{V}{\pi r^2} \\ h&=\frac{V}{\pi r^2} \end{aligned}$$

41. Computing the amount:

$$\begin{aligned} 0.86(240)&=x \\ x&=206.4 \end{aligned}$$

So 86% of 240 is 206.4

43. Let x represent the number. The equation is:

$$\begin{aligned} 2x+6&=28 \\ 2x&=22 \\ x&=11 \end{aligned}$$

The number is 11.

45. Completing the table:

	Dollars Invested at 9%	Dollars Invested at 10%
Number of	x	$x+300$
Interest on	$0.09(x)$	$0.10(x+300)$

The equation is:

$$\begin{aligned} 0.09(x)+0.10(x+300)&=125 \\ 0.09x+0.10x+30&=125 \\ 0.19x+30&=125 \\ 0.19x&=95 \\ x&=500 \\ x+300&=800 \end{aligned}$$

The man invested \$500 at 9% and \$800 at 10%.

47. Solving the inequality:

$$\begin{aligned} -2x&<4 \\ -\frac{1}{2}(-2x)&>-\frac{1}{2}(4) \\ x&>-2 \end{aligned}$$

49. Solving the inequality:

$$\begin{aligned} -\frac{a}{2}&\le -3 \\ -2\left(-\frac{a}{2}\right)&\ge -2(-3) \\ a&\ge 6 \end{aligned}$$

51. Solving the inequality:

$$\begin{aligned} -4x+5&>37 \\ -4x&>32 \\ x&<-8 \end{aligned}$$

Graphing the solution set:

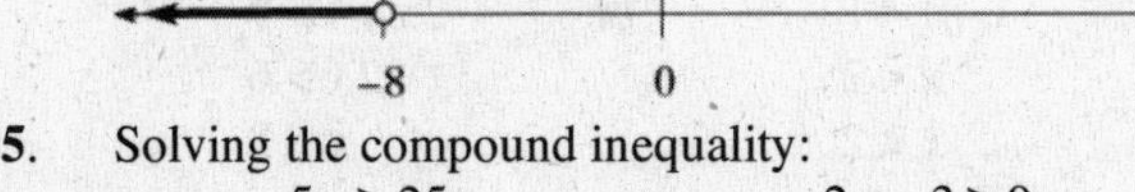

53. Solving the inequality:

$$\begin{aligned} 2(3t+1)+6&\ge 5(2t+4) \\ 6t+2+6&\ge 10t+20 \\ 6t+8&\ge 10t+20 \\ -4t+8&\ge 20 \\ -4t&\ge 12 \\ t&\le -3 \end{aligned}$$

Graphing the solution set:

-3 0

55. Solving the compound inequality:

$$\begin{aligned} -5x&\ge 25 \quad \text{or} \quad & 2x-3&\ge 9 \\ x&\le -5 & 2x&\ge 12 \\ x&\le -5 & x&\ge 6 \end{aligned}$$

Graphing the solution set:

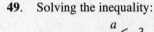

© 2012 Cengage Learning. All Rights Reserved. May not be scanned, copied or duplicated, or posted to a publicly accessible website, in whole or in part.

Chapter 2 Cumulative Review

1. Simplifying: $7-9-12=7+(-9)+(-12)=7+(-21)=-14$

3. Simplifying: $8-4\bullet5=8-20=8+(-20)=-12$

5. Simplifying: $6+3(6+2)=6+3(8)=6+24=30$

7. Simplifying: $\left(-\frac{2}{3}\right)^3=\left(-\frac{2}{3}\right)\left(-\frac{2}{3}\right)\left(-\frac{2}{3}\right)=-\frac{8}{27}$

9. Simplifying: $\frac{-30}{120}=-\frac{30\bullet1}{30\bullet4}=-\frac{1}{4}$

11. Simplifying: $-\frac{3}{4}\div\frac{15}{16}=-\frac{3}{4}\bullet\frac{16}{15}=-\frac{49}{60}=-\frac{4}{5}$

13. Simplifying: $\frac{-4(-6)}{-9}=\frac{24}{-9}=-\frac{8}{3}$

15. Simplifying: $\frac{5}{9}+\frac{1}{3}=\frac{5}{9}+\frac{1\bullet3}{3\bullet3}=\frac{5}{9}+\frac{3}{9}=\frac{8}{9}$

17. Using the associative property: $\frac{1}{5}(10x)=\left(\frac{1}{5}\bullet10\right)x=2x$

19. Simplifying: $\frac{1}{4}(8x-4)=\frac{1}{4}\bullet8x-\frac{1}{4}\bullet4=2x-1$

21. Solving the equation:

$$\begin{aligned}7x&=6x+4\\7x+(-6x)&=6x+(-6x)+4\\x&=4\end{aligned}$$

23. Solving the equation:

$$\begin{aligned}-\frac{3}{5}x&=30\\-\frac{5}{3}\left(-\frac{3}{5}x\right)&=-\frac{5}{3}(30)\\x&=-50\end{aligned}$$

25. Solving the equation:

$$\begin{aligned}3x-4&=11\\3x-4+4&=11+4\\3x&=15\\\frac{1}{3}(3x)&=\frac{1}{3}(15)\\x&=5\end{aligned}$$

27. Solving the equation:

$$\begin{aligned}4(3x-8)+5(2x+7)&=25\\12x-32+10x+35&=25\\22x+3&=25\\22x+3+(-3)&=25+(-3)\\22x&=22\\\frac{1}{22}(22x)&=\frac{1}{22}(22)\\x&=1\end{aligned}$$

29. Solving the equation:

$$\begin{aligned}3(2a-7)-4(a-3)&=15\\6a-21-4a+12&=15\\2a-9&=15\\2a-9+9&=15+9\\2a&=24\\\frac{1}{2}(2a)&=\frac{1}{2}(24)\\a&=12\end{aligned}$$

31. Solving for c:

$$\begin{aligned}a+b+c&=P\\a+b-a-b+c&=P-a-b\\c&=P-a-b\end{aligned}$$

33. Solving the inequality:

$$\begin{aligned}3(x-4)&\le6\\3x-12&\le6\\3x-12+12&\le6+12\\3x&\le18\\\frac{1}{3}(3x)&\le\frac{1}{3}(18)\\x&\le6\end{aligned}$$

Graphing the solution set:

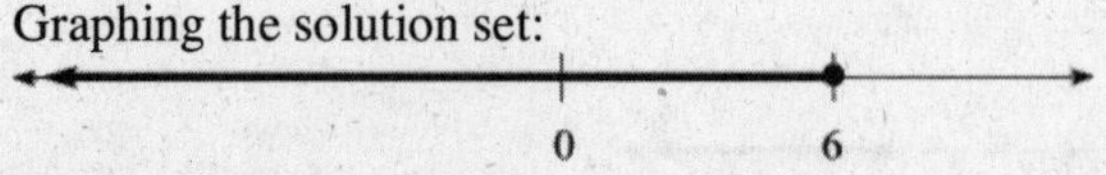

35. Solving the compound inequality:

$$\begin{aligned}x-1&<1 &\quad\text{or}\quad 2x+1&>7\\x&<2 & 2x&>6\\x&<2 & x&>3\end{aligned}$$

Graphing the solution set:

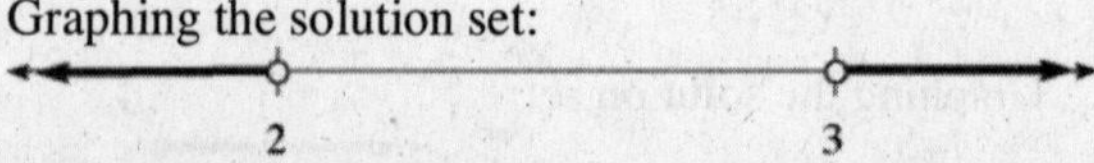

37. Translating into symbols: $x-5=12$

39. The integers are: –5, –3

41. Evaluating when $x=-2$: $x^2-8x-9=(-2)^2-8(-2)-9=4+16-9=20-9=11$

© 2012 Cengage Learning. All Rights Reserved. May not be scanned, copied or duplicated, or posted to a publicly accessible website, in whole or in part.

43. Computing the amount:

$$0.30(50) = x$$
$$15 = x$$

So 30% of 50 is 15.

45. Let x represent the number. The equation is:

$$x + 9 = 23$$
$$x = 14$$

The number is 14.

47. Let x represent the acute angle and 90° is the right angle. Since the sum of the three angles is 180°, the equation is:

$$x + 42° + 90° = 180°$$
$$x + 132° = 180°$$
$$x = 48°$$

The other two angles are 48° and 90°.

49. Let x represent her hours worked in the week. She worked 15 hours at \$8 per hour and $x - 15$ hours at \$10 per hour, so the equation is:

$$8(15) + 10(x - 15) = 150$$
$$120 + 10x - 150 = 150$$
$$10x - 30 = 150$$
$$10x = 180$$
$$x = 18$$

Carol worked 18 hours for the week.

51. The total annual circulation is: 8.3 + 24.1 = 32.4 million copies

Chapter 2 Test

1. Simplifying: $3x + 2 - 7x + 3 = 3x - 7x + 2 + 3 = -4x + 5$
2. Simplifying: $4a - 5 - a + 1 = 4a - a - 5 + 1 = 3a - 4$
3. Simplifying: $7 - 3(y + 5) - 4 = 7 - 3y - 15 - 4 = -3y + 7 - 15 - 4 = -3y - 12$
4. Simplifying: $8(2x + 1) - 5(x - 4) = 16x + 8 - 5x + 20 = 16x - 5x + 8 + 20 = 11x + 28$
5. Evaluating when $x = -5$: $2x - 3 - 7x = -5x - 3 = -5(-5) - 3 = 25 - 3 = 22$
6. Evaluating when $x = 2$ and $y = 3$: $x^2 + 2xy + y^2 = (2)^2 + 2(2)(3) + (3)^2 = 4 + 12 + 9 = 25$

7. **a.** Completing the table:

n	$(n+1)^2$
1	$(1+1)^2 = (2)^2 = 4$
2	$(2+1)^2 = (3)^2 = 9$
3	$(3+1)^2 = (4)^2 = 16$
4	$(4+1)^2 = (5)^2 = 25$

b. Completing the table:

n	$n^2 + 2$
1	$(1)^2 + 1 = 1 + 1 = 2$
2	$(2)^2 + 1 = 4 + 1 = 5$
3	$(3)^2 + 1 = 9 + 1 = 10$
4	$(4)^2 + 1 = 16 + 1 = 17$

8. Solving the equation:

$$2x - 5 = 7$$
$$2x - 5 + 5 = 7 + 5$$
$$2x = 12$$
$$\frac{1}{2}(2x) = \frac{1}{2}(12)$$
$$x = 6$$

9. Solving the equation:

$$2y + 4 = 5y$$
$$-2y + 2y + 4 = -2y + 5y$$
$$4 = 3y$$
$$\frac{1}{3}(4) = \frac{1}{3}(3y)$$
$$y = \frac{4}{3}$$

© 2012 Cengage Learning. All Rights Reserved. May not be scanned, copied or duplicated, or posted to a publicly accessible website, in whole or in part.

10. First clear the equation of fractions by multiplying by 12:

$$\frac{1}{2}x-\frac{1}{10}=\frac{1}{5}x+\frac{1}{2}$$
$$10\left(\frac{1}{2}x-\frac{1}{10}\right)=10\left(\frac{1}{5}x+\frac{1}{2}\right)$$
$$5x-1=2x+5$$
$$5x+(-2x)-1=2x+(-2x)+5$$
$$3x-1=5$$
$$3x-1+1=5+1$$
$$3x=6$$
$$\frac{1}{3}(3x)=\frac{1}{3}(6)$$
$$x=2$$

11. Solving the equation:

$$\frac{2}{5}(5x-10)=-5$$
$$2x-4=-5$$
$$2x-4+4=-5+4$$
$$2x=-1$$
$$\frac{1}{2}(2x)=\frac{1}{2}(-1)$$
$$x=-\frac{1}{2}$$

12. Solving the equation:

$$-5(2x+1)-6=19$$
$$-10x-5-6=19$$
$$-10x-11=19$$
$$-10x-11+11=19+11$$
$$-10x=30$$
$$-\frac{1}{10}(-10x)=-\frac{1}{10}(30)$$
$$x=-3$$

13. Solving the equation:

$$0.04x+0.06(100-x)=4.6$$
$$0.04x+6-0.06x=4.6$$
$$-0.02x+6=4.6$$
$$-0.02x+6+(-6)=4.6+(-6)$$
$$-0.02x=-1.4$$
$$\frac{-0.02x}{-0.02}=\frac{-1.4}{-0.02}$$
$$x=70$$

14. Solving the equation:

$$2(t-4)+3(t+5)=2t-2$$
$$2t-8+3t+15=2t-2$$
$$5t+7=2t-2$$
$$5t+(-2t)+7=2t+(-2t)-2$$
$$3t+7=-2$$
$$3t+7+(-7)=-2+(-7)$$
$$3t=-9$$
$$\frac{1}{3}(3t)=\frac{1}{3}(-9)$$
$$t=-3$$

15. Solving the equation:

$$2x-4(5x+1)=3x+17$$
$$2x-20x-4=3x+17$$
$$-18x-4=3x+17$$
$$-18x+(-3x)-4=3x+(-3x)+17$$
$$-21x-4=17$$
$$-21x-4+4=17+4$$
$$-21x=21$$
$$-\frac{1}{21}(21x)=-\frac{1}{21}(21)$$
$$x=-1$$

16. Finding the amount:

$$0.15(38)=x$$
$$5.7=x$$

So 15% of 38 is 5.7.

17. Finding the base:

$$0.12x=240$$
$$\frac{0.12x}{0.12}=\frac{240}{0.12}$$
$$x=2{,}000$$

So 12% of 2,000 is 240.

© 2012 Cengage Learning. All Rights Reserved. May not be scanned, copied or duplicated, or posted to a publicly accessible website, in whole or in part.

18. Substituting $y = -2$:

$$\begin{aligned} 2x - 3(-2) &= 12 \\ 2x + 6 &= 12 \\ 2x &= 6 \\ x &= 3 \end{aligned}$$

19. Substituting $x = -6$:

$$\begin{aligned} 2(-6) - 3y &= 12 \\ -12 - 3y &= 12 \\ -3y &= 24 \\ y &= -8 \end{aligned}$$

20. Solving for y:

$$\begin{aligned} 2x + 5y &= 20 \\ 5y &= -2x + 20 \\ y &= -\frac{2}{5}x + 4 \end{aligned}$$

21. Solving for v:

$$\begin{aligned} x + vt + 16t^2 &= h \\ vt &= h - x - 16t^2 \\ v &= \frac{h - x - 16t^2}{t} \end{aligned}$$

22. Completing the table:

	Ten Years Ago	Now
Dave	$2x - 10$	$2x$
Rick	$x - 10$	x

The equation is:

$$\begin{aligned} 2x - 10 + x - 10 &= 40 \\ 3x - 20 &= 40 \\ 3x &= 60 \\ x &= 20 \\ 2x &= 40 \end{aligned}$$

Rick is 20 years old and Dave is 40 years old.

23. Let w represent the width and $2w$ represent the length. Using the perimeter formula:

$$\begin{aligned} 2(w) + 2(2w) &= 60 \\ 2w + 4w &= 60 \\ 6w &= 60 \\ w &= 10 \\ 2w &= 20 \end{aligned}$$

The width is 10 inches and the length is 20 inches.

24. Completing the table:

	Dimes	Quarters
Number	$x + 7$	x
Value (cents)	$10(x + 7)$	$25(x)$

The equation is:

$$\begin{aligned} 10(x + 7) + 25(x) &= 350 \\ 10x + 70 + 25x &= 350 \\ 35x + 70 &= 350 \\ 35x &= 280 \\ x &= 8 \\ x + 7 &= 15 \end{aligned}$$

He has 8 quarters and 15 dimes in his collection.

25. Completing the table:

	Dollars Invested at 7%	Dollars Invested at 9%
Number of	x	$x + 600$
Interest on	$0.07(x)$	$0.09(x + 600)$

The equation is:

$$\begin{aligned} 0.07(x) + 0.09(x + 600) &= 182 \\ 0.07x + 0.09x + 54 &= 182 \\ 0.16x + 54 &= 182 \\ 0.16x &= 128 \\ x &= 800 \\ x + 600 &= 1,400 \end{aligned}$$

She has \$800 invested at 7% and \$1,400 invested at 9%.

© 2012 Cengage Learning. All Rights Reserved. May not be scanned, copied or duplicated, or posted to a publicly accessible website, in whole or in part.

26. Solving the inequality:

$$2x+3<5$$
$$2x+3+(-3)<5+(-3)$$
$$2x<2$$
$$\frac{1}{2}(2x)<\frac{1}{2}(2)$$
$$x<1$$

Graphing the solution set:

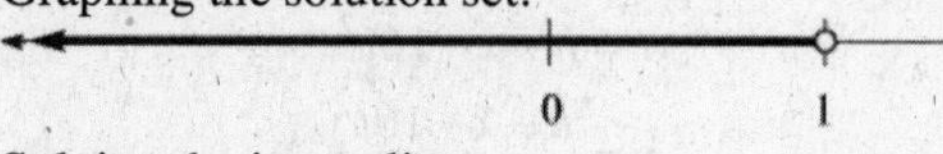

27. Solving the inequality:

$$-5a>20$$
$$-\frac{1}{5}(-5a)<-\frac{1}{5}(20)$$
$$a<-4$$

Graphing the solution set:

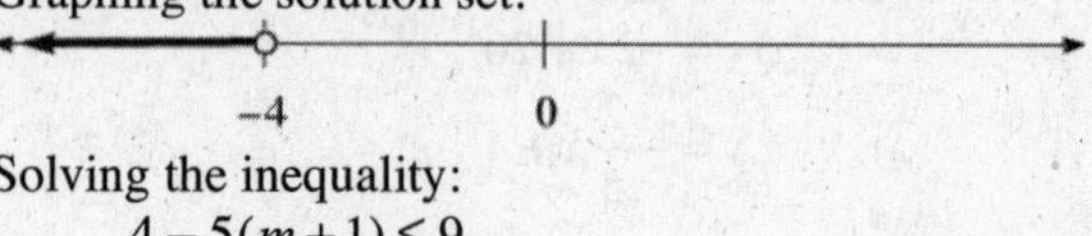

28. Solving the inequality:

$$0.4-0.2x\geq 1$$
$$-0.4+0.4-0.2x\geq -0.4+1$$
$$-0.2x\geq 0.6$$
$$\frac{-0.2x}{-0.2}\leq\frac{0.6}{-0.2}$$
$$x\leq -3$$

Graphing the solution set:

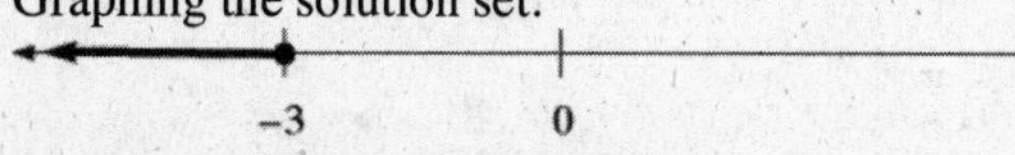

29. Solving the inequality:

$$4-5(m+1)\leq 9$$
$$4-5m-5\leq 9$$
$$-5m-1\leq 9$$
$$-5m-1+1\leq 9+1$$
$$-5m\leq 10$$
$$-\frac{1}{5}(-5m)\geq -\frac{1}{5}(10)$$
$$m\geq -2$$

Graphing the solution set:

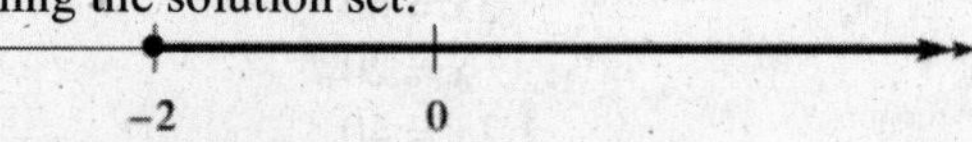

30. Solving the compound inequality:

$$3-4x\geq -5 \quad \text{or} \quad 2x\geq 10$$
$$-4x\geq -8 \qquad x\geq 5$$
$$x\leq 2 \qquad x\geq 5$$

Graphing the solution set:

31. Solving the compound inequality:

$$-7<2x-1<9$$
$$-6<2x<10$$
$$-3<x<5$$

Graphing the solution set:

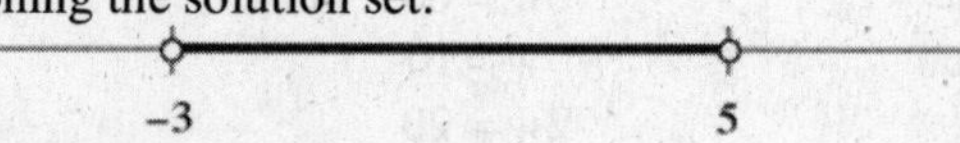

32. The inequality is $x>-4$.

33. The inequality is $-3\leq x\leq 1$.

34. The inequality is $x<-4$ or $x>1$.

35. Finding the percent:

$$p\bullet 168=6.9$$
$$p=\frac{6.9}{168}\approx 4.1\%$$

36. Finding the percent:

$$p\bullet 168=5.4$$
$$p=\frac{5.4}{168}\approx 3.2\%$$

© 2012 Cengage Learning. All Rights Reserved. May not be scanned, copied or duplicated, or posted to a publicly accessible website, in whole or in part.

Chapter 3
Linear Equations and Inequalities in Two Variables

Getting Ready for Chapter 3

1. Simplifying: $2(2)+(-1)=4-1=3$

2. Simplifying: $2(5)-1=10-1=9$

3. Simplifying: $-\frac{1}{3}(3)+2=-1+2=1$

4. Simplifying: $\frac{3}{2}(2)-3=3-3=0$

5. Simplifying: $-\frac{1}{2}(-3x+6)=\frac{3}{2}x-3$

6. Simplifying: $-\frac{2}{3}(x-3)=-\frac{2}{3}x+2$

7. Solving the equation:

$$\begin{aligned}6+3y&=6\\3y&=0\\y&=0\end{aligned}$$

8. Solving the equation:

$$\begin{aligned}2x-10&=20\\2x&=30\\x&=15\end{aligned}$$

9. Solving the equation:

$$\begin{aligned}7&=2x-1\\8&=2x\\x&=4\end{aligned}$$

10. Solving the equation:

$$\begin{aligned}3&=2x-1\\4&=2x\\x&=2\end{aligned}$$

11. Substituting $x=0$:

$$\begin{aligned}3(0)-5y&=10\\-5y&=10\\y&=-2\end{aligned}$$

12. Substituting $x=6$:

$$\begin{aligned}y&=\frac{2}{3}(6)+5\\y&=4+5\\y&=9\end{aligned}$$

13. Evaluating when $x=-2$ and $y=3$: $\frac{y-1}{x+4}=\frac{3-1}{-2+4}=\frac{2}{2}=1$

14. Evaluating when $x=-2$ and $y=3$: $\frac{y+7}{x-3}=\frac{3+7}{-2-3}=\frac{10}{-5}=-2$

15. Evaluating when $x=-2$ and $y=3$: $\frac{4-y}{-1-x}=\frac{4-3}{-1-(-2)}=\frac{1}{1}=1$

16. Evaluating when $x=-2$ and $y=3$: $\frac{3-y}{x+8}=\frac{3-3}{-2+8}=\frac{0}{6}=0$

17. Solving for y:

$$\begin{aligned}7x-2y&=14\\-2y&=-7x+14\\y&=\frac{7}{2}x-7\end{aligned}$$

18. Solving for y:

$$\begin{aligned}-2x-5y&=4\\-5y&=2x+4\\y&=-\frac{2}{5}x-\frac{4}{5}\end{aligned}$$

© 2012 Cengage Learning. All Rights Reserved. May not be scanned, copied or duplicated, or posted to a publicly accessible website, in whole or in part.

19. Solving for y:

$$y-3=-2(x+4)$$
$$y-3=-2x-8$$
$$y=-2x-5$$

20. Solving for y:

$$y+1=-\frac{2}{3}x+2$$
$$y=-\frac{2}{3}x+1$$

3.1 Paired Data and Graphing Ordered Pairs

1. Graphing the ordered pair:

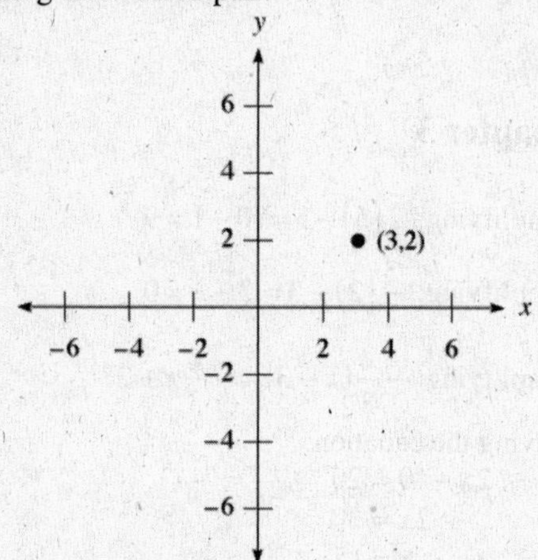

3. Graphing the ordered pair:

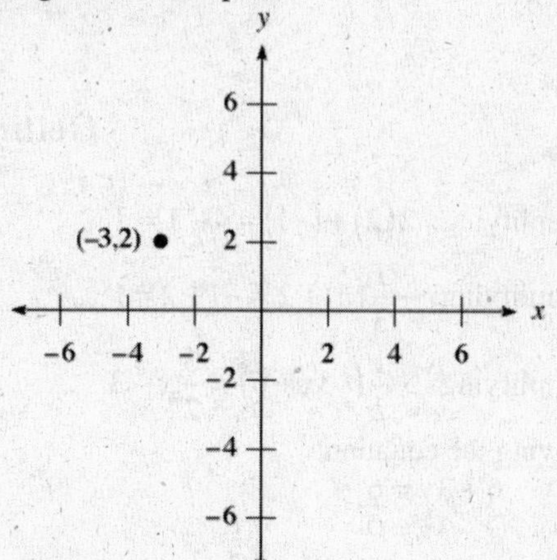

5. Graphing the ordered pair:

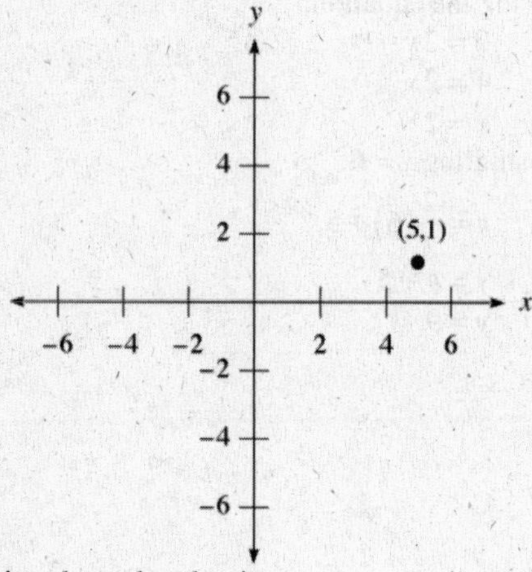

7. Graphing the ordered pair:

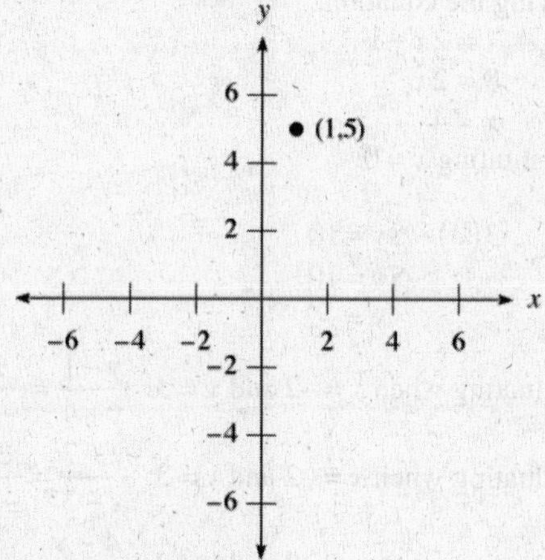

9. Graphing the ordered pair:

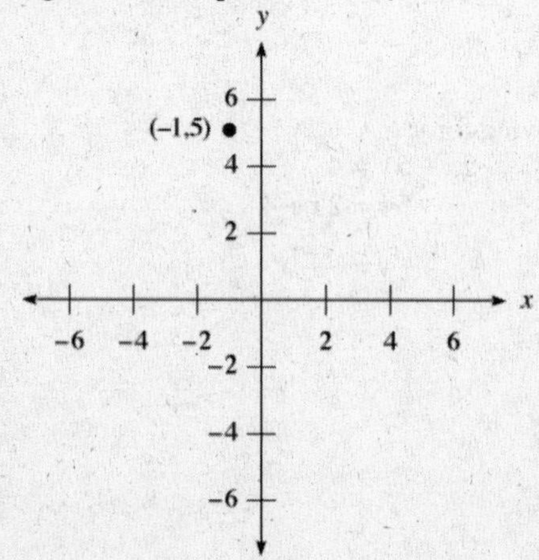

11. Graphing the ordered pair:

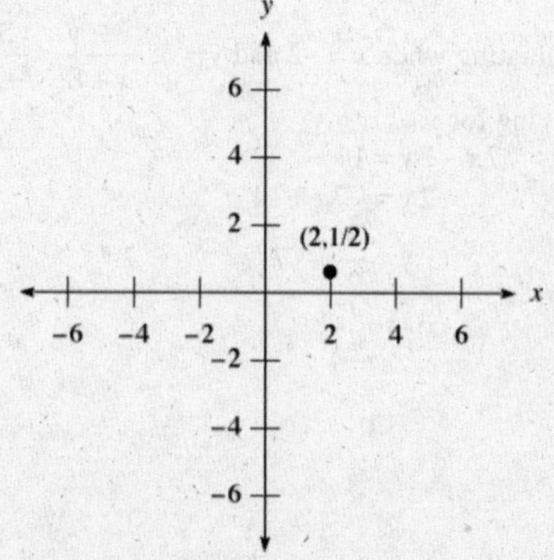

© 2012 Cengage Learning. All Rights Reserved. May not be scanned, copied or duplicated, or posted to a publicly accessible website, in whole or in part.

13. Graphing the ordered pair:

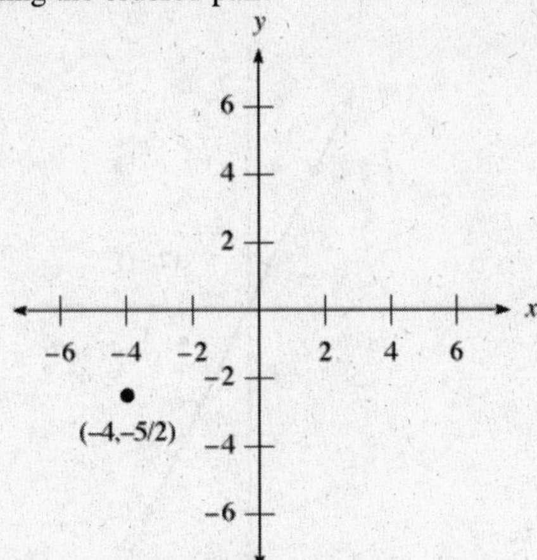

15. Graphing the ordered pair:

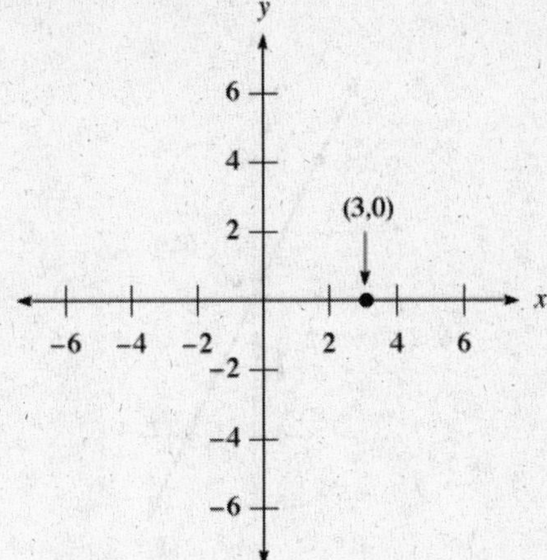

17. Graphing the ordered pair:

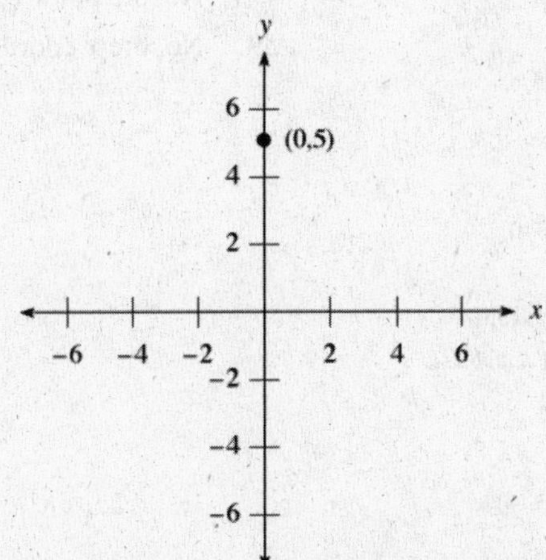

19. The coordinates are $(-4,4)$.

21. The coordinates are $(-4,2)$.

23. The coordinates are $(-3,0)$.

25. The coordinates are $(2,-2)$.

27. The coordinates are $(-5,-5)$.

29. Graphing the line:

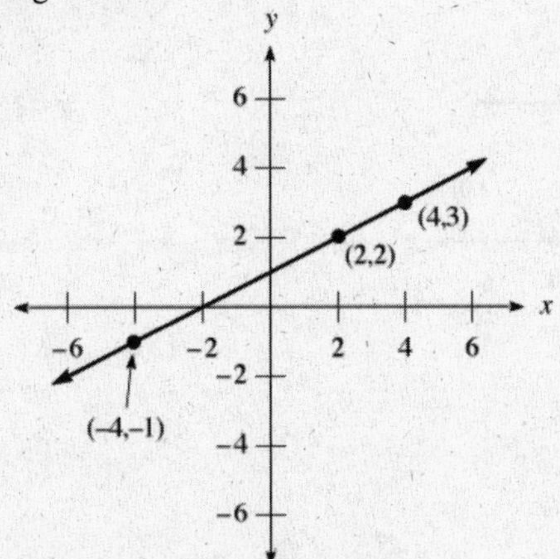

Yes, the point $(2,2)$ lies on the line.

31. Graphing the line:

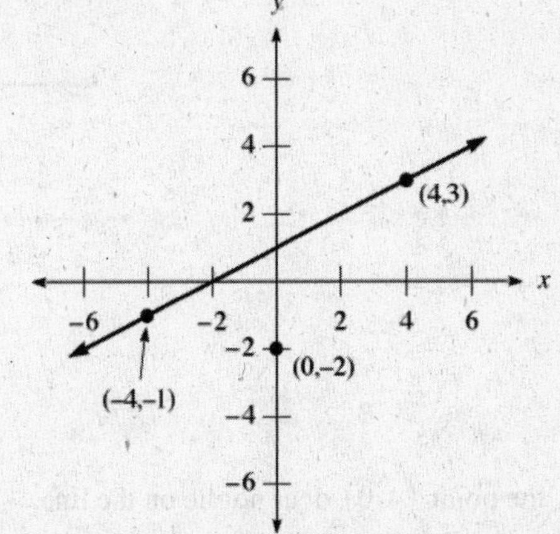

No, the point $(0,-2)$ does not lie on the line.

© 2012 Cengage Learning. All Rights Reserved. May not be scanned, copied or duplicated, or posted to a publicly accessible website, in whole or in part.

33. Graphing the line:

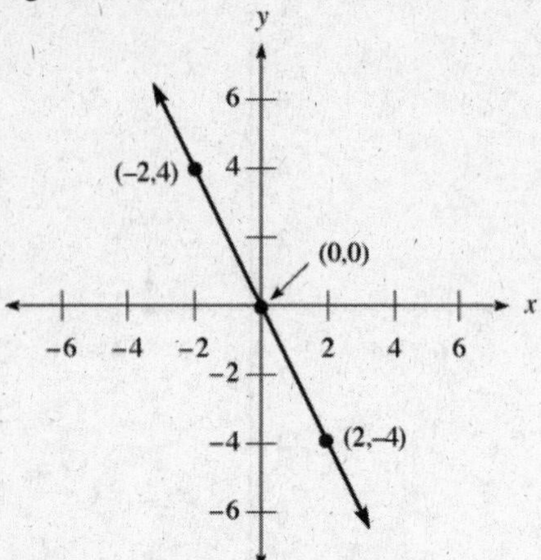

Yes, the point $(0,0)$ lies on the line.

35. Graphing the line:

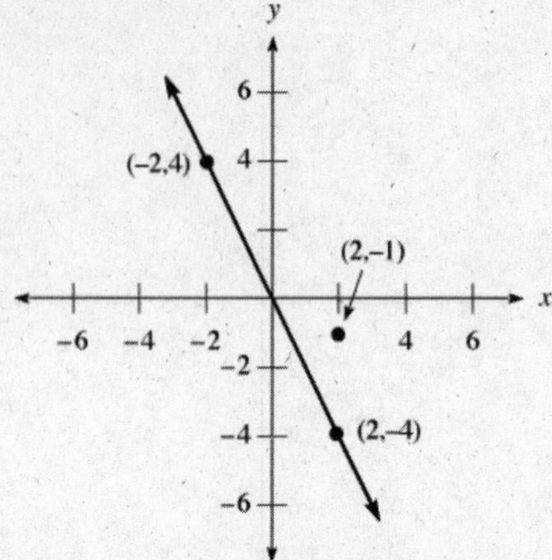

No, the point $(2,-1)$ does not lie on the line.

37. Graphing the line:

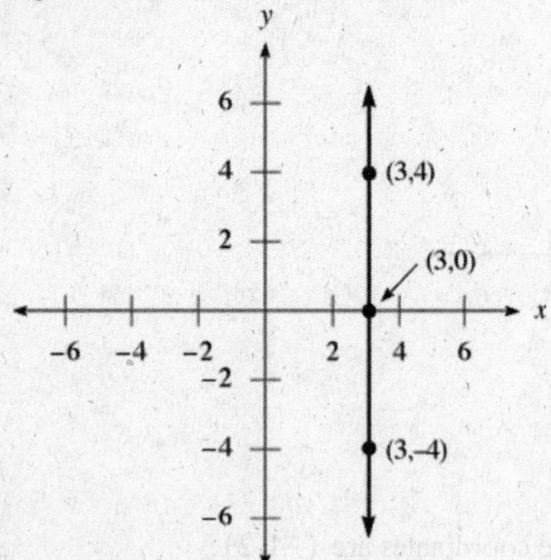

Yes, the point $(3,0)$ lies on the line.

39. No, the x-coordinate of every point on this line is 3.

41. Graphing the line:

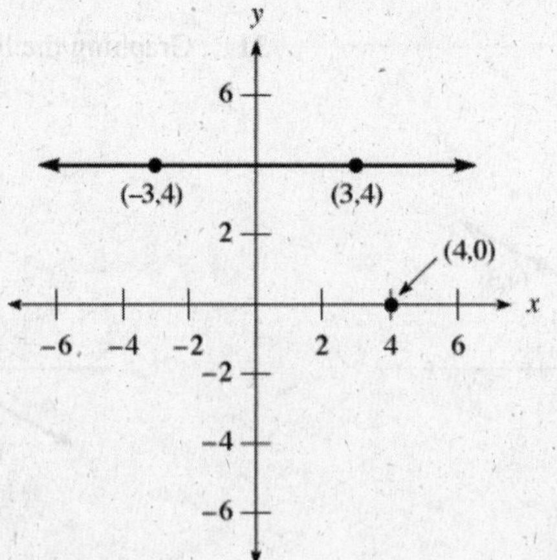

No, the point $(4,0)$ does not lie on the line.

43. No, the y-coordinate of every point on this line is 4.

45.
- **a.** The energy used is 60 watts.
- **b.** The energy used is 20 watts.
- **c.** The energy used is 100 watts.

© 2012 Cengage Learning. All Rights Reserved. May not be scanned, copied or duplicated, or posted to a publicly accessible website, in whole or in part.

47. **a.** Three ordered pairs on the graph are (5,40), (10,80), and (20,160).
b. She will earn \$320 for working 40 hours.
c. If her check is \$240, she worked 30 hours that week.
d. No. She should be paid \$280 for working 35 hours, not \$260.

49. Point A is (6 – 5,2) = (1,2), and point B is (6,2 + 5) = (6,7).

51. Point A is (7 – 5,2) = (2,2), point B is (2,2 + 3) = (2,5), and point C is (2 + 5,5) = (7,5).

53. **a.** Substituting $y = 4$:
$$\begin{aligned}2x+3(4)&=6\\2x+12&=6\\2x&=-6\\x&=-3\end{aligned}$$
b. Substituting $y = -2$:
$$\begin{aligned}2x+3(-2)&=6\\2x-6&=6\\2x&=12\\x&=6\end{aligned}$$
c. Substituting $x = 3$:
$$\begin{aligned}2(3)+3y&=6\\6+3y&=6\\3y&=0\\y&=0\end{aligned}$$
d. Substituting $x = 9$:
$$\begin{aligned}2(9)+3y&=6\\18+3y&=6\\3y&=-12\\y&=-4\end{aligned}$$

55. **a.** Substituting $y = 7$:
$$\begin{aligned}2x-1&=7\\2x&=8\\x&=4\end{aligned}$$
b. Substituting $y = 3$:
$$\begin{aligned}2x-1&=3\\2x&=4\\x&=2\end{aligned}$$
c. Substituting $x = 0$: $y=2(0)-1=0-1=-1$
d. Substituting $x = 5$: $y=2(5)-1=10-1=9$

57. Simplifying: $\frac{x}{5}+\frac{3}{5}=\frac{x+3}{5}$

59. Simplifying: $\frac{2}{7}-\frac{a}{7}=\frac{2-a}{7}$

61. Simplifying: $\frac{1}{14}-\frac{y}{7}=\frac{1}{14}-\frac{y}{7}\bullet\frac{2}{2}=\frac{1-2y}{14}$

63. Simplifying: $\frac{1}{2}+\frac{3}{x}=\frac{1}{2}\bullet\frac{x}{x}+\frac{3}{x}\bullet\frac{2}{2}=\frac{x+6}{2x}$

65. Simplifying: $\frac{5+x}{6}-\frac{5}{6}=\frac{5+x-5}{6}=\frac{x}{6}$

67. Simplifying: $\frac{4}{x}+\frac{1}{2}=\frac{4}{x}\bullet\frac{2}{2}+\frac{1}{2}\bullet\frac{x}{x}=\frac{8+x}{2x}$

3.2 Solutions to Linear Equations in Two Variables

1. Substituting $x = 0$, $y = 0$, and $y = -6$:
$$\begin{aligned}2(0)+y&=6\\0+y&=6\\y&=6\end{aligned}\qquad\begin{aligned}2x+0&=6\\2x&=6\\x&=3\end{aligned}\qquad\begin{aligned}2x+(-6)&=6\\2x&=12\\x&=6\end{aligned}$$
The ordered pairs are $(0,6)$, $(3,0)$, and $(6,-6)$.

3. Substituting $x = 0$, $y = 0$, and $x = -4$:
$$\begin{aligned}3(0)+4y&=12\\0+4y&=12\\4y&=12\\y&=3\end{aligned}\qquad\begin{aligned}3x+4(0)&=12\\3x+0&=12\\3x&=12\\x&=4\end{aligned}\qquad\begin{aligned}3(-4)+4y&=12\\-12+4y&=12\\4y&=24\\y&=6\end{aligned}$$
The ordered pairs are $(0,3)$, $(4,0)$, and $(-4,6)$.

5. Substituting $x = 1$, $y = 0$, and $x = 5$:
$$\begin{aligned}y&=4(1)-3\\y&=4-3\\y&=1\end{aligned}\qquad\begin{aligned}0&=4x-3\\3&=4x\\x&=\frac{3}{4}\end{aligned}\qquad\begin{aligned}y&=4(5)-3\\y&=20-3\\y&=17\end{aligned}$$
The ordered pairs are $(1,1)$, $\left(\frac{3}{4},0\right)$, and $(5,17)$.

© 2012 Cengage Learning. All Rights Reserved. May not be scanned, copied or duplicated, or posted to a publicly accessible website, in whole or in part.

7. Substituting $x = 2$, $y = 6$, and $x = 0$:

$$y = 7(2) - 1 \quad\quad 6 = 7x - 1 \quad\quad y = 7(0) - 1$$
$$y = 14 - 1 \quad\quad 7 = 7x \quad\quad y = 0 - 1$$
$$y = 13 \quad\quad x = 1 \quad\quad y = -1$$

The ordered pairs are $(2,13)$, $(1,6)$, and $(0,-1)$.

9. Substituting $y = 4$, $y = -3$, and $y = 0$ results (in each case) in $x = -5$. The ordered pairs are $(-5,4)$, $(-5,-3)$, and $(-5,0)$.

11. Completing the table:

x	y
1	3
−3	−9
4	12
6	18

13. Completing the table:

x	y
0	0
−1/2	−2
−3	−12
3	12

15. Completing the table:

x	y
2	3
3	2
5	0
9	−4

17. Completing the table:

x	y
2	0
3	2
1	−2
−3	−10

19. Completing the table:

x	y
0	−1
−1	−7
−3	−19
3/2	8

21. Substituting each ordered pair into the equation:

$$(2,3):\ 2(2) - 5(3) = 4 - 15 = -11 \neq 10$$
$$(0,-2):\ 2(0) - 5(-2) = 0 + 10 = 10$$
$$\left(\frac{5}{2},1\right):\ 2\left(\frac{5}{2}\right) - 5(1) = 5 - 5 = 0 \neq 10$$

Only the ordered pair $(0,-2)$ is a solution.

23. Substituting each ordered pair into the equation:

$$(1,5):\ 7(1) - 2 = 7 - 2 = 5$$
$$(0,-2):\ 7(0) - 2 = 0 - 2 = -2$$
$$(-2,-16):\ 7(-2) - 2 = -14 - 2 = -16$$

All the ordered pairs $(1,5)$, $(0,-2)$ and $(-2,-16)$ are solutions.

25. Substituting each ordered pair into the equation:

$$(1,6):\ 6(1) = 6$$
$$(-2,-12):\ 6(-2) = -12$$
$$(0,0):\ 6(0) = 0$$

All the ordered pairs $(1,6)$, $(-2,-12)$ and $(0,0)$ are solutions.

© 2012 Cengage Learning. All Rights Reserved. May not be scanned, copied or duplicated, or posted to a publicly accessible website, in whole or in part.

27. Substituting each ordered pair into the equation:

$(1,1):\ 1+1=2\neq 0$

$(2,-2):\ 2+(-2)=0$

$(3,3):\ 3+3=6\neq 0$

Only the ordered pair $(2,-2)$ is a solution.

29. Since $x = 3$, the ordered pair $(5,3)$ cannot be a solution. The ordered pairs $(3,0)$ and $(3,-3)$ are solutions.

31. Substituting $y = 120{,}000$:

$$\begin{aligned} 120{,}000 &= 40{,}000x - 80{,}200{,}000 \\ 80{,}320{,}000 &= 40{,}000x \\ x &= 2008 \end{aligned}$$

Substituting $x = 2009$:

$$\begin{aligned} y &= 40{,}000(2009) - 80{,}200{,}000 \\ y &= 80{,}360{,}00 - 80{,}200{,}000 \\ y &= 160{,}000 \end{aligned}$$

Substituting $y = 200{,}000$:

$$\begin{aligned} 200{,}000 &= 40{,}000x - 80{,}200{,}000 \\ 80{,}400{,}000 &= 40{,}000x \\ x &= 2010 \end{aligned}$$

The ordered pairs are (2008, 120,000), (2009, 160,000) and (2010, 200,000).

33. Substituting $w = 3$:

$$\begin{aligned} 2l + 2(3) &= 30 \\ 2l + 6 &= 30 \\ 2l &= 24 \\ l &= 12 \end{aligned}$$

The length is 12 inches.

35.
a. This is correct, since; $y = 12(5) = \$60$

b. This is not correct, since: $y = 12(9) = \$108$. Her check should be for $108.

c. This is not correct, since: $y = 12(7) = \$84$. Her check should be for $84.

d. This is correct, since: $y = 12(14) = \$168$

37.
a. Substituting $t = 5$: $V = -45{,}000(5) + 600{,}000 = -225{,}000 + 600{,}000 = \$375{,}000$

b. Solving when $V = 330{,}000$:

$$\begin{aligned} -45{,}000t + 600{,}000 &= 330{,}000 \\ -45{,}000t &= -270{,}000 \\ t &= 6 \end{aligned}$$

The crane will be worth $330,000 at the end of 6 years.

c. Substituting $t = 9$: $V = -45{,}000(9) + 600{,}000 = -405{,}000 + 600{,}000 = \$195{,}000$

No, the crane will be worth $195,000 after 9 years.

d. The crane cost $600,000 (the value when $t = 0$).

39. Substituting $x = 4$:

$$\begin{aligned} 3(4) + 2y &= 6 \\ 12 + 2y &= 6 \\ 2y &= -6 \\ y &= -3 \end{aligned}$$

41. Substituting $x = 0$: $y = -\frac{1}{3}(0) + 2 = 0 + 2 = 2$

43. Substituting $x = 2$: $y = \frac{3}{2}(2) - 3 = 3 - 3 = 0$

45. Solving for y:

$$\begin{aligned} 5x + y &= 4 \\ y &= -5x + 4 \end{aligned}$$

47. Solving for y:

$$\begin{aligned} 3x - 2y &= 6 \\ -2y &= -3x + 6 \\ y &= \frac{3}{2}x - 3 \end{aligned}$$

49. Simplifying: $\frac{11(-5) - 17}{2(-6)} = \frac{-55 - 17}{-12} = \frac{-72}{-12} = 6$

51. Simplifying: $\frac{13(-6) + 18}{4(-5)} = \frac{-78 + 18}{-20} = \frac{-60}{-20} = 3$

© 2012 Cengage Learning. All Rights Reserved. May not be scanned, copied or duplicated, or posted to a publicly accessible website, in whole or in part.

53. Simplifying: $\dfrac{7^2-5^2}{(7-5)^2}=\dfrac{49-25}{2^2}=\dfrac{24}{4}=6$

55. Simplifying: $\dfrac{-3\bullet 4^2-3\bullet 2^4}{-3(8)}=\dfrac{-3\bullet 16-3\bullet 16}{-3(8)}=\dfrac{-48-48}{-24}=\dfrac{-96}{-24}=4$

3.3 Graphing Linear Equations in Two Variables

1. The ordered pairs are $(0,4)$, $(2,2)$, and $(4,0)$:

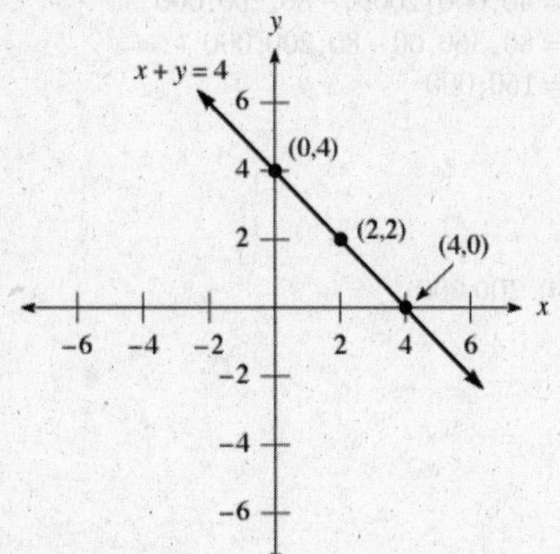

3. The ordered pairs are $(0,3)$, $(2,1)$, and $(4,-1)$:

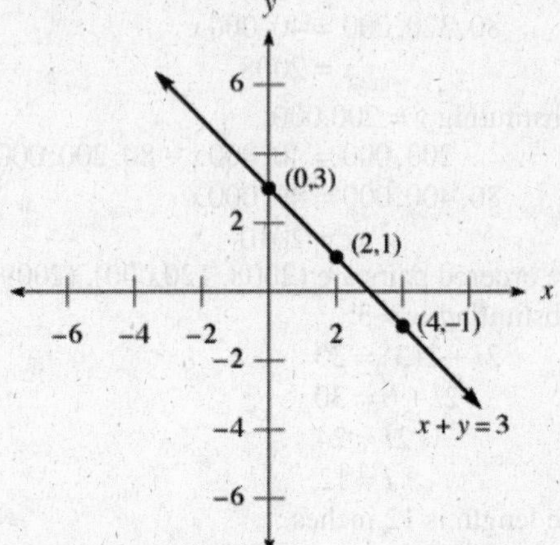

5. The ordered pairs are $(0,0)$, $(-2,-4)$, and $(2,4)$:

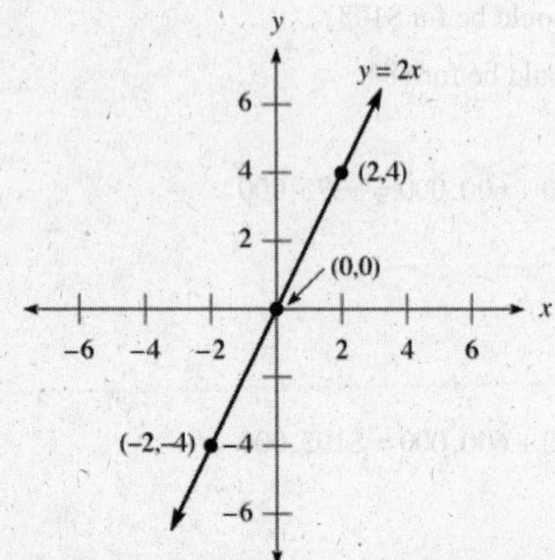

7. The ordered pairs are $(-3,-1)$, $(0,0)$, and $(3,1)$:

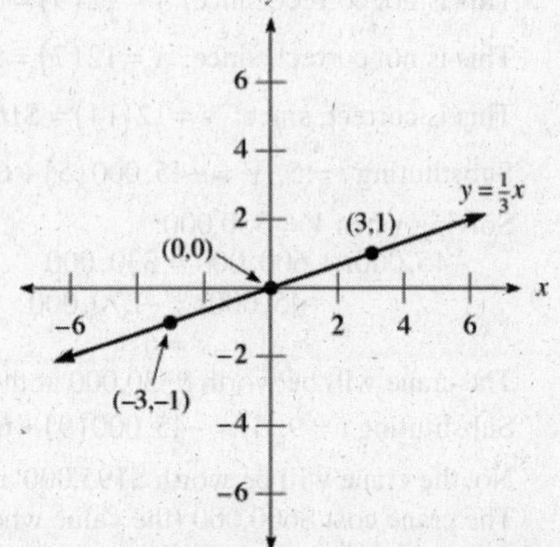

9. The ordered pairs are $(0,1)$, $(-1,-1)$, and $(1,3)$:

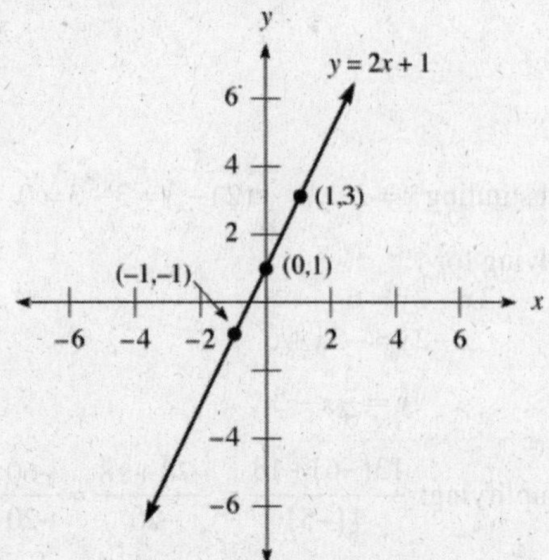

11. The ordered pairs are $(0,4)$, $(-1,4)$, and $(2,4)$:

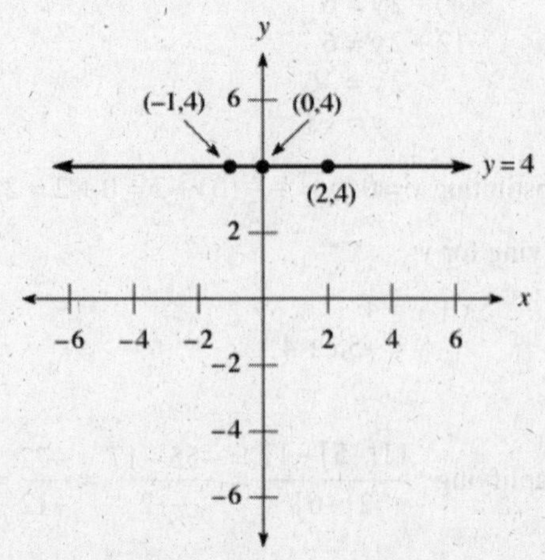

© 2012 Cengage Learning. All Rights Reserved. May not be scanned, copied or duplicated, or posted to a publicly accessible website, in whole or in part.

13. The ordered pairs are $(-2,2)$, $(0,3)$, and $(2,4)$:

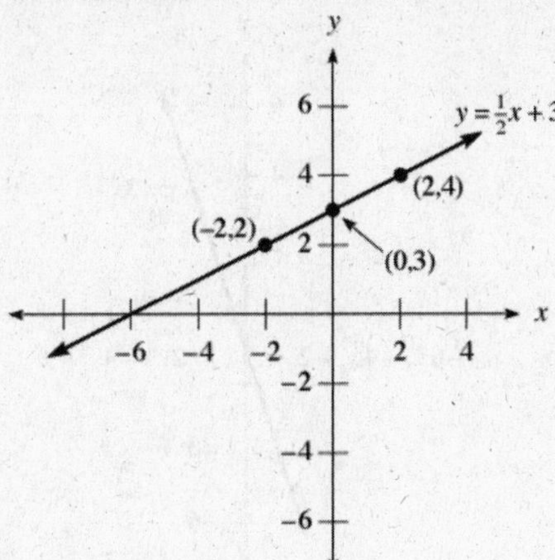

15. The ordered pairs are $(-3,3)$, $(0,1)$, and $(3,-1)$:

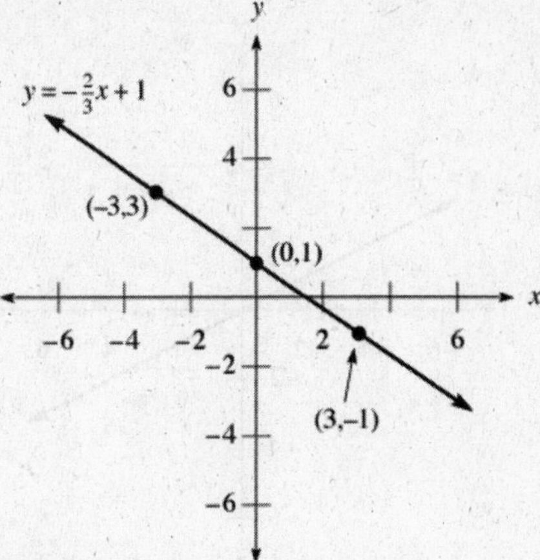

17. Solving for *y*:

$$2x+y=3$$
$$y=-2x+3$$

The ordered pairs are $(-1,5)$, $(0,3)$, and $(1,1)$:

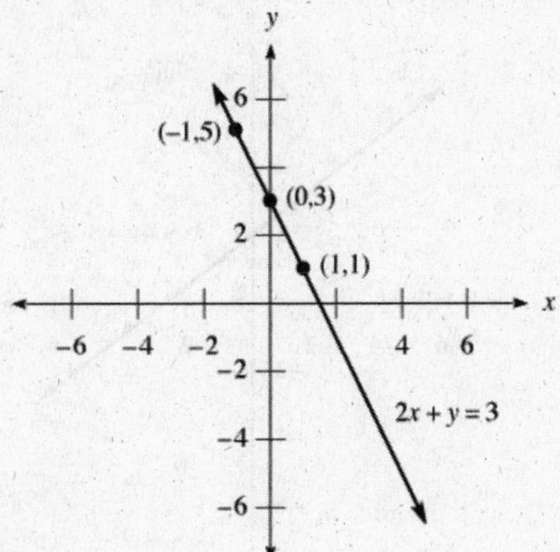

19. Solving for *y*:

$$3x+2y=6$$
$$2y=-3x+6$$
$$y=-\frac{3}{2}x+3$$

The ordered pairs are $(0,3)$, $(2,0)$, and $(4,-3)$:

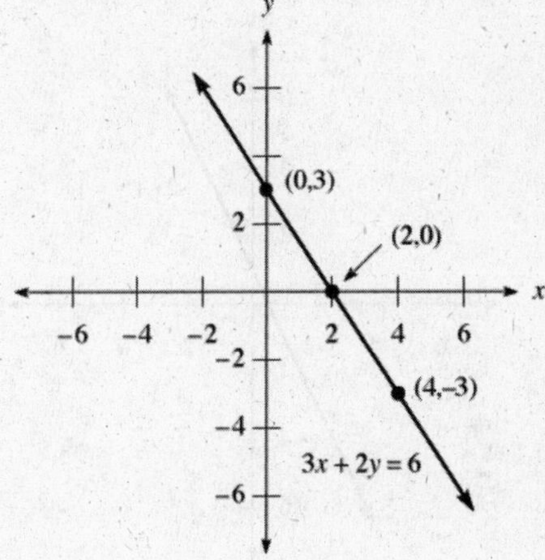

21. Solving for *y*:

$$-x+2y=6$$
$$2y=x+6$$
$$y=\frac{1}{2}x+3$$

The ordered pairs are $(-2,2)$, $(0,3)$, and $(2,4)$:

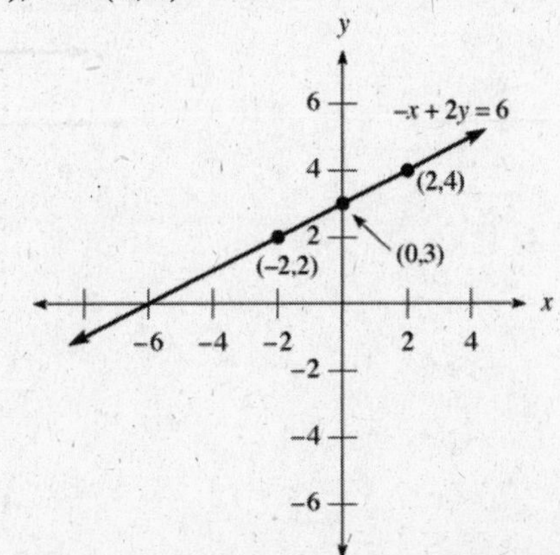

© 2012 Cengage Learning. All Rights Reserved. May not be scanned, copied or duplicated, or posted to a publicly accessible website, in whole or in part.

23. Three solutions are $(-4,2)$, $(0,0)$, and $(4,-2)$:

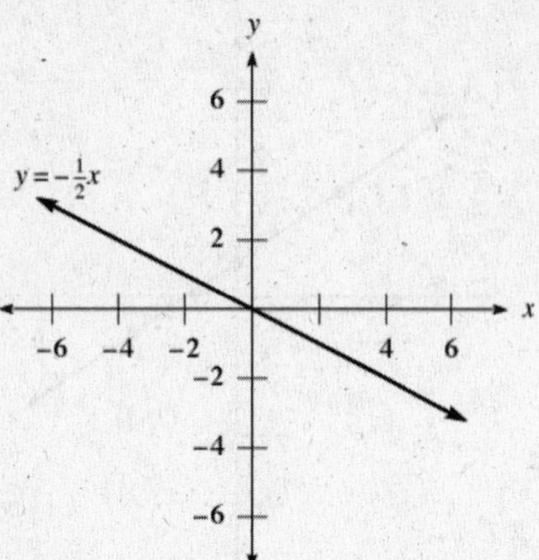

25. Three solutions are $(-1,-4)$, $(0,-1)$, and $(1,2)$:

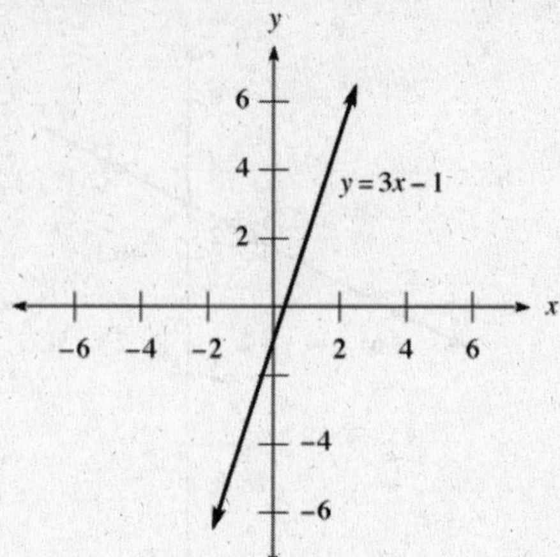

27. Solving for *y*:

$$-2x+y=1$$
$$y=2x+1$$

Three solutions are $(-2,-3)$, $(0,1)$, and $(2,5)$:

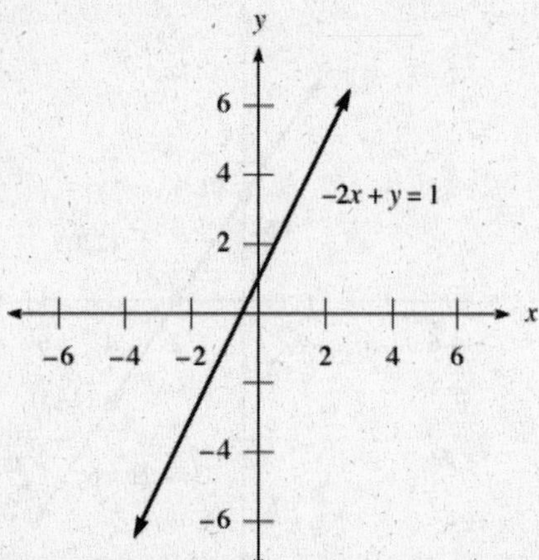

29. Solving for *y*:

$$3x+4y=8$$
$$4y=-3x+8$$
$$y=-\frac{3}{4}x+2$$

Three solutions are $(-4,5)$, $(0,2)$, and $(4,-1)$:

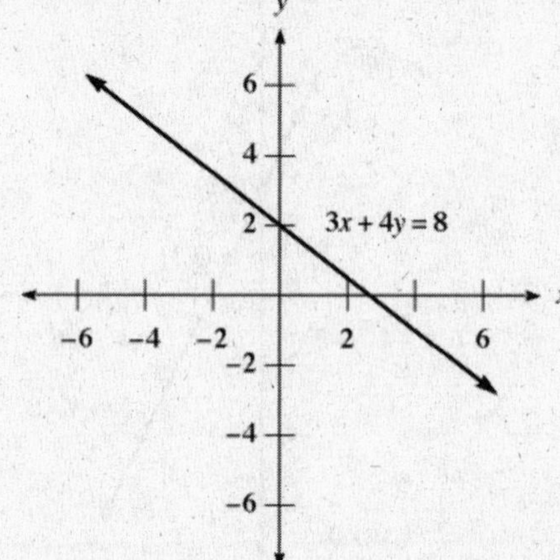

31. Three solutions are $(-2,-4)$, $(-2,0)$, and $(-2,4)$:

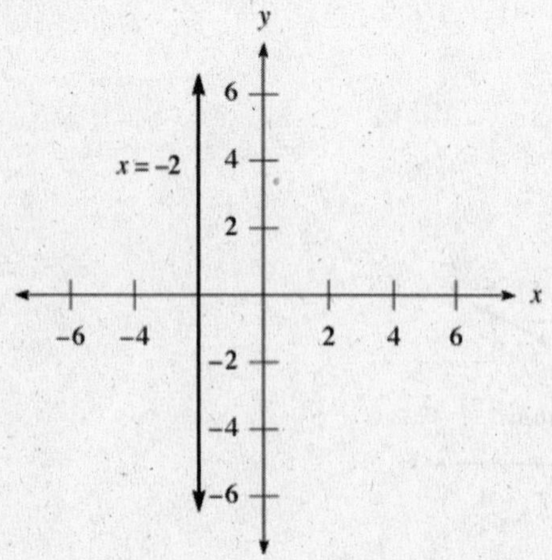

33. Three solutions are $(-4,2)$, $(0,2)$, and $(4,2)$:

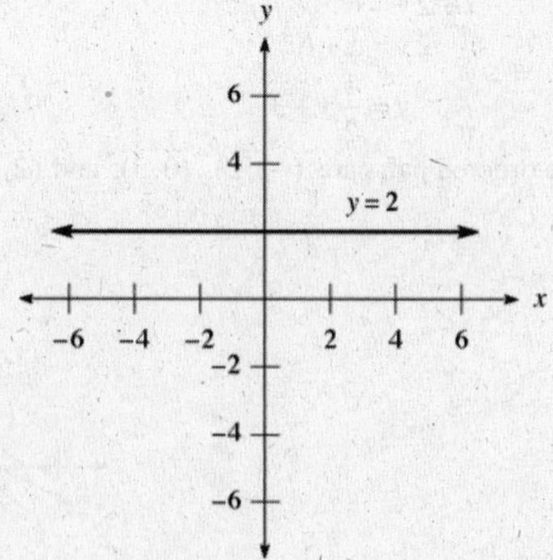

© 2012 Cengage Learning. All Rights Reserved. May not be scanned, copied or duplicated, or posted to a publicly accessible website, in whole or in part.

35. Graphing the equation:

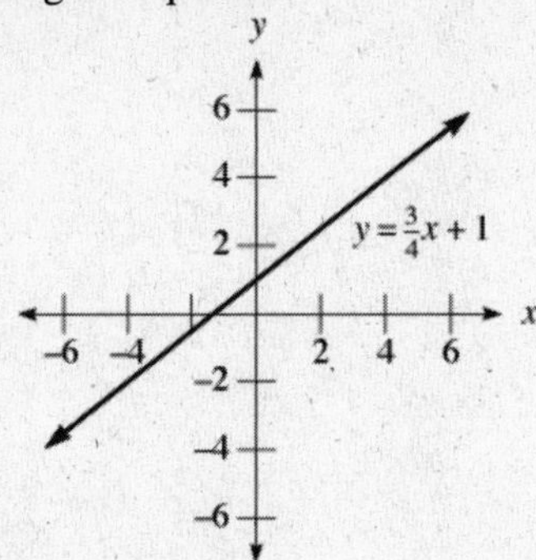

37. Graphing the equation:

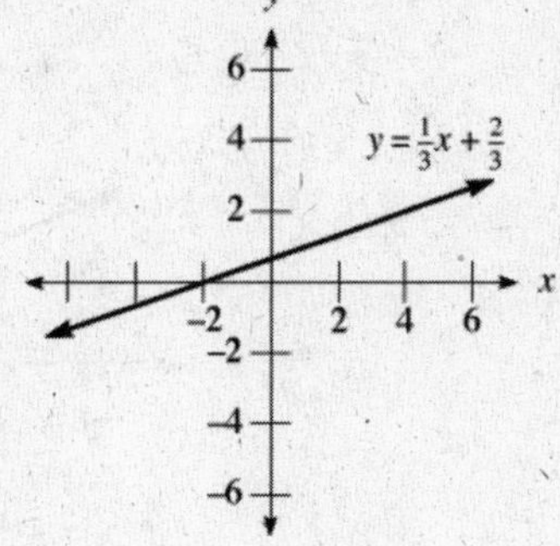

39. Graphing the equation:

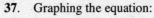

41. Completing the table:

Equation	H, V, and/or O
$x = 3$	V
$y = 3$	H
$y = 3x$	O
$y = 0$	O,H

43. Completing the table:

Equation	H, V, and/or O
$x = -\frac{3}{5}$	V
$y = -\frac{3}{5}$	H
$y = -\frac{3}{5}x$	O
$x = 0$	O,V

45. **a.** Solving the equation:

$$2x + 5 = 10$$
$$2x = 5$$
$$x = \frac{5}{2}$$

b. Substituting $y = 0$:

$$2x + 5(0) = 10$$
$$2x = 10$$
$$x = 5$$

c. Substituting $x = 0$:

$$2(0) + 5y = 10$$
$$5y = 10$$
$$y = 2$$

© 2012 Cengage Learning. All Rights Reserved. May not be scanned, copied or duplicated, or posted to a publicly accessible website, in whole or in part.

d. Graphing the line:

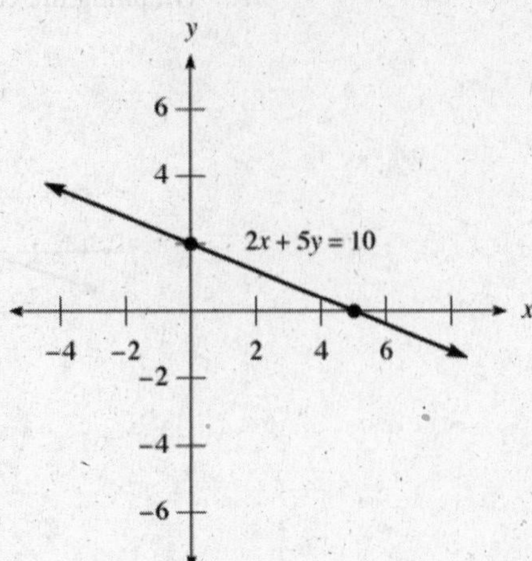

e. Solving for y:

$$2x+5y=10$$
$$5y=-2x+10$$
$$y=-\frac{2}{5}x+2$$

47. **a.** Yes, (2000, 7500) is a point on the graph.

b. No, (2004, 15000) is not a point on the graph.

c. Yes, (2007, 15000) is a point on the graph.

49. **a.** Substituting $y = 0$:

$$3x-2(0)=6$$
$$3x-0=6$$
$$3x=6$$
$$x=2$$

b. Substituting $x = 0$:

$$3(0)-2y=6$$
$$0-2y=6$$
$$-2y=6$$
$$y=-3$$

51. **a.** Substituting $y = 0$:

$$-x+2(0)=4$$
$$-x+0=4$$
$$-x=4$$
$$x=-4$$

b. Substituting $x = 0$:

$$-(0)+2y=4$$
$$0+2y=4$$
$$2y=4$$
$$y=2$$

53. **a.** Substituting $y = 0$:

$$0=-\frac{1}{3}x+2$$
$$-\frac{1}{3}x=-2$$
$$x=6$$

b. Substituting $x = 0$:

$$y=-\frac{1}{3}(0)+2$$
$$y=2$$

55. Simplifying: $\frac{1}{2}(4x+10)=\frac{1}{2}\bullet 4x+\frac{1}{2}\bullet 10=2x+5$

57. Simplifying: $\frac{2}{3}(3x-9)=\frac{2}{3}\bullet 3x-\frac{2}{3}\bullet 9=2x-6$

59. Simplifying: $\frac{3}{4}(4x+10)=\frac{3}{4}\bullet 4x+\frac{3}{4}\bullet 10=3x+\frac{15}{2}$

61. Simplifying: $\frac{3}{5}(10x+15)=\frac{3}{5}\bullet 10x+\frac{3}{5}\bullet 15=6x+9$

63. Simplifying: $5\left(\frac{2}{5}x+10\right)=5\bullet\frac{2}{5}x+5\bullet 10=2x+50$

65. Simplifying: $4\left(\frac{3}{2}x-7\right)=4\bullet\frac{3}{2}x-4\bullet 7=6x-28$

67. Simplifying: $\frac{3}{4}(2x+12y)=\frac{3}{4}\bullet 2x+\frac{3}{4}\bullet 12y=\frac{3}{2}x+9y$

69. Simplifying: $\frac{1}{2}(5x-10y)+6=\frac{1}{2}\bullet 5x-\frac{1}{2}\bullet 10y+6=\frac{5}{2}x-5y+6$

© 2012 Cengage Learning. All Rights Reserved. May not be scanned, copied or duplicated, or posted to a publicly accessible website, in whole or in part.

3.4 More on Graphing: Intercepts

1. To find the x-intercept, let $y = 0$:

$$2x + 0 = 4$$
$$2x = 4$$
$$x = 2$$

To find the y-intercept, let $x = 0$:

$$2(0) + y = 4$$
$$0 + y = 4$$
$$y = 4$$

Graphing the line:

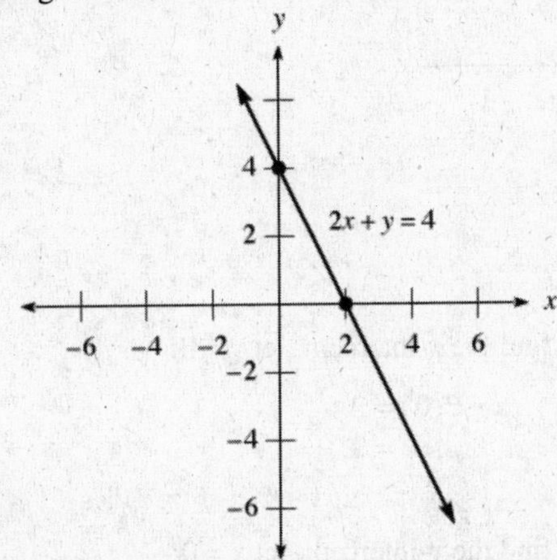

3. To find the x-intercept, let $y = 0$:

$$-x + 0 = 3$$
$$-x = 3$$
$$x = -3$$

To find the y-intercept, let $x = 0$:

$$-0 + y = 3$$
$$y = 3$$

Graphing the line:

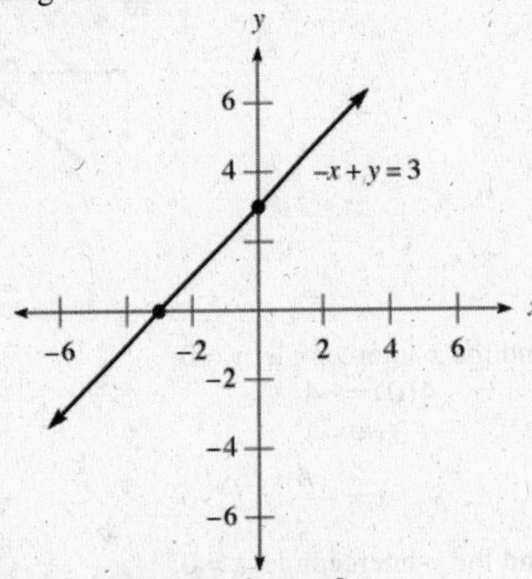

5. To find the x-intercept, let $y = 0$:

$$-x + 2(0) = 2$$
$$-x = 2$$
$$x = -2$$

To find the y-intercept, let $x = 0$:

$$-0 + 2y = 2$$
$$2y = 2$$
$$y = 1$$

Graphing the line:

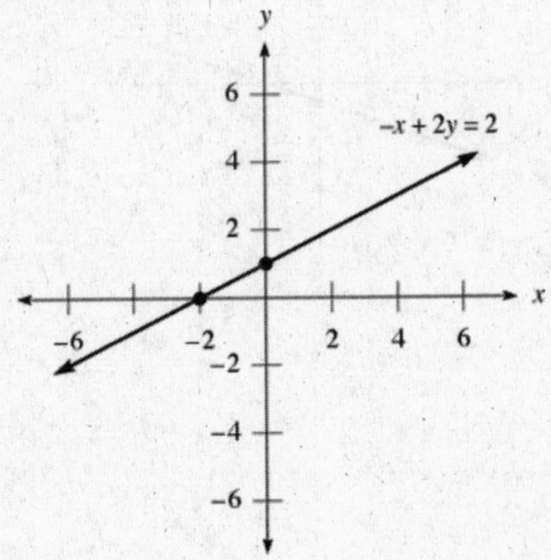

7. To find the x-intercept, let $y = 0$:

$$5x + 2(0) = 10$$
$$5x = 10$$
$$x = 2$$

To find the y-intercept, let $x = 0$:

$$5(0) + 2y = 10$$
$$2y = 10$$
$$y = 5$$

Graphing the line:

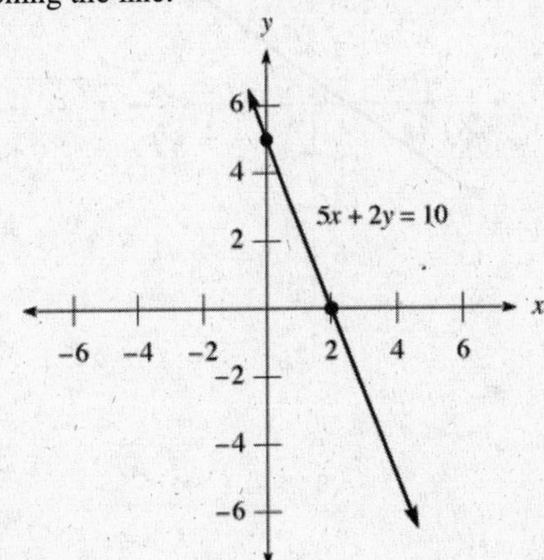

© 2012 Cengage Learning. All Rights Reserved. May not be scanned, copied or duplicated, or posted to a publicly accessible website, in whole or in part.

9. To find the x-intercept, let $y = 0$:

$$-4x + 5(0) = 20$$
$$-4x = 20$$
$$x = -5$$

To find the y-intercept, let $x = 0$:

$$-4(0) + 5y = 20$$
$$5y = 20$$
$$y = 4$$

Graphing the line:

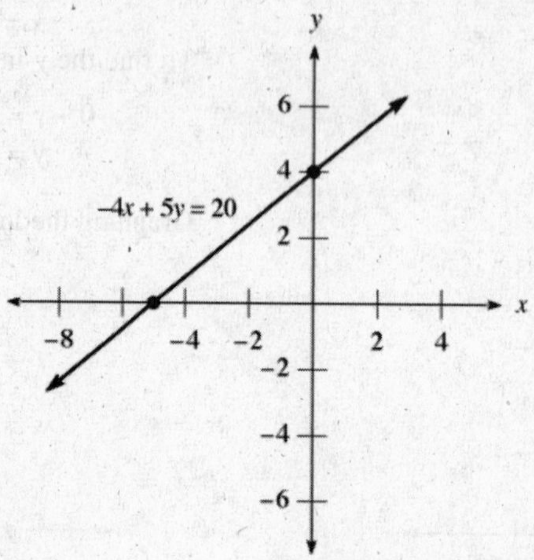

11. To find the x-intercept, let $y = 0$:

$$3x - 4(0) = -4$$
$$3x = -4$$
$$x = -\frac{4}{3}$$

To find the y-intercept, let $x = 0$:

$$3(0) - 4y = -4$$
$$-4y = -4$$
$$y = 1$$

Graphing the line:

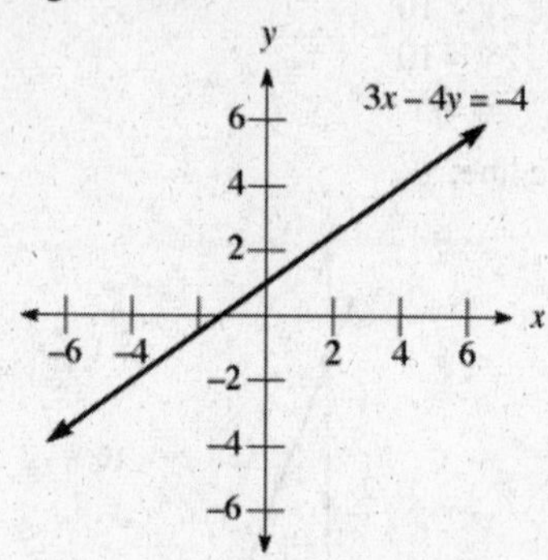

13. To find the x-intercept, let $y = 0$:

$$x - 3(0) = 2$$
$$x = 2$$

To find the y-intercept, let $x = 0$:

$$0 - 3y = 2$$
$$-3y = 2$$
$$y = -\frac{2}{3}$$

Graphing the line:

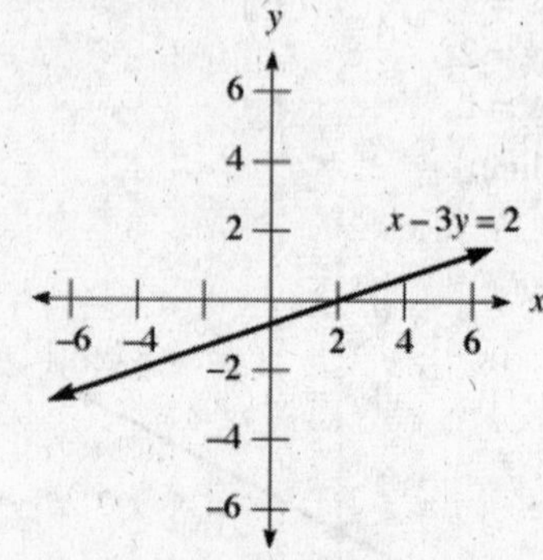

© 2012 Cengage Learning. All Rights Reserved. May not be scanned, copied or duplicated, or posted to a publicly accessible website, in whole or in part.

15. To find the x-intercept, let $y = 0$:

$$2x - 3(0) = -2$$
$$2x = -2$$
$$x = -1$$

To find the y-intercept, let $x = 0$:

$$2(0) - 3y = -2$$
$$-3y = -2$$
$$y = \frac{2}{3}$$

Graphing the line:

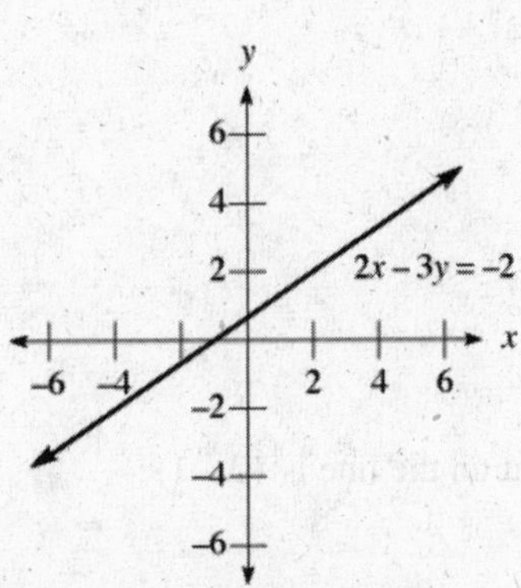

17. To find the x-intercept, let $y = 0$:

$$2x - 6 = 0$$
$$2x = 6$$
$$x = 3$$

To find the y-intercept, let $x = 0$:

$$y = 2(0) - 6$$
$$y = -6$$

Graphing the line:

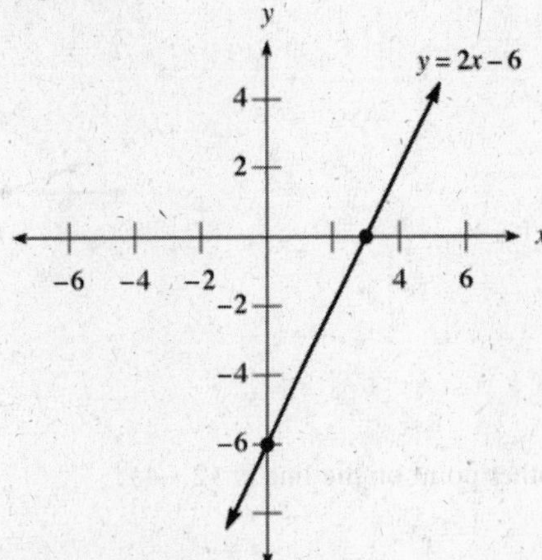

19. To find the x-intercept, let $y = 0$:

$$2x - 1 = 0$$
$$2x = 1$$
$$x = \frac{1}{2}$$

To find the y-intercept, let $x = 0$:

$$y = 2(0) - 1$$
$$y = -1$$

Graphing the line:

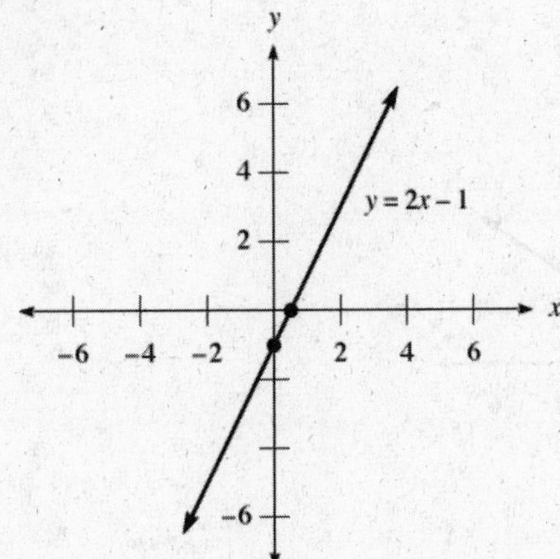

21. To find the x-intercept, let $y = 0$:

$$\frac{1}{2}x + 3 = 0$$
$$\frac{1}{2}x = -3$$
$$x = -6$$

To find the y-intercept, let $x = 0$:

$$y = \frac{1}{2}(0) + 3$$
$$y = 3$$

Graphing the line:

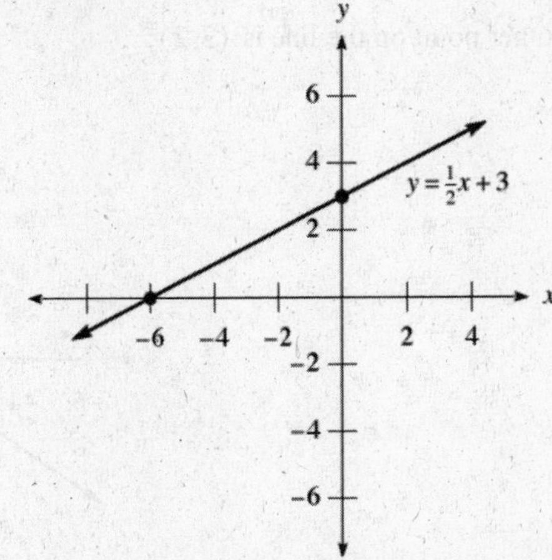

© 2012 Cengage Learning. All Rights Reserved. May not be scanned, copied or duplicated, or posted to a publicly accessible website, in whole or in part.

23. To find the x-intercept, let $y = 0$:

$$-\frac{1}{3}x - 2 = 0$$
$$-\frac{1}{3}x = 2$$
$$x = -6$$

To find the y-intercept, let $x = 0$:

$$y = -\frac{1}{3}(0) - 2$$
$$y = -2$$

Graphing the line:

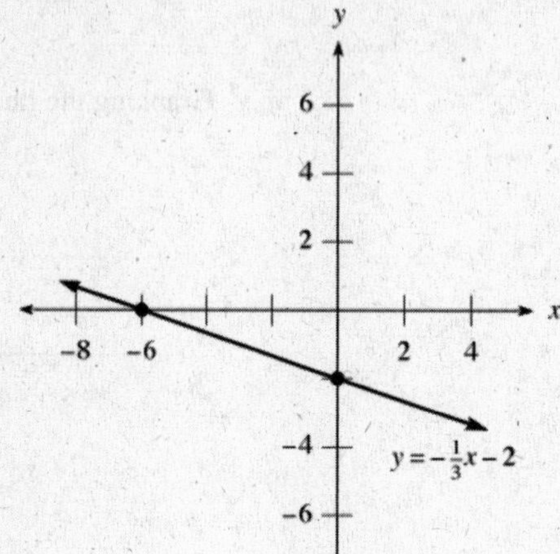

25. Another point on the line is $(2,-4)$:

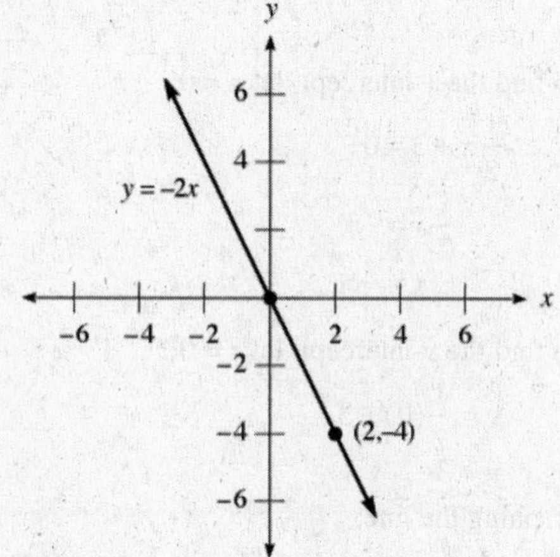

27. Another point on the line is $(3,-1)$:

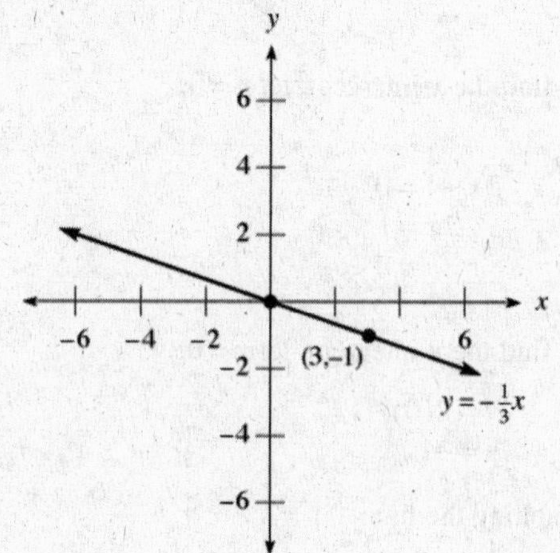

29. Another point on the line is $(3,2)$:

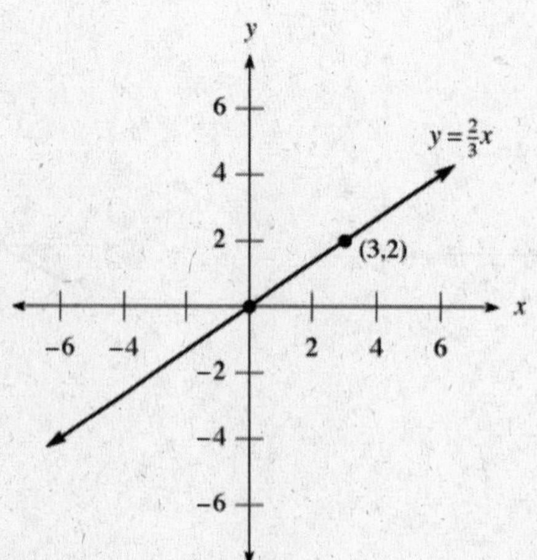

© 2012 Cengage Learning. All Rights Reserved. May not be scanned, copied or duplicated, or posted to a publicly accessible website, in whole or in part.

31. Completing the table:

Equation	x-intercept	y-intercept
$3x+4y=12$	4	3
$3x+4y=4$	$4/3$	1
$3x+4y=3$	1	$3/4$
$3x+4y=2$	$2/3$	$1/2$

33. Completing the table:

Equation	x-intercept	y-intercept
$x-3y=2$	2	$-2/3$
$y=\frac{1}{3}x-\frac{2}{3}$	2	$-2/3$
$x-3y=0$	0	0
$y=\frac{1}{3}x$	0	0

35. **a.** Solving the equation:

$$2x-3=-3$$
$$2x=0$$
$$x=0$$

b. Substituting $y=0$:

$$2x-3(0)=-3$$
$$2x-0=-3$$
$$2x=-3$$
$$x=-\frac{3}{2}$$

c. Substituting $x=0$:

$$2(0)-3y=-3$$
$$0-3y=-3$$
$$-3y=-3$$
$$y=1$$

d. Graphing the equation:

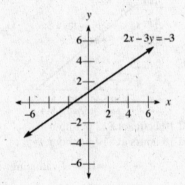

e. Solving for y:

$$2x-3y=-3$$
$$-3y=-2x-3$$
$$y=\frac{2}{3}x+1$$

37. Graphing the line:

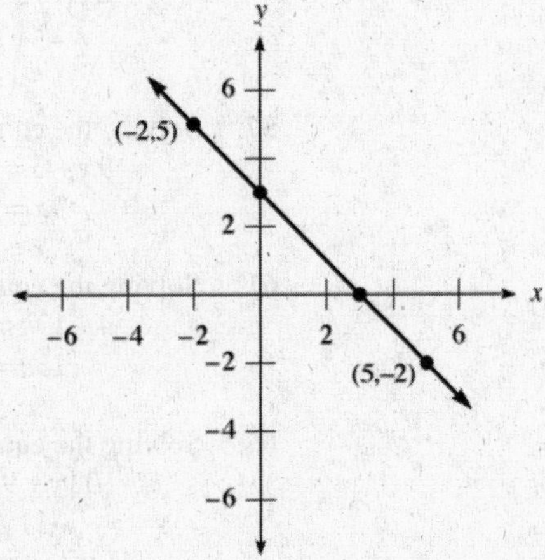

The x- and y-intercepts are both 3.

© 2012 Cengage Learning. All Rights Reserved. May not be scanned, copied or duplicated, or posted to a publicly accessible website, in whole or in part.

39. The x-intercept is -1 and the y-intercept is -3.

41. Completing the table:

x	y
–2	1
0	–1
–1	0
1	–2

43. Finding the x-intercept:

$$0=\frac{1}{175}x+\frac{17}{7}$$
$$\frac{1}{175}x=-\frac{17}{7}$$
$$x=-425$$

Finding the y-intercept:

$$y=\frac{1}{175}(0)+\frac{17}{7}$$
$$y=\frac{17}{7}$$

45. **a.** The equation is $10x+12y=480$.

b. Let $y = 0$:

$$10x+12(0)=480$$
$$10x=480$$
$$x=48$$

Let $x = 0$:

$$10(0)+12y=480$$
$$12y=480$$
$$y=40$$

The x-intercept is 48 and the y-intercept is 40.

c. Graphing the equation in the first quadrant:

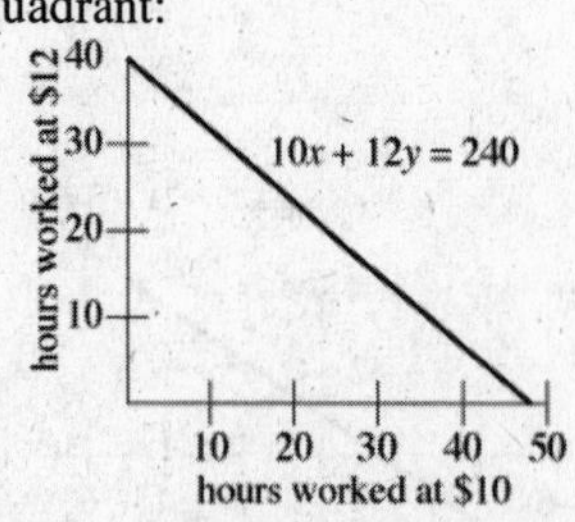

d. When $x = 36$, $y = 10$. She worked 10 hours at $12 per hour.

e. When $y = 25$, $x = 18$. She worked 18 hours at $10 per hour.

47. **a.** Evaluating: $\frac{5-2}{3-1}=\frac{3}{2}$

b. Evaluating: $\frac{2-5}{1-3}=\frac{-3}{-2}=\frac{3}{2}$

49. **a.** Evaluating when $x = 3$ and $y = 5$: $\frac{y-2}{x-1}=\frac{5-2}{3-1}=\frac{3}{2}$

b. Evaluating when $x = 3$ and $y = 5$: $\frac{2-y}{1-x}=\frac{2-5}{1-3}=\frac{-3}{-2}=\frac{3}{2}$

51. Solving the equation:

$$-12y-4=-148$$
$$-12y=-144$$
$$y=12$$

53. Solving the equation:

$$-5y-4=51$$
$$-5y=55$$
$$y=-11$$

55. Solving the equation:

$$11x-12=-78$$
$$11x=-66$$
$$x=-6$$

57. Solving the equation:

$$9x+3=66$$
$$9x=63$$
$$x=7$$

59. Solving the equation:

$$-9c-6=12$$
$$-9c=18$$
$$c=-2$$

61. Solving the equation:

$$4+13c=-9$$
$$13c=-13$$
$$c=-1$$

63. Solving the equation:

$$3y-12=30$$
$$3y=42$$
$$y=14$$

65. Solving the equation:

$$-11y+9=75$$
$$-11y=66$$
$$y=-6$$

© 2012 Cengage Learning. All Rights Reserved. May not be scanned, copied or duplicated, or posted to a publicly accessible website, in whole or in part.

3.5 The Slope of a Line

1. The slope is given by: $m = \frac{4-1}{4-2} = \frac{3}{2}$

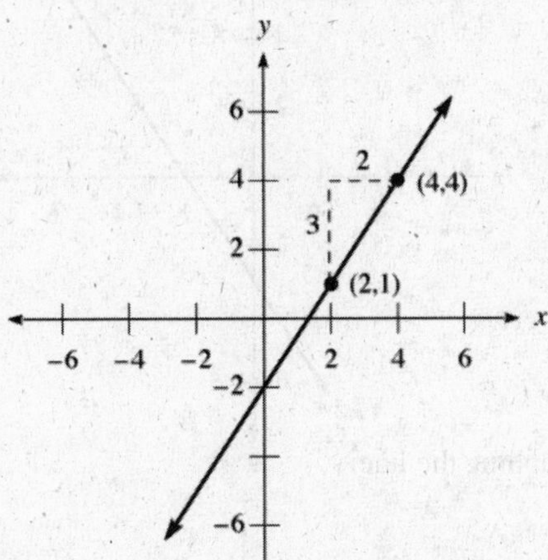

3. The slope is given by: $m = \frac{2-4}{5-1} = \frac{-2}{4} = -\frac{1}{2}$

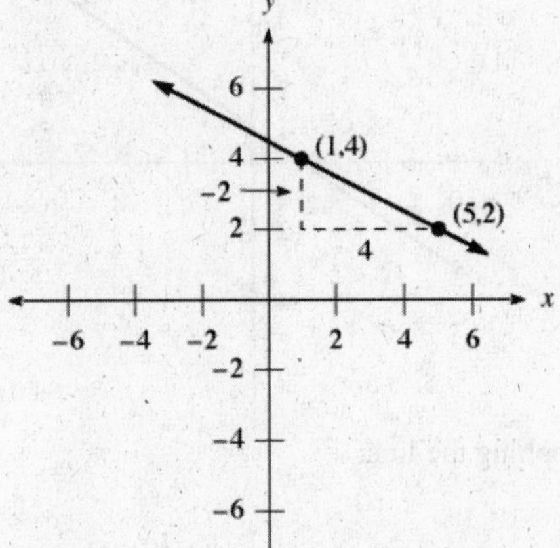

5. The slope is given by: $m = \frac{2-(-3)}{4-1} = \frac{5}{3}$

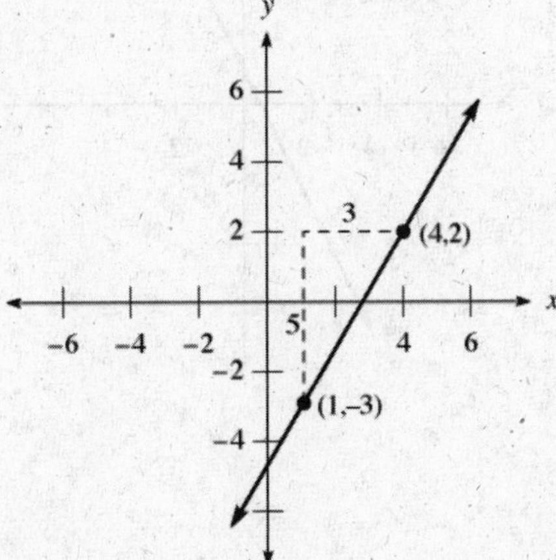

7. The slope is given by: $m = \frac{3-(-2)}{1-(-3)} = \frac{5}{4}$

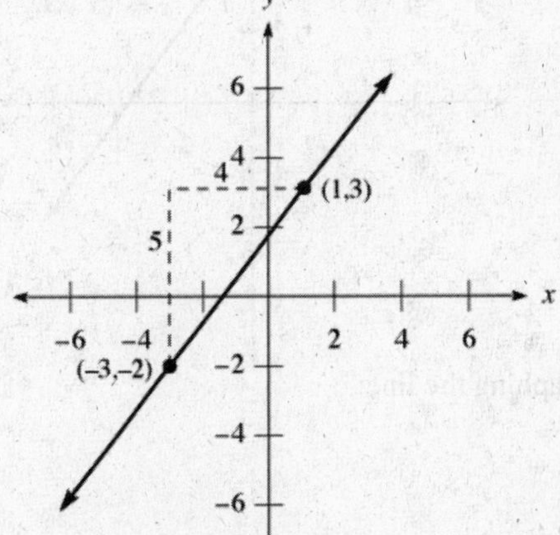

9. The slope is given by: $m = \frac{-2-2}{3-(-3)} = \frac{-4}{6} = -\frac{2}{3}$

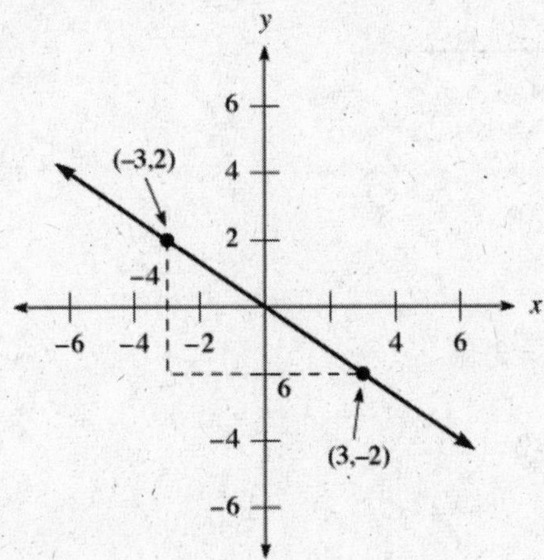

11. The slope is given by: $m = \frac{-2-(-5)}{3-2} = \frac{3}{1} = 3$

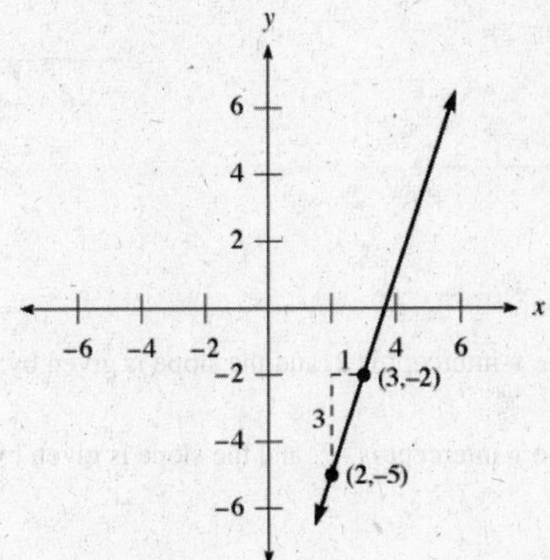

© 2012 Cengage Learning. All Rights Reserved. May not be scanned, copied or duplicated, or posted to a publicly accessible website, in whole or in part.

13. Graphing the line:

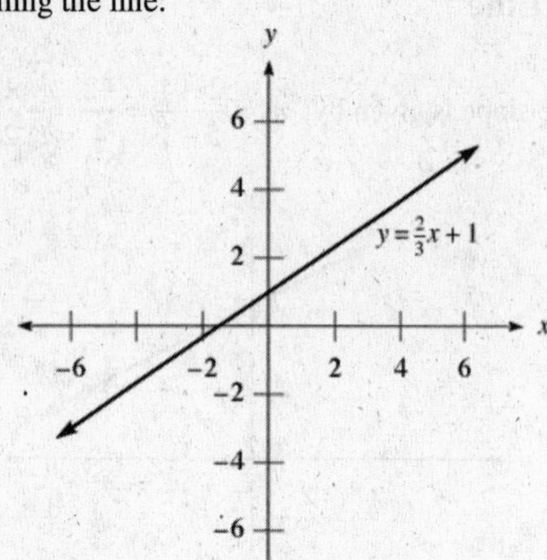

15. Graphing the line:

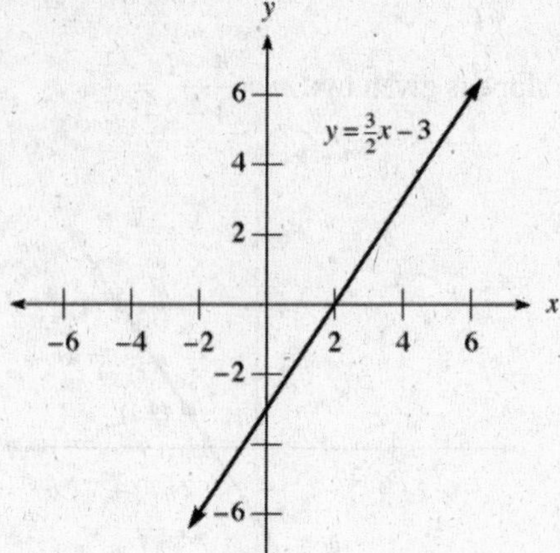

17. Graphing the line:

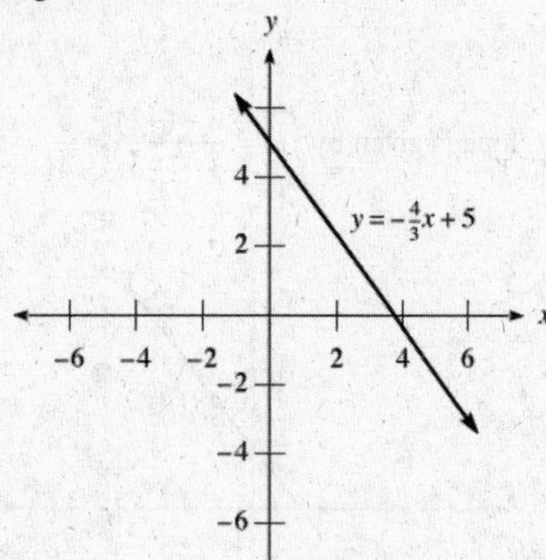

19. Graphing the line:

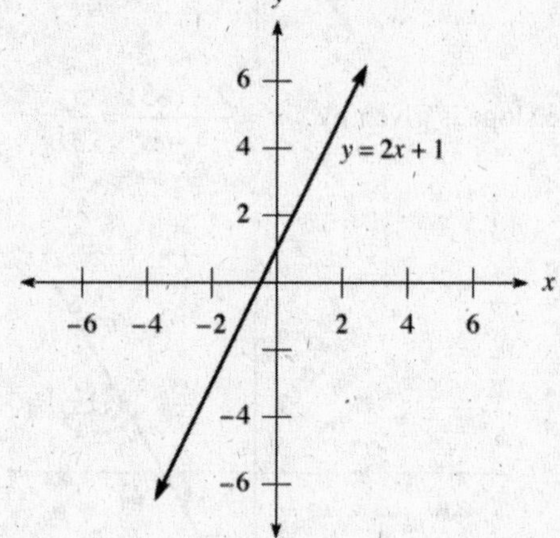

21. Graphing the line:

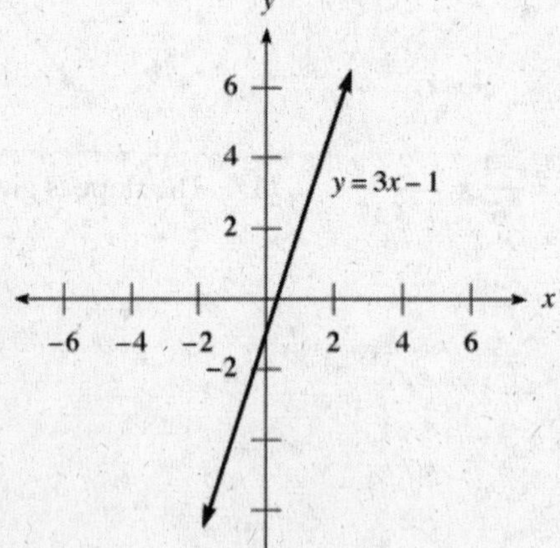

23. The y-intercept is 2, and the slope is given by: $m=\frac{5-(-1)}{1-(-1)}=\frac{6}{2}=3$

25. The y-intercept is –2, and the slope is given by: $m=\frac{2-0}{2-1}=\frac{2}{1}=2$

© 2012 Cengage Learning. All Rights Reserved. May not be scanned, copied or duplicated, or posted to a publicly accessible website, in whole or in part.

27. The slope is given by: $m = \frac{0-(-2)}{3-0} = \frac{2}{3}$

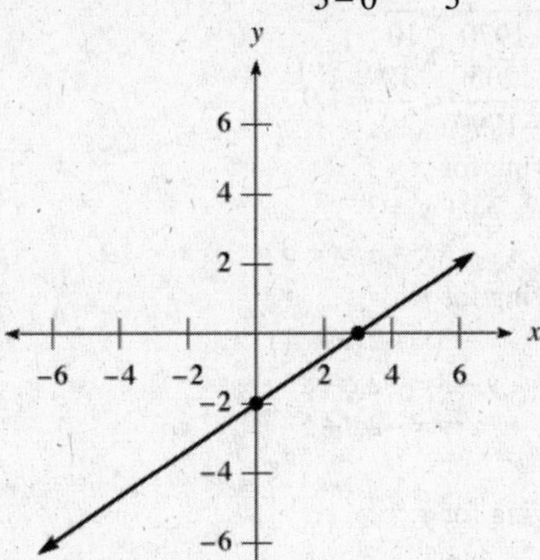

29. The slope is given by: $m = \frac{0-2}{4-0} = \frac{-2}{4} = -\frac{1}{2}$

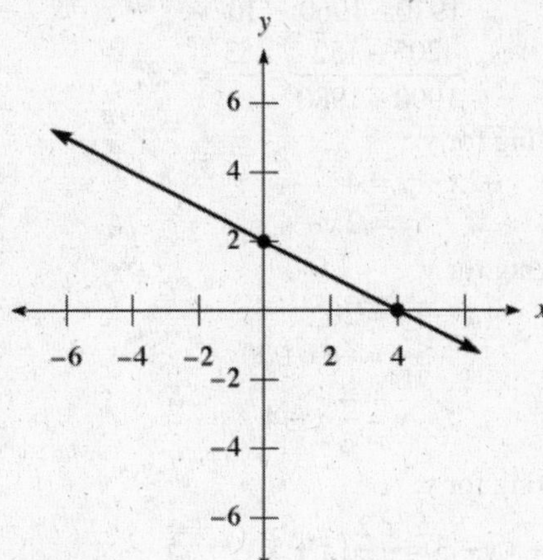

31. Graphing the line:

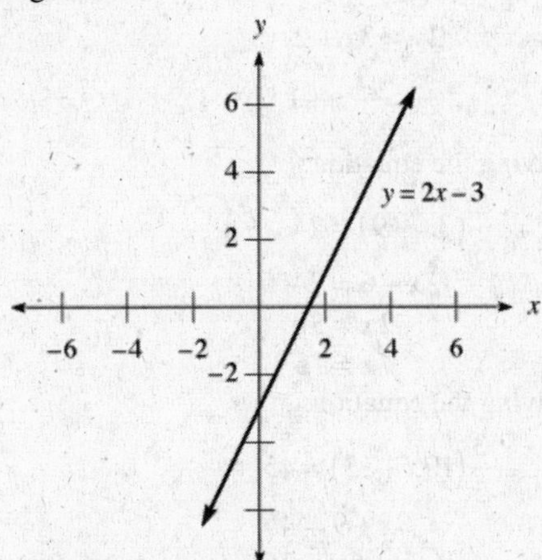

The slope is 2 and the y-intercept is –3.

33. Graphing the line:

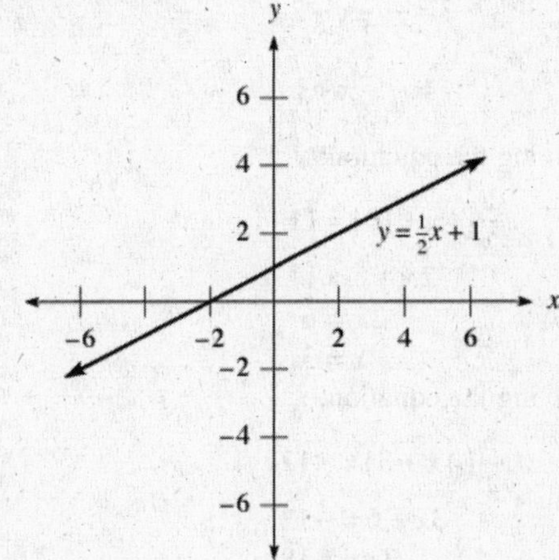

The slope is $\frac{1}{2}$ and the y-intercept is 1.

35. Using the slope formula:

$$\frac{y-2}{6-4} = 2$$
$$\frac{y-2}{2} = 2$$
$$y-2 = 4$$
$$y = 6$$

37. The slopes are given in the table:

Equation	Slope
$x = 3$	undefined
$y = 3$	0
$y = 3x$	3

39. The slopes are given in the table:

Equation	Slope
$y = -\frac{2}{3}$	0
$x = -\frac{2}{3}$	undefined
$y = -\frac{2}{3}x$	$-\frac{2}{3}$

41. Finding the slope: $m = \frac{2,399.2 - 1,936.5}{2008 - 2005} = \frac{462.7}{3} \approx 154.23$

© 2012 Cengage Learning. All Rights Reserved. May not be scanned, copied or duplicated, or posted to a publicly accessible website, in whole or in part.

43. Finding the slopes:

$A: \frac{121-88}{1970-1960}=\frac{33}{10}=3.3$ $\quad B: \frac{152-121}{1980-1970}=\frac{31}{10}=3.1$

$C: \frac{205-152}{1990-1980}=\frac{53}{10}=5.3$ $\quad D: \frac{224-205}{2000-1990}=\frac{19}{10}=1.9$

45. Solving for y:

$$\begin{aligned} -2x+y&=4 \\ y&=2x+4 \end{aligned}$$

47. Solving for y:

$$\begin{aligned} 3x+y&=3 \\ y&=-3x+3 \end{aligned}$$

49. Solving for y:

$$\begin{aligned} 4x-5y&=20 \\ -5y&=-4x+20 \\ y&=\frac{4}{5}x-4 \end{aligned}$$

51. Solving for y:

$$\begin{aligned} y-3&=-2(x+4) \\ y-3&=-2x-8 \\ y&=-2x-5 \end{aligned}$$

53. Solving for y:

$$\begin{aligned} y-3&=-\frac{2}{3}(x+3) \\ y-3&=-\frac{2}{3}x-2 \\ y&=-\frac{2}{3}x+1 \end{aligned}$$

55. Solving for y:

$$\begin{aligned} \frac{y-1}{x}&=\frac{3}{2} \\ 2y-2&=3x \\ 2y&=3x+2 \\ y&=\frac{3}{2}x+1 \end{aligned}$$

57. Solving the equation:

$$\begin{aligned} \frac{1}{2}(4x+10)&=11 \\ 2x+5&=11 \\ 2x&=6 \\ x&=3 \end{aligned}$$

59. Solving the equation:

$$\begin{aligned} \frac{2}{3}(3x-9)&=24 \\ 2x-6&=24 \\ 2x&=30 \\ x&=15 \end{aligned}$$

61. Solving the equation:

$$\begin{aligned} \frac{3}{4}(4x+8)&=-12 \\ 3x+6&=-12 \\ 3x&=-18 \\ x&=-6 \end{aligned}$$

63. Solving the equation:

$$\begin{aligned} \frac{3}{5}(10x+15)&=45 \\ 6x+9&=45 \\ 6x&=36 \\ x&=6 \end{aligned}$$

65. Solving the equation:

$$\begin{aligned} 5\left(\frac{2}{5}x+10\right)&=-28 \\ 2x+50&=-28 \\ 2x&=-78 \\ x&=-39 \end{aligned}$$

67. Solving the equation:

$$\begin{aligned} 4\left(\frac{3}{2}x-7\right)&=-4 \\ 6x-28&=-4 \\ 6x&=24 \\ x&=4 \end{aligned}$$

69. Solving the equation:

$$\begin{aligned} \frac{3}{4}(2x+12)&=24 \\ \frac{3}{2}x+9&=24 \\ \frac{3}{2}x&=15 \\ x&=\frac{2}{3}(15)=10 \end{aligned}$$

71. Solving the equation:

$$\begin{aligned} \frac{1}{2}(5x-10)+6&=-49 \\ \frac{5}{2}x-5+6&=-49 \\ \frac{5}{2}x+1&=-49 \\ \frac{5}{2}x&=-50 \\ x&=\frac{2}{5}(-50)=-20 \end{aligned}$$

© 2012 Cengage Learning. All Rights Reserved. May not be scanned, copied or duplicated, or posted to a publicly accessible website, in whole or in part.

3.6 Finding the Equation of a Line

1. The slope-intercept form is $y=\frac{2}{3}x+1$.

3. The slope-intercept form is $y=\frac{3}{2}x-1$.

5. The slope-intercept form is $y=-\frac{2}{5}x+3$.

7. The slope-intercept form is $y=2x-4$.

9. Solving for *y*:

$$-2x+y=4$$
$$y=2x+4$$

The slope is 2 and the *y*-intercept is 4:

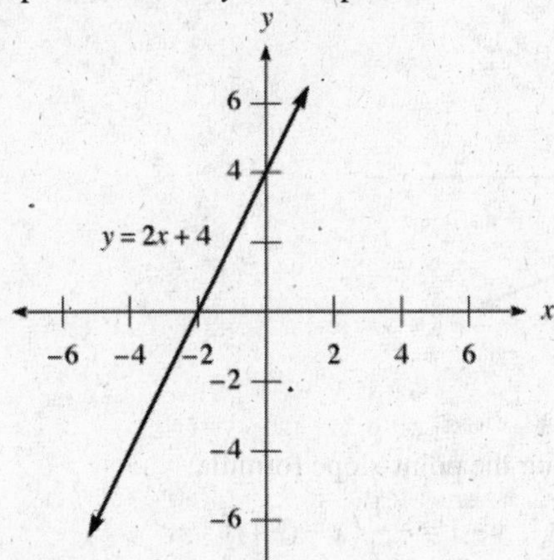

11. Solving for *y*:

$$3x+y=3$$
$$y=-3x+3$$

The slope is –3 and the *y*-intercept is 3:

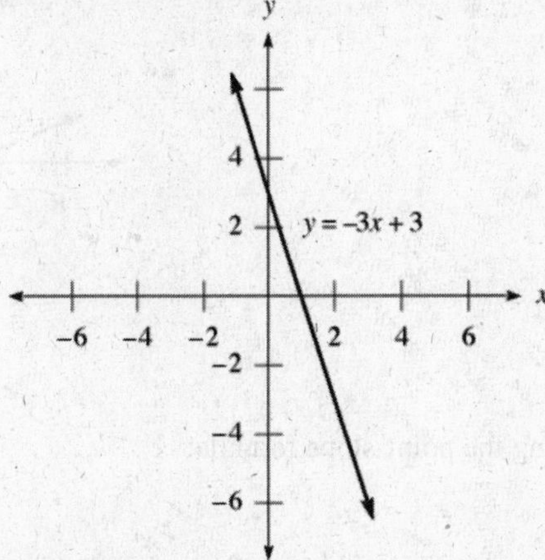

13. Solving for *y*:

$$3x+2y=6$$
$$2y=-3x+6$$
$$y=-\frac{3}{2}x+3$$

The slope is $-\frac{3}{2}$ and the *y*-intercept is 3:

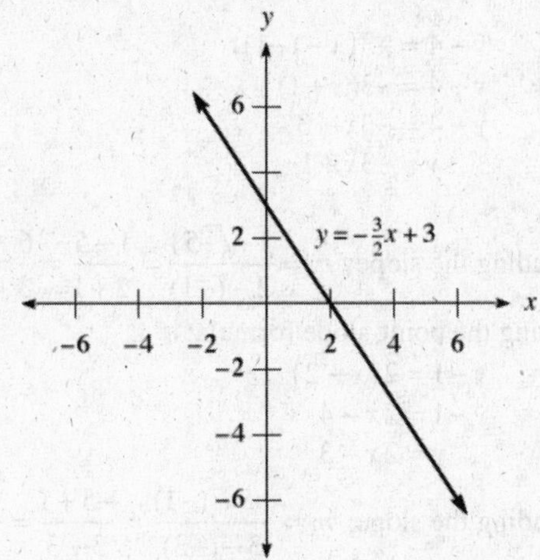

15. Solving for *y*:

$$4x-5y=20$$
$$-5y=-4x+20$$
$$y=\frac{4}{5}x-4$$

The slope is $\frac{4}{5}$ and the *y*-intercept is –4:

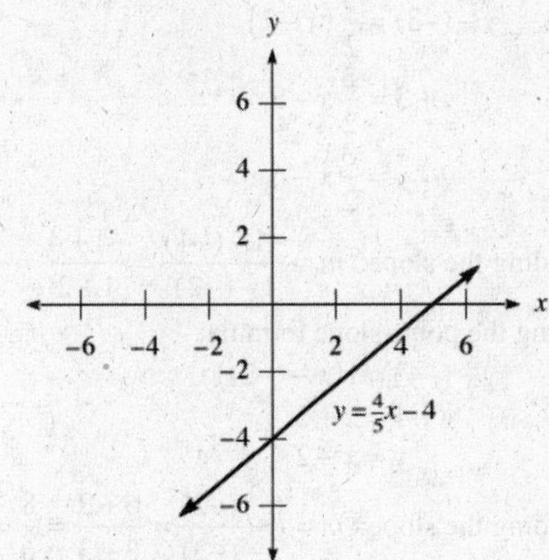

© 2012 Cengage Learning. All Rights Reserved. May not be scanned, copied or duplicated, or posted to a publicly accessible website, in whole or in part.

17. Solving for y:

$$\begin{aligned} -2x-5y&=10 \\ -5y&=2x+10 \\ y&=-\frac{2}{5}x-2 \end{aligned}$$

The slope is $-\frac{2}{5}$ and the y-intercept is –2:

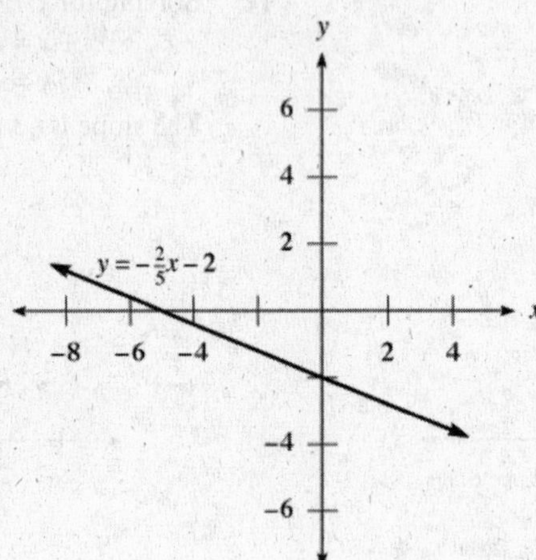

19. Using the point-slope formula:

$$\begin{aligned} y-(-5)&=2(x-(-2)) \\ y+5&=2(x+2) \\ y+5&=2x+4 \\ y&=2x-1 \end{aligned}$$

21. Using the point-slope formula:

$$\begin{aligned} y-1&=-\frac{1}{2}(x-(-4)) \\ y-1&=-\frac{1}{2}(x+4) \\ y-1&=-\frac{1}{2}x-2 \\ y&=-\frac{1}{2}x-1 \end{aligned}$$

23. Using the point-slope formula:

$$\begin{aligned} y-(-3)&=\frac{3}{2}(x-2) \\ y+3&=\frac{3}{2}x-3 \\ y&=\frac{3}{2}x-6 \end{aligned}$$

25. Using the point-slope formula:

$$\begin{aligned} y-4&=-3(x-(-1)) \\ y-4&=-3(x+1) \\ y-4&=-3x-3 \\ y&=-3x+1 \end{aligned}$$

27. Finding the slope: $m=\frac{-1-(-4)}{1-(-2)}=\frac{-1+4}{1+2}=\frac{3}{3}=1$

Using the point-slope formula:

$$\begin{aligned} y-(-4)&=1(x-(-2)) \\ y+4&=x+2 \\ y&=x-2 \end{aligned}$$

29. Finding the slope: $m=\frac{1-(-5)}{2-(-1)}=\frac{1+5}{2+1}=\frac{6}{3}=2$

Using the point-slope formula:

$$\begin{aligned} y-1&=2(x-2) \\ y-1&=2x-4 \\ y&=2x-3 \end{aligned}$$

31. Finding the slope: $m=\frac{6-(-2)}{3-(-3)}=\frac{6+2}{3+3}=\frac{8}{6}=\frac{4}{3}$

Using the point-slope formula:

$$\begin{aligned} y-6&=\frac{4}{3}(x-3) \\ y-6&=\frac{4}{3}x-4 \\ y&=\frac{4}{3}x+2 \end{aligned}$$

33. Finding the slope: $m=\frac{-5-(-1)}{3-(-3)}=\frac{-5+1}{3+3}=\frac{-4}{6}=-\frac{2}{3}$

Using the point-slope formula:

$$\begin{aligned} y-(-5)&=-\frac{2}{3}(x-3) \\ y+5&=-\frac{2}{3}x+2 \\ y&=-\frac{2}{3}x-3 \end{aligned}$$

© 2012 Cengage Learning. All Rights Reserved. May not be scanned, copied or duplicated, or posted to a publicly accessible website, in whole or in part.

35. The y-intercept is 3, and the slope is: $m = \frac{3-0}{0-(-1)} = \frac{3}{1} = 3$. The slope-intercept form is $y = 3x + 3$.

37. The y-intercept is –1, and the slope is: $m = \frac{0-(-1)}{4-0} = \frac{0+1}{4} = \frac{1}{4}$. The slope-intercept form is $y = \frac{1}{4}x - 1$.

39. a. Solving the equation:
$$-2x + 1 = 6$$
$$-2x = 5$$
$$x = -\frac{5}{2}$$

b. Writing in slope-intercept form:
$$-2x + y = 6$$
$$y = 2x + 6$$

c. Substituting $x = 0$:
$$-2(0) + y = 6$$
$$0 + y = 6$$
$$y = 6$$

d. Finding the slope:
$$-2x + y = 6$$
$$y = 2x + 6$$
The slope is 2.

e. Graphing the line:

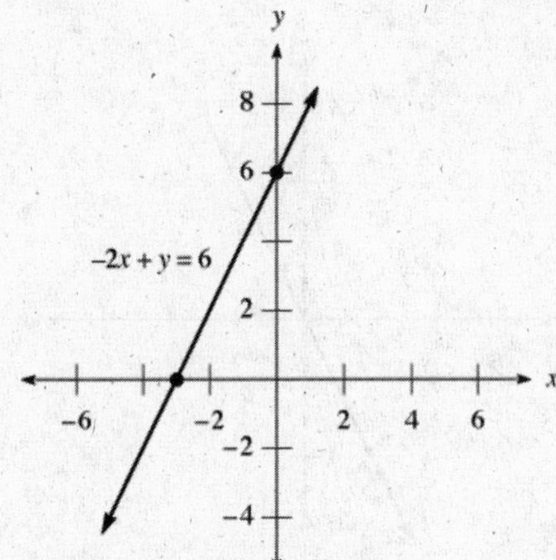

41. The slope is given by: $m = \frac{0-2}{3-0} = -\frac{2}{3}$. Since $b = 2$, the equation is $y = -\frac{2}{3}x + 2$.

43. The slope is given by: $m = \frac{0-(-5)}{-2-0} = -\frac{5}{2}$. Since $b = -5$, the equation is $y = -\frac{5}{2}x - 5$.

45. Its equation is $x = 3$.

47. The slope of the line must be 4, so any line of the form $y = 4x + b$ will be parallel.

49. Finding the slope between (2,3) and (4,1): $m = \frac{1-3}{4-2} = -\frac{2}{2} = -1$. This will be the slope of any parallel line.

51. Finding the slope between (4,0) and (0,–2): $m = \frac{-2-0}{0-4} = \frac{-2}{-4} = \frac{1}{2}$. So the slope of any perpendicular line is –2.

53. Using the points (2000, 65.4) and (2005, 104), the slope is: $m = \frac{104 - 65.4}{2005 - 2000} = \frac{38.6}{5} = 7.72$

Using the point-slope formula:
$$y - 65.4 = 7.72(x - 2000)$$
$$y - 65.4 = 7.72x - 15{,}440$$
$$y = 7.72x - 15{,}374.6$$

55. a. After 5 years, the copier is worth \$6,000.
b. The copier is worth \$12,000 after 3 years.
c. The slope of the line is –3000.
d. The copier is decreasing in value by \$3000 per year.
e. The equation is $V = -3000t + 21{,}000$.

© 2012 Cengage Learning. All Rights Reserved. May not be scanned, copied or duplicated, or posted to a publicly accessible website, in whole or in part.

57. Graphing the line:

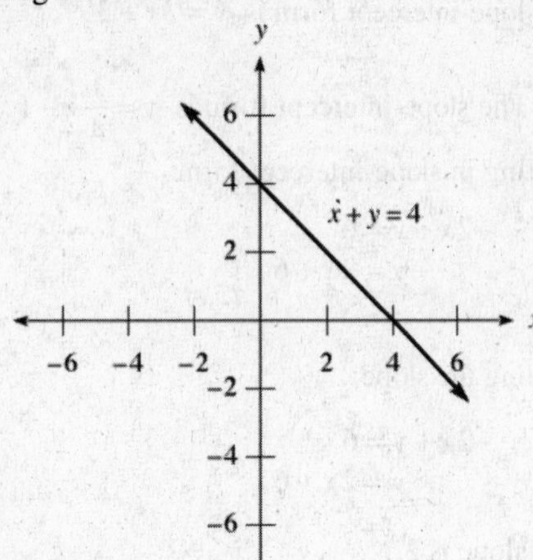

59. Graphing the line:

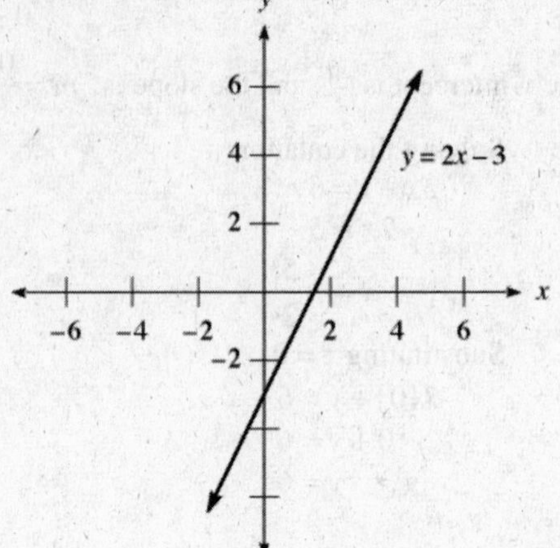

61. Graphing the line:

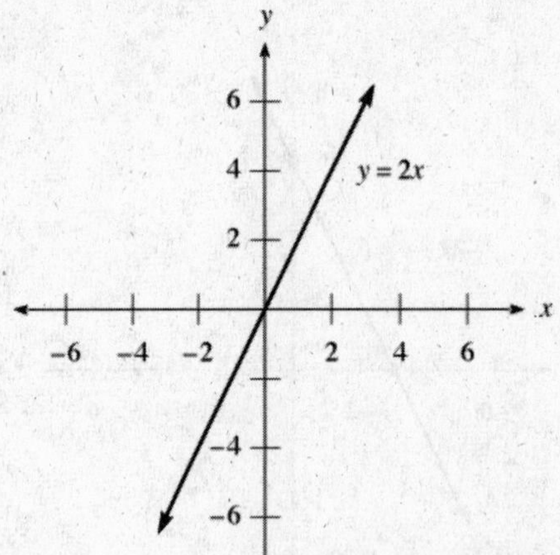

63. Substituting $y = 13$:

$$\begin{aligned} 3x - 5 &= 13 \\ 3x &= 18 \\ x &= 6 \end{aligned}$$

65. Substituting $y = -11$:

$$\begin{aligned} 3x - 5 &= -11 \\ 3x &= -6 \\ x &= -2 \end{aligned}$$

67. Substituting $x = 0$: $y = \frac{3}{7}(0) + 4 = 0 + 4 = 4$

69. Substituting $x = -35$: $y = \frac{3}{7}(-35) + 4 = -15 + 4 = -11$

71. Substituting $x = 5$: $y = 3(5) - 5 = 15 - 5 = 10$

73. Substituting $x = -11$: $y = 3(-11) - 5 = -33 - 5 = -38$

75. Substituting $y = -9$:

$$\begin{aligned} \frac{x-6}{2} &= -9 \\ x - 6 &= -18 \\ x &= -12 \end{aligned}$$

77. Substituting $y = -12$:

$$\begin{aligned} \frac{x-6}{2} &= -12 \\ x - 6 &= -24 \\ x &= -18 \end{aligned}$$

© 2012 Cengage Learning. All Rights Reserved. May not be scanned, copied or duplicated, or posted to a publicly accessible website, in whole or in part.

3.7 Linear Inequalities in Two Variables

1. Checking the point $(0,0)$:

$2(0)-3(0)=0-0<6$ (true)

Graphing the linear inequality:

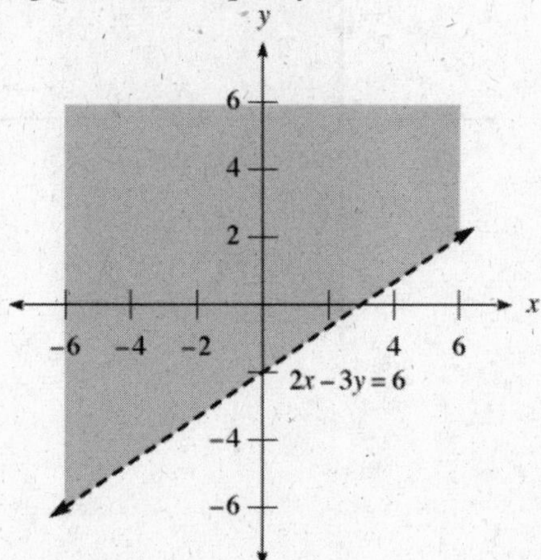

3. Checking the point $(0,0)$:

$0-2(0)=0-0\leq 4$ (true)

Graphing the linear inequality:

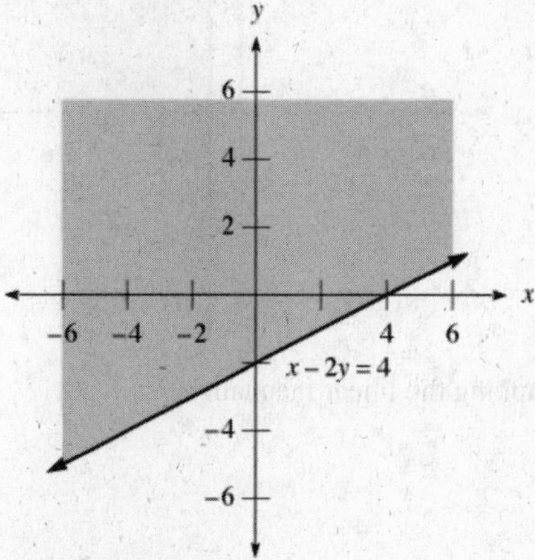

5. Checking the point $(0,0)$:

$0-0\leq 2$ (true)

Graphing the linear inequality:

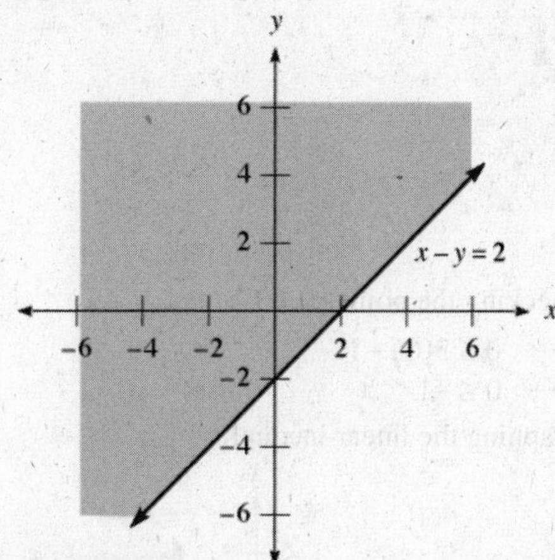

7. Checking the point $(0,0)$:

$3(0)-4(0)=0-0\geq 12$ (false)

Graphing the linear inequality:

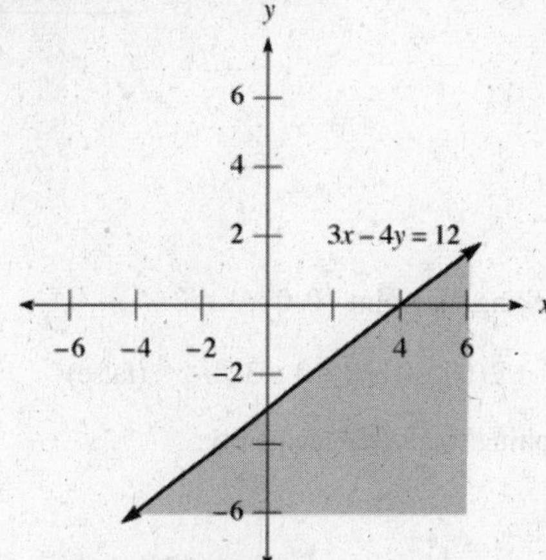

9. Checking the point $(0,0)$:

$5(0)-0=0-0\leq 5$ (true)

Graphing the linear inequality:

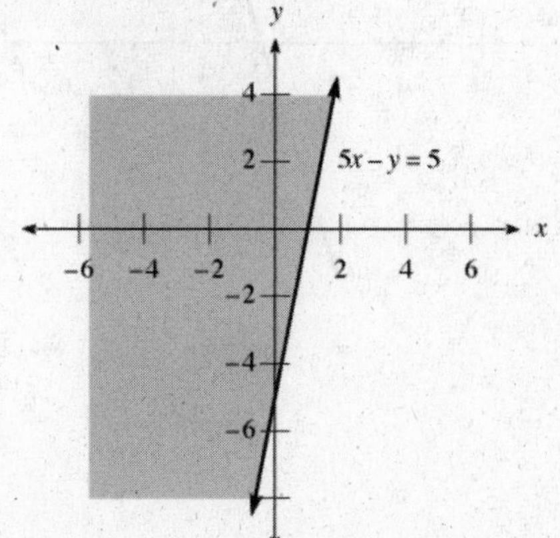

11. Checking the point $(0,0)$:

$2(0)+6(0)=0+0\leq 12$ (true)

Graphing the linear inequality:

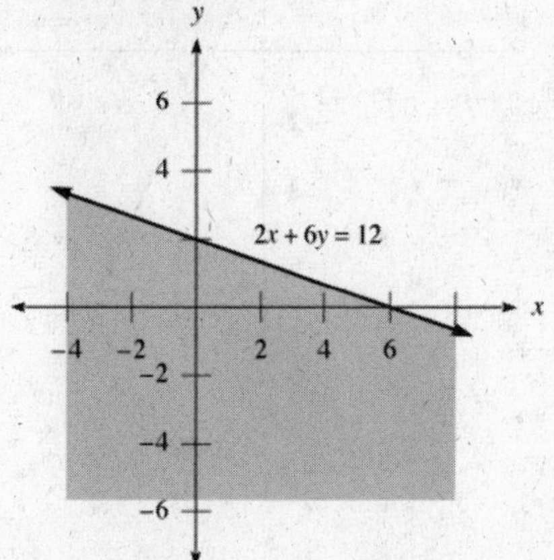

© 2012 Cengage Learning. All Rights Reserved. May not be scanned, copied or duplicated, or posted to a publicly accessible website, in whole or in part.

13. Graphing the linear inequality:

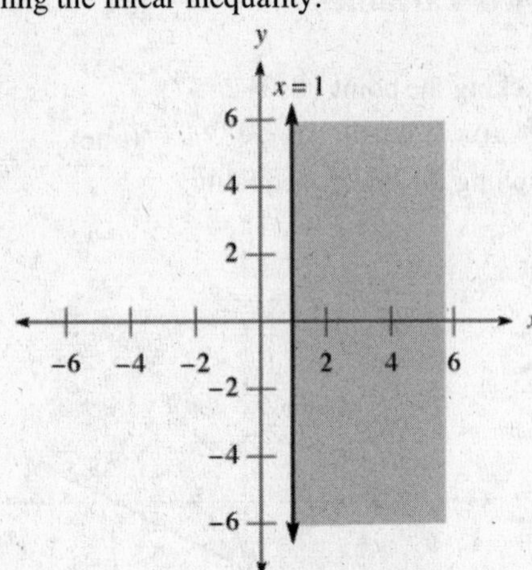

15. Graphing the linear inequality:

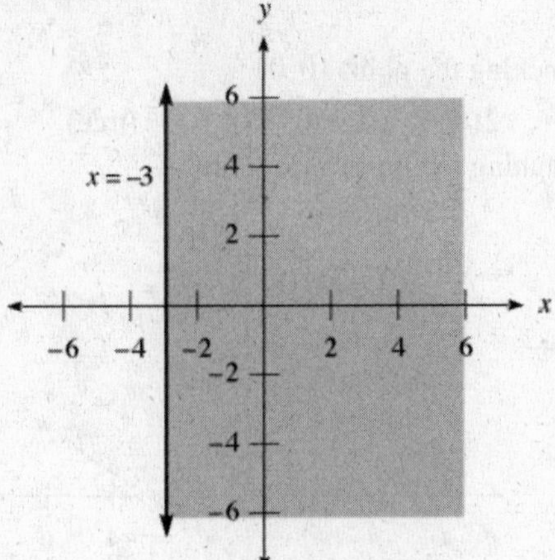

17. Graphing the linear inequality:

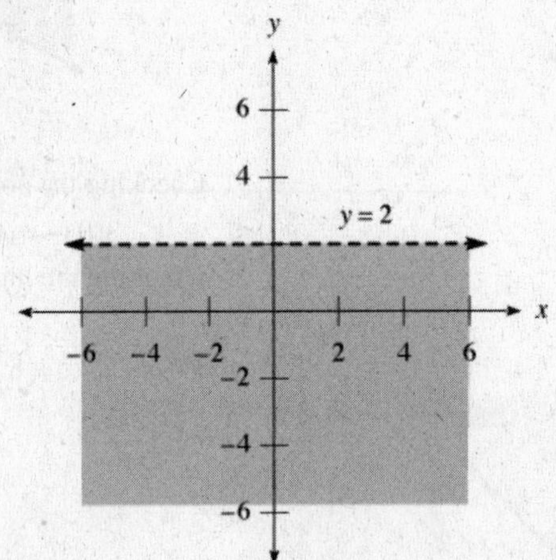

19. Checking the point $(0,0)$:

$2(0)+0=0+0>3$ (false)

Graphing the linear inequality:

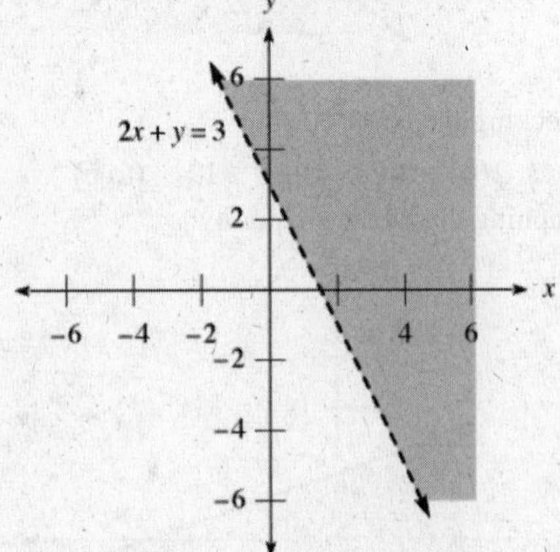

21. Checking the point $(0,0)$:

$0 \le 3(0)-1$

$0 \le -1$ (false)

Graphing the linear inequality:

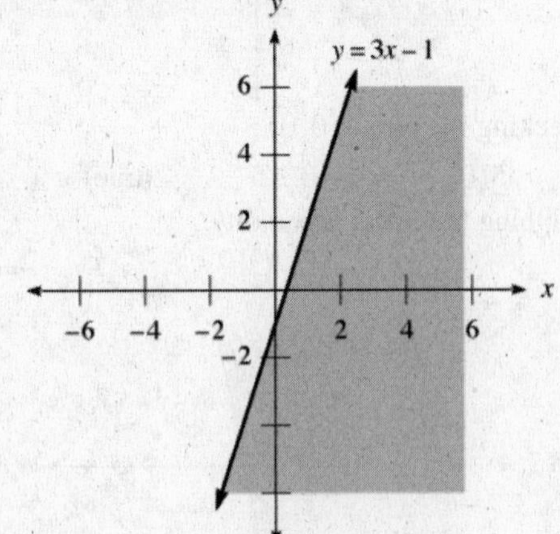

© 2012 Cengage Learning. All Rights Reserved. May not be scanned, copied or duplicated, or posted to a publicly accessible website, in whole or in part.

23. Checking the point $(0,0)$:

$$0 \le -\frac{1}{2}(0)+2$$
$$0 \le 2 \quad \text{(true)}$$

Graphing the linear inequality:

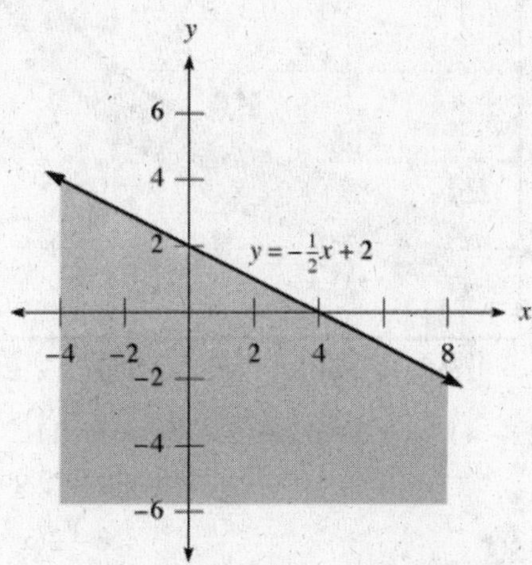

25. **a.** Solving the inequality:

$$4+3y<12$$
$$3y<8$$
$$y<\frac{8}{3}$$

b. Solving the inequality:

$$4-3y<12$$
$$-3y<8$$
$$y \ge -\frac{8}{3}$$

c. Solving for *y*:

$$4x+3y=12$$
$$3y=-4x+12$$
$$y=-\frac{4}{3}x+4$$

d. Graphing the inequality:

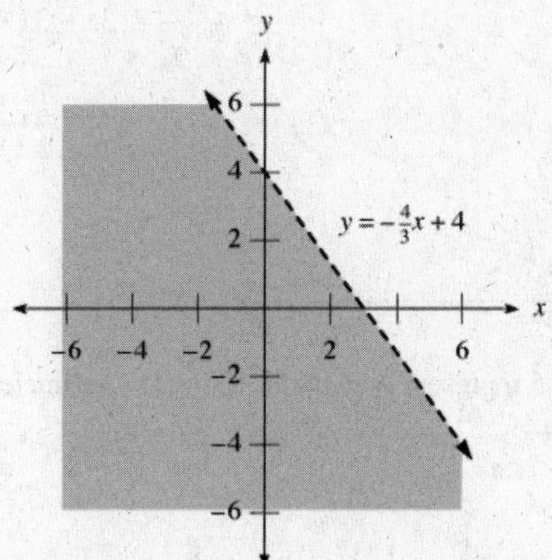

27. **a.** The slope is $\frac{2}{5}$ and the *y*-intercept is 2, so the equation is $y=\frac{2}{5}x+2$.

b. Since the shading is below the line, the inequality is $y<\frac{2}{5}x+2$.

c. Since the shading is above the line, the inequality is $y>\frac{2}{5}x+2$.

© 2012 Cengage Learning. All Rights Reserved. May not be scanned, copied or duplicated, or posted to a publicly accessible website, in whole or in part.

29. Sketching the graph:

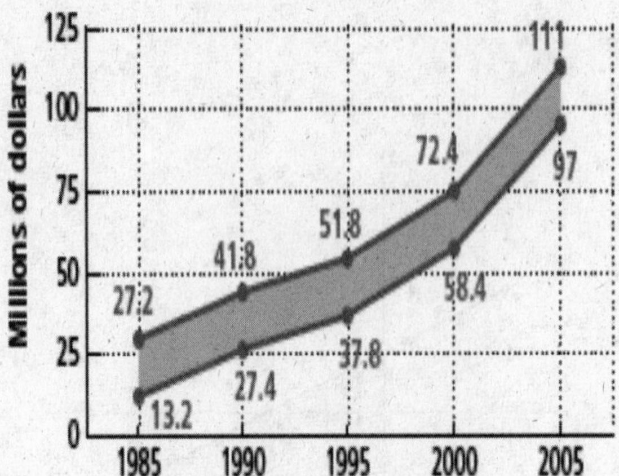

31. Simplifying the expression: $7-3(2x-4)-8=7-6x+12-8=-6x+11$

33. Solving the equation:

$$-\frac{3}{2}x=12$$
$$-\frac{2}{3}\left(-\frac{3}{2}x\right)=-\frac{2}{3}(12)$$
$$x=-8$$

35. Solving the equation:

$$8-2(x+7)=2$$
$$8-2x-14=2$$
$$-2x-6=2$$
$$-2x=8$$
$$x=-4$$

37. Solving for w:

$$2l+2w=P$$
$$2w=P-2l$$
$$w=\frac{P-2l}{2}$$

39. Solving the inequality:

$$3-2x>5$$
$$3-3-2x>5-3$$
$$-2x>2$$
$$-\frac{1}{2}(-2x)<-\frac{1}{2}(2)$$
$$x<-1$$

Graphing the solution set:

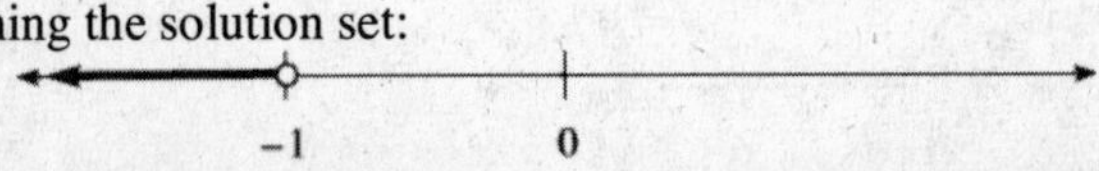

41. Solving for y:

$$3x-2y\le 12$$
$$3x-3x-2y\le -3x+12$$
$$-2y\le -3x+12$$
$$-\frac{1}{2}(-2y)\ge -\frac{1}{2}(-3x+12)$$
$$y\ge \frac{3}{2}x-6$$

43. Let w represent the width and $3w+5$ represent the length. Using the perimeter formula:

$$2(w)+2(3w+5)=26$$
$$2w+6w+10=26$$
$$8w+10=26$$
$$8w=16$$
$$w=2$$
$$3w+5=3(2)+5=11$$

The width is 2 inches and the length is 11 inches.

© 2012 Cengage Learning. All Rights Reserved. May not be scanned, copied or duplicated, or posted to a publicly accessible website, in whole or in part.

Chapter 3 Review

1. Graphing the ordered pair:

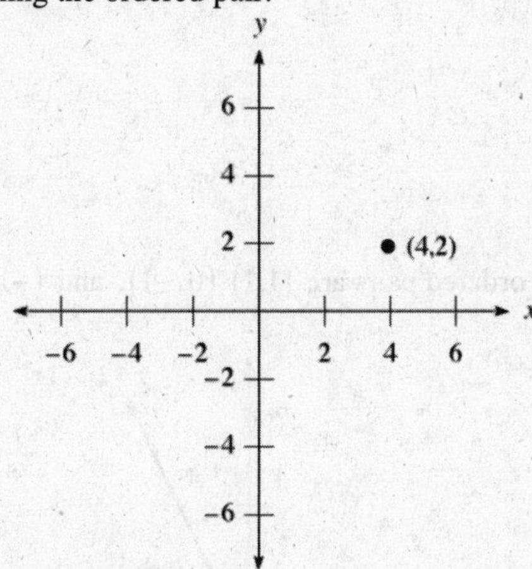

3. Graphing the ordered pair:

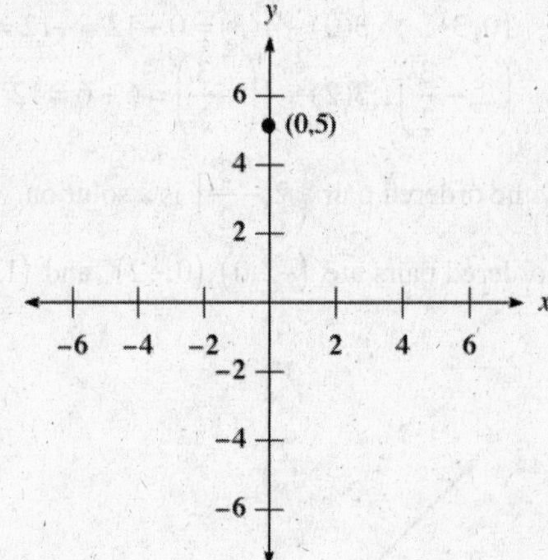

5. Graphing the ordered pair:

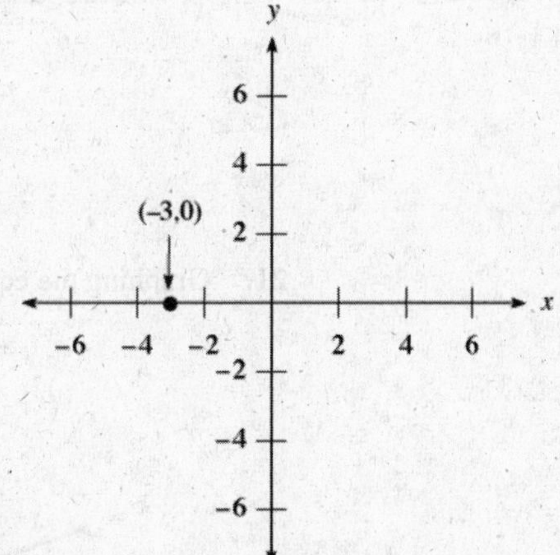

7. Substituting $x = 4$, $x = 0$, $y = 3$, and $y = 0$:

$$\begin{aligned} 3(4)+y&=6 \\ 12+y&=6 \\ y&=-6 \end{aligned} \qquad \begin{aligned} 3(0)+y&=6 \\ 0+y&=6 \\ y&=6 \end{aligned} \qquad \begin{aligned} 3x+3&=6 \\ 3x&=3 \\ x&=1 \end{aligned} \qquad \begin{aligned} 3x+0&=6 \\ 3x&=6 \\ x&=2 \end{aligned}$$

The ordered pairs are $(4,-6),(0,6),(1,3)$, and $(2,0)$.

9. Substituting $x = 4$, $y = -2$, and $y = 3$:

$$\begin{aligned} y&=2(4)-6 \\ y&=8-6 \\ y&=2 \end{aligned} \qquad \begin{aligned} -2&=2x-6 \\ 4&=2x \\ x&=2 \end{aligned} \qquad \begin{aligned} 3&=2x-6 \\ 9&=2x \\ x&=\frac{9}{2} \end{aligned}$$

The ordered pairs are $(4,2),(2,-2)$, and $\left(\frac{9}{2},3\right)$.

11. Substituting $x = 2$, $x = -1$, and $x = -3$ results (in each case) in $y = -3$. The ordered pairs are $(2,-3),(-1,-3)$, and $(-3,-3)$.

© 2012 Cengage Learning. All Rights Reserved. May not be scanned, copied or duplicated, or posted to a publicly accessible website, in whole or in part.

13. Substituting each ordered pair into the equation:

$$\left(-2,\frac{9}{2}\right):\ 3(-2)-4\left(\frac{9}{2}\right)=-6-18=-24\neq 12$$

$$(0,3):\quad 3(0)-4(3)=0-12=-12\neq 12$$

$$\left(2,-\frac{3}{2}\right):\ 3(2)-4\left(-\frac{3}{2}\right)=6+6=12$$

Only the ordered pair $\left(2,-\frac{3}{2}\right)$ is a solution.

15. The ordered pairs are $(-2,0),(0,-2)$, and $(1,-3)$:

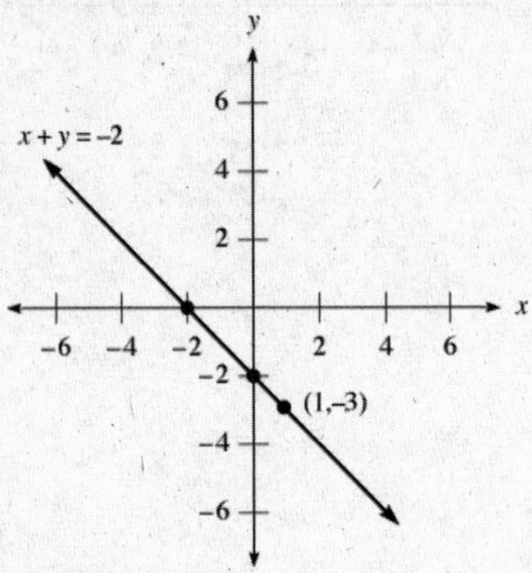

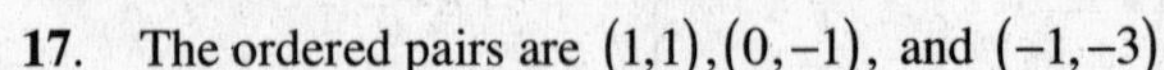

17. The ordered pairs are $(1,1),(0,-1)$, and $(-1,-3)$:

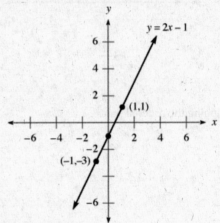

19. Graphing the equation:

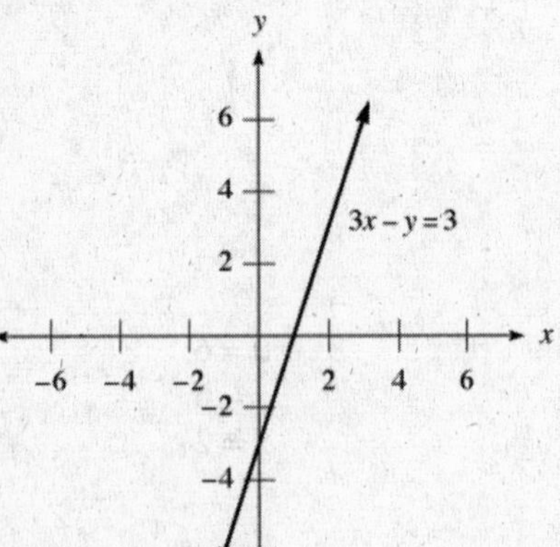

21. Graphing the equation:

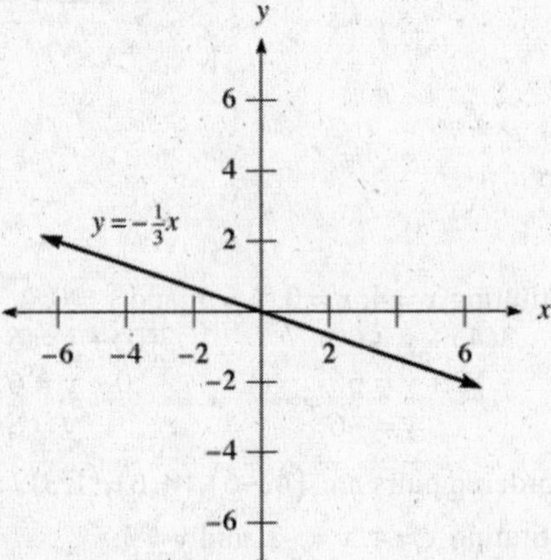

23. Graphing the equation:

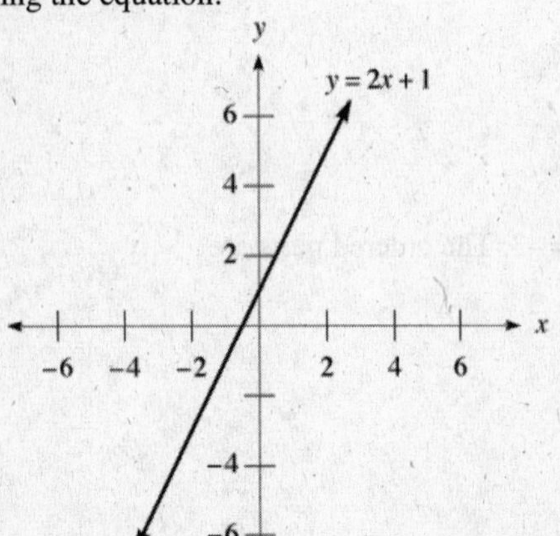

25. Graphing the equation:

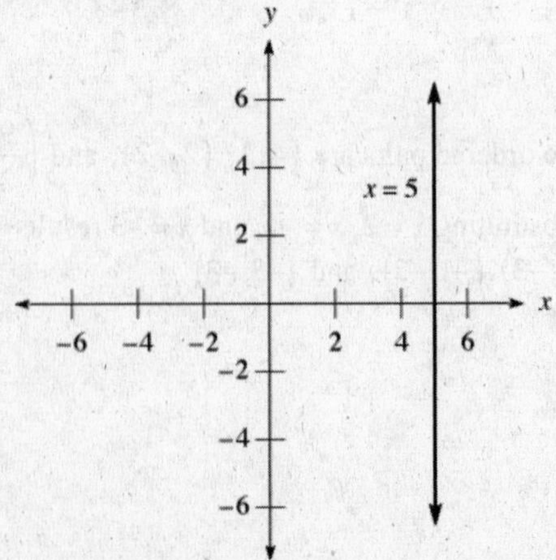

© 2012 Cengage Learning. All Rights Reserved. May not be scanned, copied or duplicated, or posted to a publicly accessible website, in whole or in part.

27. Graphing the equation:

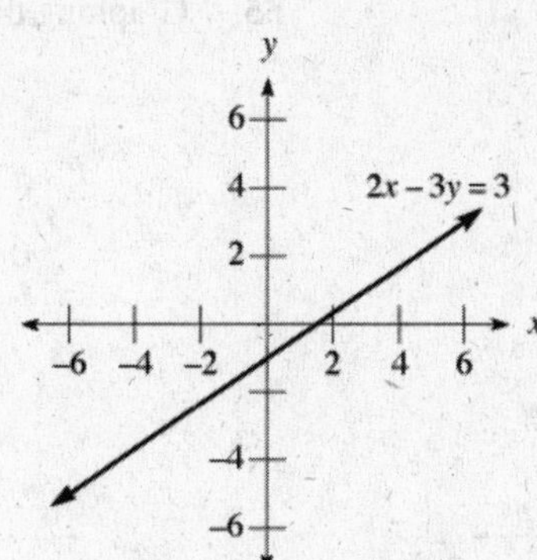

29. To find the x-intercept, let $y = 0$:

$$3x - 0 = 6$$
$$3x = 6$$
$$x = 2$$

To find the y-intercept, let $x = 0$:

$$3(0) - y = 6$$
$$-y = 6$$
$$y = -6$$

31. To find the x-intercept, let $y = 0$:

$$0 = x - 3$$
$$x = 3$$

To find the y-intercept, let $x = 0$:

$$y = 0 - 3$$
$$y = -3$$

33. Since the equation is $y = -5$, there is no x-intercept and the y-intercept is -5.

35. The slope is given by: $m = \dfrac{5-3}{3-2} = \dfrac{2}{1} = 2$

37. The slope is given by: $m = \dfrac{-8-(-4)}{-3-(-1)} = \dfrac{-8+4}{-3+1} = \dfrac{-4}{-2} = 2$

39. Using the slope formula:

$$\frac{9-3}{x-3} = 2$$
$$\frac{6}{x-3} = 2$$
$$6 = 2(x-3)$$
$$6 = 2x - 6$$
$$12 = 2x$$
$$x = 6$$

41. Using the slope formula:

$$y - 4 = -2\left(x - (-1)\right)$$
$$y - 4 = -2(x+1)$$
$$y - 4 = -2x - 2$$
$$y = -2x + 2$$

43. Using the point-slope formula:

$$y - (-2) = -\frac{3}{4}\left(x - 3\right)$$
$$y + 2 = -\frac{3}{4}x + \frac{9}{4}$$
$$y = -\frac{3}{4}x + \frac{1}{4}$$

45. The slope-intercept form is $y = 3x + 2$.

47. The slope-intercept form is $y = -\dfrac{1}{3}x + \dfrac{3}{4}$.

49. The equation is in slope-intercept form, so $m = 4$ and $b = -1$.

51. Solving for y:

$$6x + 3y = 9$$
$$3y = -6x + 9$$
$$y = -2x + 3$$

The equation is in slope-intercept form, so $m = -2$ and $b = 3$.

© 2012 Cengage Learning. All Rights Reserved. May not be scanned, copied or duplicated, or posted to a publicly accessible website, in whole or in part.

53. Checking the point $(0,0)$:

$0-0<3$ (true)

Graphing the linear inequality:

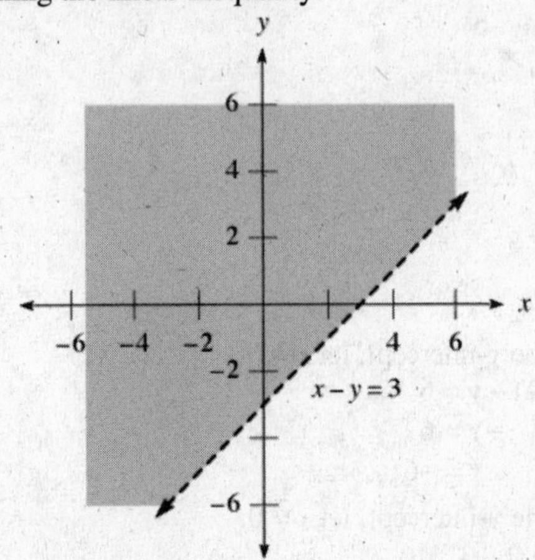

55. Graphing the linear inequality:

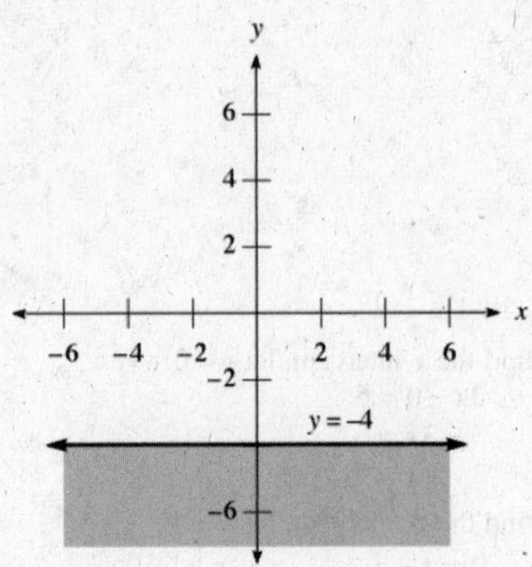

Chapter 3 Cumulative Review

1. Simplifying using order of operations: $7-2\bullet 6=7-12=-5$

3. Simplifying using order of operations: $\frac{3}{8}(16)-\frac{2}{3}(9)=3(2)-2(3)=6-6=0$

5. Simplifying using order of operations: $(4-9)(-3-8)=(-5)(-11)=55$

7. Simplifying using order of operations: $\frac{18+(-34)}{7-7}=\frac{-16}{0}$, which is undefined

9. Simplifying: $\frac{75}{135}=\frac{3\bullet 5\bullet 5}{3\bullet 3\bullet 3\bullet 5}=\frac{5}{3\bullet 3}=\frac{5}{9}$

11. Using the associative property: $5(8x)=(5\bullet 8)x=40x$

13. Solving the equation:

$$\begin{aligned}5x+6-4x-3&=7\\x+3&=7\\x&=4\end{aligned}$$

15. Solving the equation:

$$\begin{aligned}-4x+7&=5x-11\\-9x+7&=-11\\-9x&=-18\\x&=2\end{aligned}$$

17. Solving the equation:

$$\begin{aligned}3&=-\frac{1}{4}(5x+2)\\3&=-\frac{5}{4}x-\frac{1}{2}\\4(3)&=4\left(-\frac{5}{4}x-\frac{1}{2}\right)\\12&=-5x-2\\14&=-5x\\x&=-\frac{14}{5}\end{aligned}$$

19. Solving the inequality:

$$\begin{aligned}-4x&>28\\-\frac{1}{4}(-4x)&<-\frac{1}{4}(28)\\x&<-7\end{aligned}$$

Graphing the solution set:

−7 0

© 2012 Cengage Learning. All Rights Reserved. May not be scanned, copied or duplicated, or posted to a publicly accessible website, in whole or in part.

21. Graphing the equation:

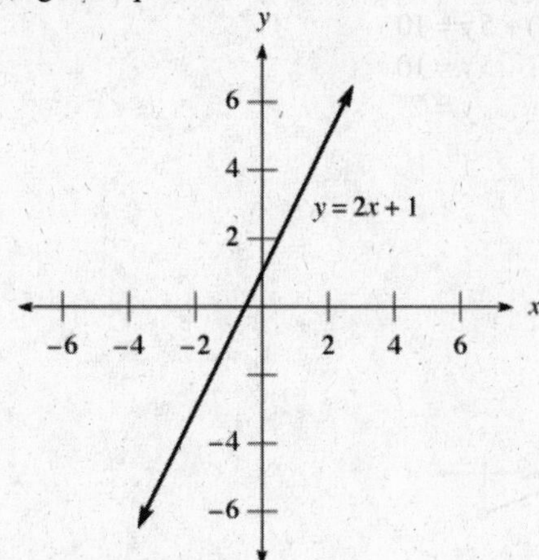

23. Graphing the equation:

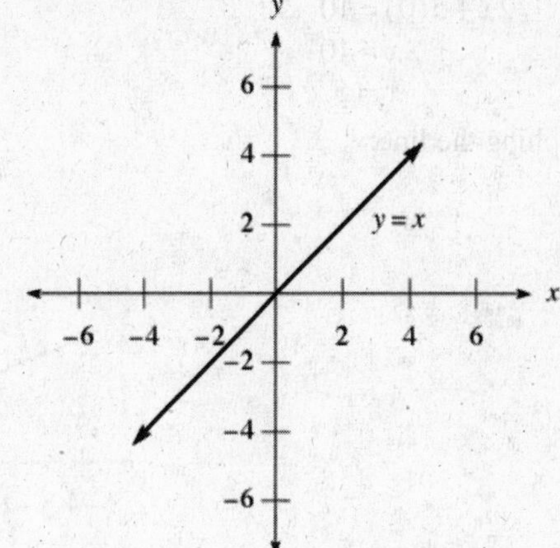

25. Checking the point $(0,0)$:

$$0-2(0)\le 4$$
$$0\le 4 \quad \text{(true)}$$

Graphing the linear inequality:

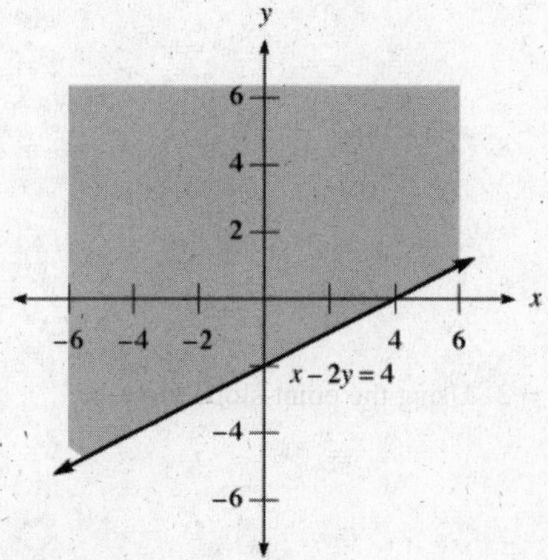

27. Graphing the line which passes through $(-3,-2)$ and $(2,3)$:

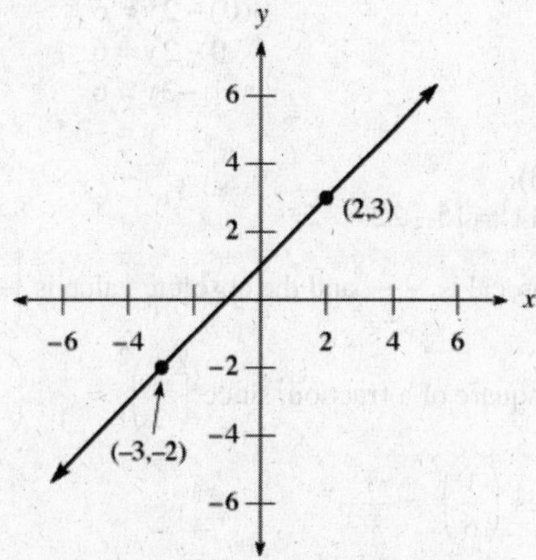

The point $(4,7)$ does not lie on the line, while $(1,2)$ does lie on the line.

© 2012 Cengage Learning. All Rights Reserved. May not be scanned, copied or duplicated, or posted to a publicly accessible website, in whole or in part.

29. To find the x-intercept, let $y = 0$:

$$\begin{aligned} 2x + 5(0) &= 10 \\ 2x &= 10 \\ x &= 5 \end{aligned}$$

To find the y-intercept, let $x = 0$:

$$\begin{aligned} 2(0) + 5y &= 10 \\ 5y &= 10 \\ y &= 2 \end{aligned}$$

Graphing the line:

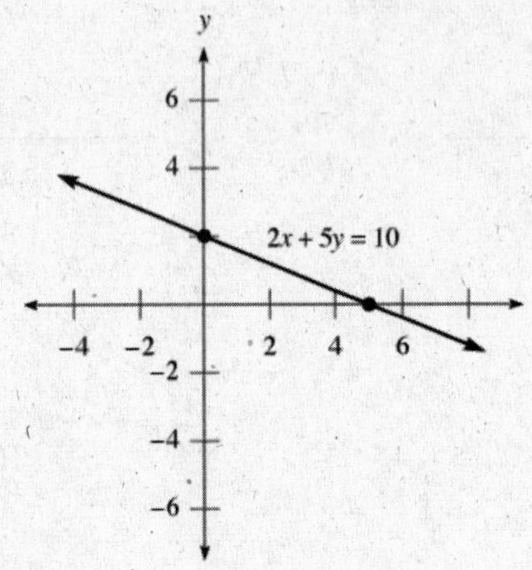

31. The slope is given by: $m = \dfrac{7-3}{5-2} = \dfrac{4}{3}$

33. Solving for y:

$$\begin{aligned} 2x + 3y &= 6 \\ 3y &= -2x + 6 \\ y &= -\frac{2}{3}x + 2 \end{aligned}$$

The slope of the line is $-\dfrac{2}{3}$.

35. First find the slope: $m = \dfrac{7-3}{4-2} = \dfrac{4}{2} = 2$. Using the point-slope formula:

$$\begin{aligned} y - 3 &= 2(x - 2) \\ y - 3 &= 2x - 4 \\ y &= 2x - 1 \end{aligned}$$

37. Substituting $y = 3$ and $x = 0$:

$$\begin{aligned} 3x - 2(3) &= 6 \\ 3x - 6 &= 6 \\ 3x &= 12 \\ x &= 4 \end{aligned}$$

$$\begin{aligned} 3(0) - 2y &= 6 \\ 0 - 2y &= 6 \\ -2y &= 6 \\ y &= -3 \end{aligned}$$

The ordered pairs are (4,3) and (0,–3).

39. Simplifying the expression: $15 - 2(11) = 15 - 22 = -7$

41. The opposite of $-\dfrac{2}{3}$ is $\dfrac{2}{3}$, the reciprocal is $-\dfrac{3}{2}$, and the absolute value is $\left|-\dfrac{2}{3}\right| = \dfrac{2}{3}$.

43. Each number in the sequence is the square of a fraction, since $\left(\dfrac{1}{2}\right)^2 = \dfrac{1}{4}$, $\left(\dfrac{1}{3}\right)^2 = \dfrac{1}{9}$, $\left(\dfrac{1}{4}\right)^2 = \dfrac{1}{16}$, and $\left(\dfrac{1}{5}\right)^2 = \dfrac{1}{25}$.

So the next number in the sequence is $\left(\dfrac{1}{6}\right)^2 = \dfrac{1}{36}$.

45. Evaluating when $x = 2$: $x^2 + 6x - 7 = (2)^2 + 6(2) - 7 = 4 + 12 - 7 = 9$

© 2012 Cengage Learning. All Rights Reserved. May not be scanned, copied or duplicated, or posted to a publicly accessible website, in whole or in part.

47. Let w represent the width and $w + 4$ represent the length. Using the perimeter formula:

$$\begin{aligned} 2(w)+2(w+4) &= 28 \\ 2w+2w+8 &= 28 \\ 4w+8 &= 28 \\ 4w &= 20 \\ w &= 5 \\ w+4 = 5+4 &= 9 \end{aligned}$$

The width is 5 in. and the length is 9 in.

49. The slope is: $m = \dfrac{13.7-14.5}{1996-1994} = \dfrac{-0.8}{2} = -0.4 = -\dfrac{2}{5}$

The poverty rate is decreasing by 0.4% per year.

Chapter 3 Test

1. Graphing the ordered pair:

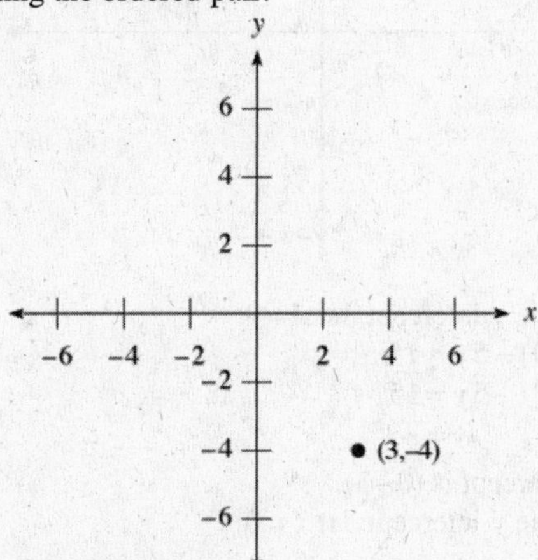

2. Graphing the ordered pair:

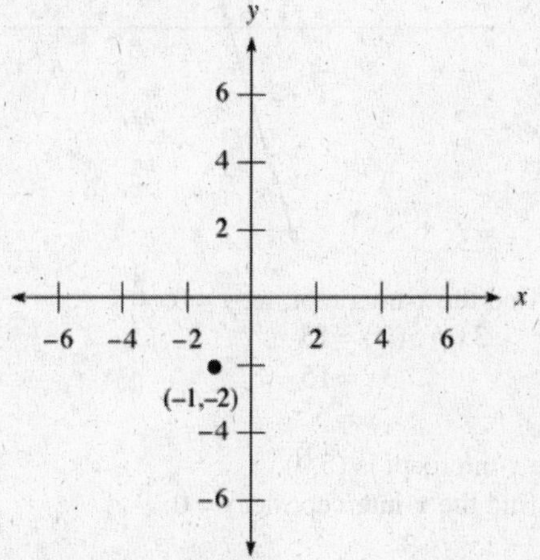

3. Graphing the ordered pair:

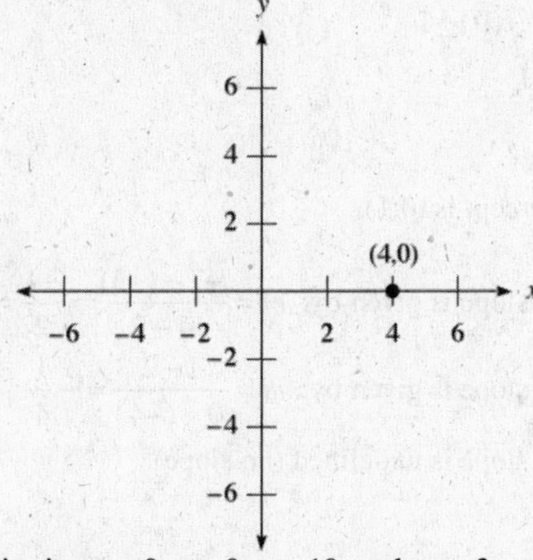

4. Graphing the ordered pair:

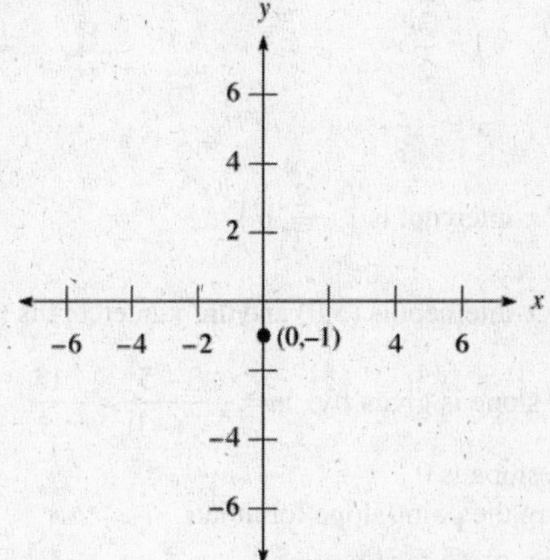

5. Substituting $x = 0$, $y = 0$, $x = 10$, and $y = -3$:

$$\begin{aligned} 2(0)-5y &= 10 \\ 0-5y &= 10 \\ -5y &= 10 \\ y &= -2 \end{aligned} \qquad \begin{aligned} 2x-5(0) &= 10 \\ 2x-0 &= 10 \\ 2x &= 10 \\ x &= 5 \end{aligned} \qquad \begin{aligned} 2(10)-5y &= 10 \\ 20-5y &= 10 \\ -5y &= -10 \\ y &= 2 \end{aligned} \qquad \begin{aligned} 2x-5(-3) &= 10 \\ 2x+15 &= 10 \\ 2x &= -5 \\ x &= -\frac{5}{2} \end{aligned}$$

The ordered pairs are $(0,-2), (5,0), (10,2)$, and $\left(-\dfrac{5}{2},-3\right)$.

© 2012 Cengage Learning. All Rights Reserved. May not be scanned, copied or duplicated, or posted to a publicly accessible website, in whole or in part.

6. Substituting each ordered pair into the equation:

$(2,5)$: $4(2)-3=8-3=5$
$(0,-3)$: $4(0)-3=0-3=-3$
$(3,0)$: $4(3)-3=12-3=9\neq 0$
$(-2,11)$: $4(-2)-3=-8-3=-11\neq 11$

The ordered pairs $(2,5)$ and $(0,-3)$ are solutions.

7. Graphing the line:

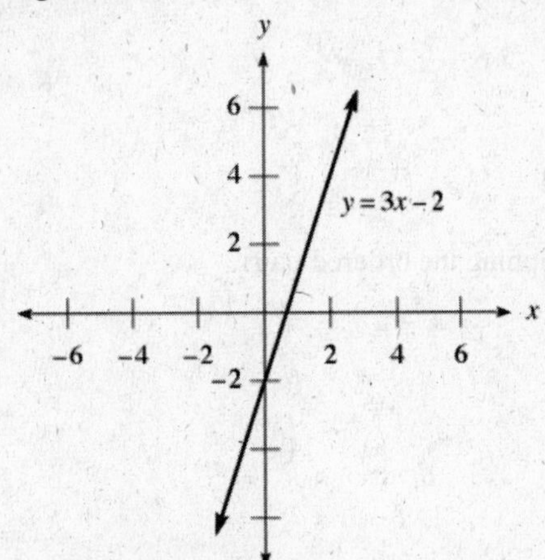

8. Graphing the line:

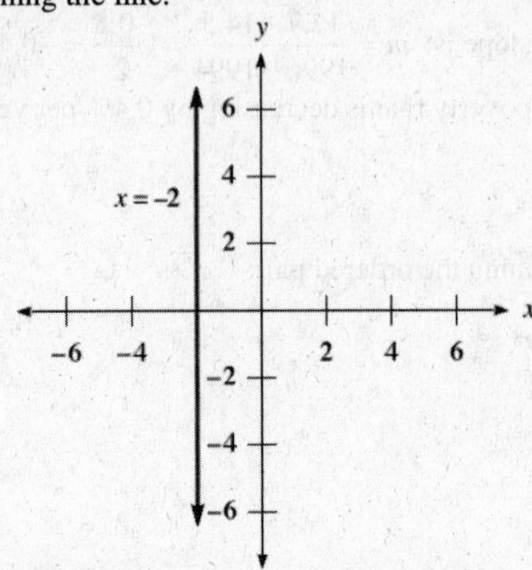

9. To find the x-intercept, let $y=0$:

$$3x-5(0)=15$$
$$3x=15$$
$$x=5$$

The x-intercept is (5,0).

To find the y-intercept, let $x=0$:

$$3(0)-5y=15$$
$$-5y=15$$
$$y=-3$$

The y-intercept is (0,–3).

10. To find the x-intercept, let $y=0$:

$$0=\frac{3}{2}x+1$$
$$-1=\frac{3}{2}x$$
$$x=-\frac{2}{3}$$

The x-intercept is $\left(-\frac{2}{3},0\right)$.

To find the y-intercept, let $x=0$:

$$y=\frac{3}{2}(0)+1$$
$$y=1$$

The y-intercept is (0,1).

11. The x-intercept is (3,0) and the y-intercept is (0,–2).

12. The slope is given by: $m=\frac{-7-(-3)}{4-2}=\frac{-4}{2}=-2$

13. The slope is given by: $m=\frac{-8-5}{2-(-3)}=-\frac{13}{5}$

14. The slope is given by: $m=\frac{1-2}{0-(-4)}=-\frac{1}{4}$

15. The slope is 0.

16. The slope is undefined (no slope).

17. Using the point-slope formula:

$$y-5=3(x-(-5))$$
$$y-5=3(x+5)$$
$$y-5=3x+15$$
$$y=3x+20$$

18. The slope-intercept form is $y=4x+8$.

© 2012 Cengage Learning. All Rights Reserved. May not be scanned, copied or duplicated, or posted to a publicly accessible website, in whole or in part.

19. First find the slope: $m = \dfrac{4-1}{-2-(-3)} = 3$. Using the point-slope formula:

$$\begin{aligned} y-1 &= 3(x-(-3)) \\ y-1 &= 3(x+3) \\ y-1 &= 3x+9 \\ y &= 3x+10 \end{aligned}$$

20. First find the slope: $m = \dfrac{4-0}{3-1} = \dfrac{4}{2} = 2$. Using the point-slope formula:

$$\begin{aligned} y-4 &= 2(x-3) \\ y-4 &= 2x-6 \\ y &= 2x-2 \end{aligned}$$

21. The slope is $m = \dfrac{4}{3}$ and the y-intercept is 4, so the equation is $y = \dfrac{4}{3}x+4$.

22. Checking the point $(0,0)$:

$$\begin{aligned} 0 &< 0+4 \\ 0 &< 4 \quad \text{(true)} \end{aligned}$$

Graphing the linear inequality:

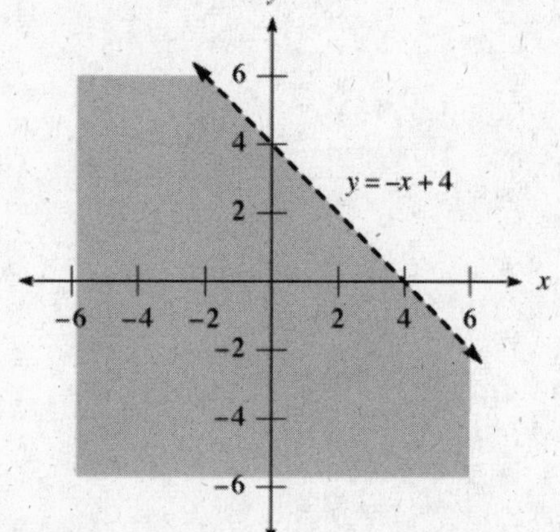

23. Checking the point $(0,0)$:

$$\begin{aligned} 3(0)-4(0) &\geq 12 \\ 0 &\geq 12 \quad \text{(false)} \end{aligned}$$

Graphing the linear inequality:

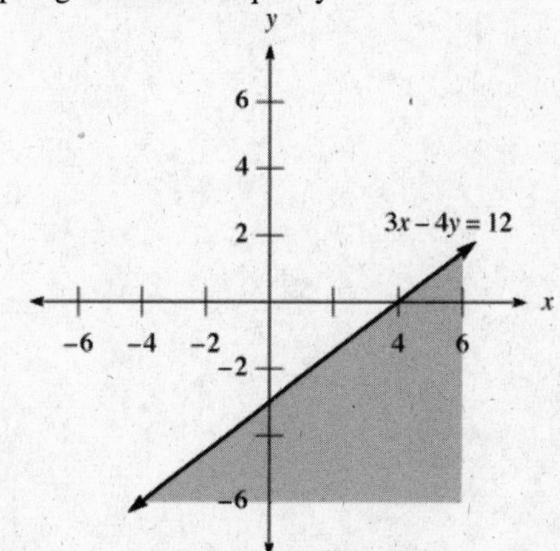

24. The points are (1986, 37), (1989, 26), (1996, 24), (1998, 27), (2000, 24), (2004, 33), and (2007, 29).

© 2012 Cengage Learning. All Rights Reserved. May not be scanned, copied or duplicated, or posted to a publicly accessible website, in whole or in part.

Chapter 4
Systems of Linear Equations

Getting Ready for Chapter 4

1. Simplifying: $(3x+2y)+(-3x+y)=3x+2y-3x+y=3y$
2. Simplifying: $(6x-7y)+(x+7y)=6x-7y+x+7y=7x$
3. Simplifying: $(5x+9y)+(-5x-12y)=5x+9y-5x-12y=-3y$
4. Simplifying: $-2(3x+4y)=-6x-8y$
5. Simplifying: $4(3x+4y)=12x+16y$
6. Simplifying: $9\left(\frac{2}{3}x-\frac{1}{4}y\right)=\frac{18}{3}x-\frac{9}{4}y=6x-\frac{9}{4}y$
7. Solving the equation:
$$\begin{aligned}4x-(2x-1)&=5\\4x-2x+1&=5\\2x+1&=5\\2x&=4\\x&=2\end{aligned}$$
8. Solving the equation:
$$\begin{aligned}5x-3(4x-7)&=-7\\5x-12x+21&=-7\\-7x+21&=-7\\-7x&=-28\\x&=4\end{aligned}$$
9. Solving the equation:
$$\begin{aligned}3(3y-1)-3y&=-27\\9y-3-3y&=-27\\6y-3&=-27\\6y&=-24\\y&=-4\end{aligned}$$
10. Solving the equation:
$$\begin{aligned}2x+9\left(-\frac{5}{3}x-\frac{8}{3}\right)&=2\\2x-15x-24&=2\\-13x-24&=2\\-13x&=26\\x&=-2\end{aligned}$$
11. Substituting $x=3$:
$$\begin{aligned}3+2y&=7\\2y&=4\\y&=2\end{aligned}$$
12. Substituting $x=-4$:
$$\begin{aligned}2(-4)-3y&=4\\-8-3y&=4\\-3y&=12\\y&=-4\end{aligned}$$
13. Substituting $x=3$:
$$\begin{aligned}3+3y&=3\\3y&=0\\y&=0\end{aligned}$$
14. Substituting $x=3$:
$$\begin{aligned}-2(3)+2y&=4\\-6+2y&=4\\2y&=10\\y&=5\end{aligned}$$

© 2012 Cengage Learning. All Rights Reserved. May not be scanned, copied or duplicated, or posted to a publicly accessible website, in whole or in part.

4.1 Solving Linear Systems by Graphing

1. Graphing both lines:

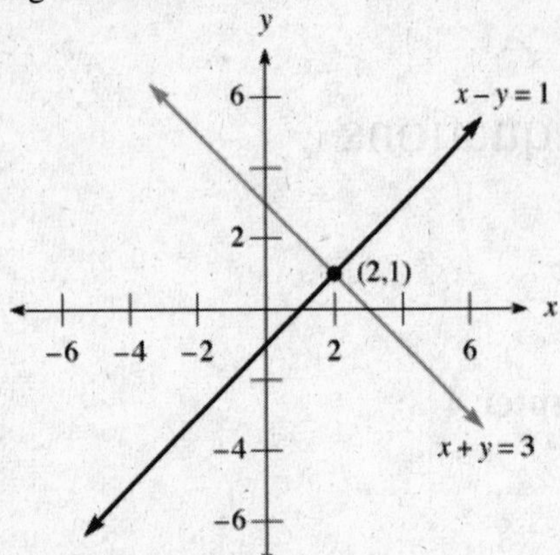

The intersection point is $(2,1)$.

3. Graphing both lines:

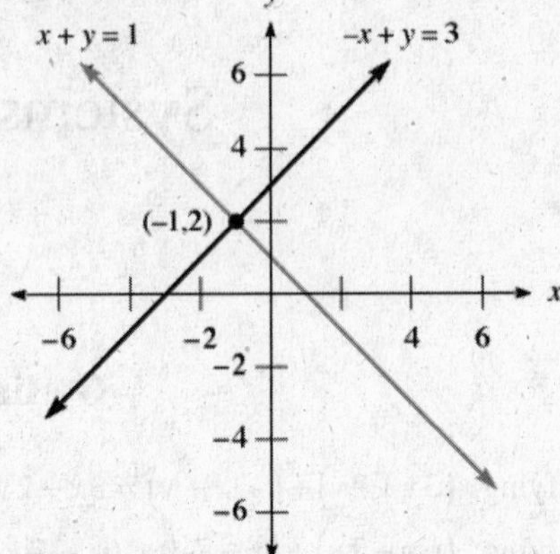

The intersection point is $(-1,2)$.

5. Graphing both lines:

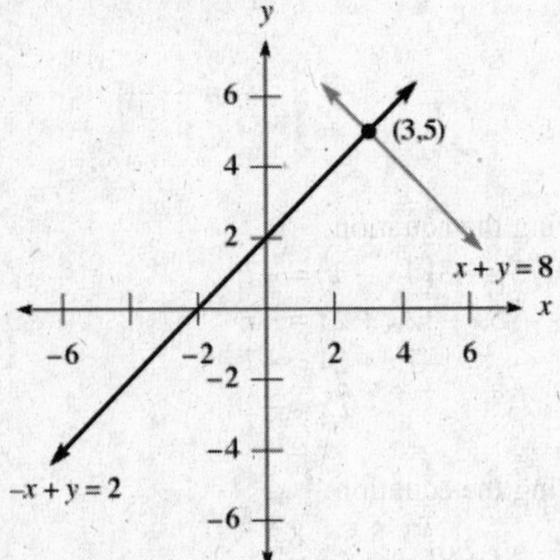

The intersection point is $(3,5)$.

7. Graphing both lines:

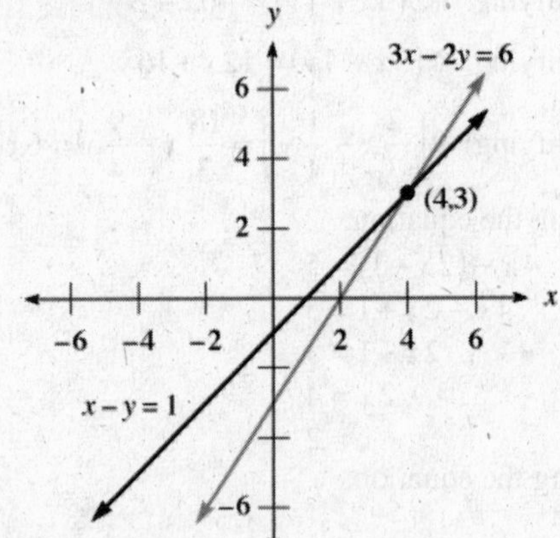

The intersection point is $(4,3)$.

9. Graphing both lines:

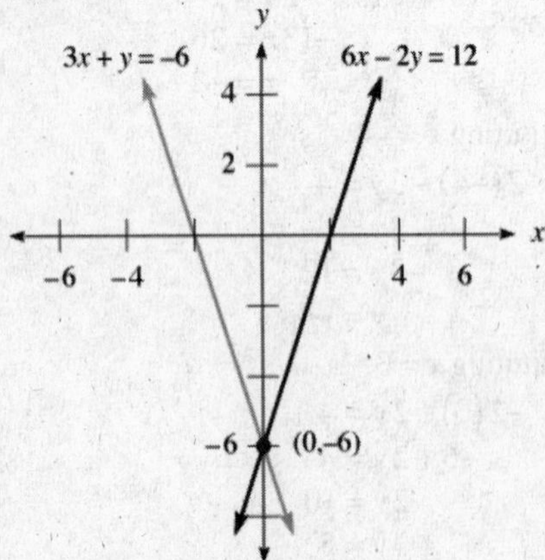

The intersection point is $(0,-6)$.

11. Graphing both lines:

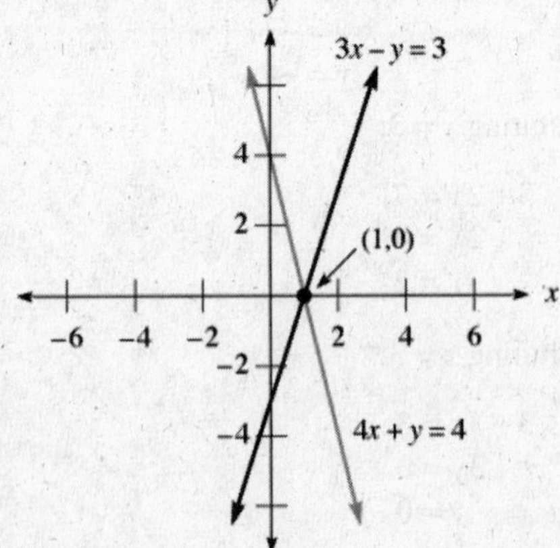

The intersection point is $(1,0)$.

© 2012 Cengage Learning. All Rights Reserved. May not be scanned, copied or duplicated, or posted to a publicly accessible website, in whole or in part.

13. Graphing both lines:

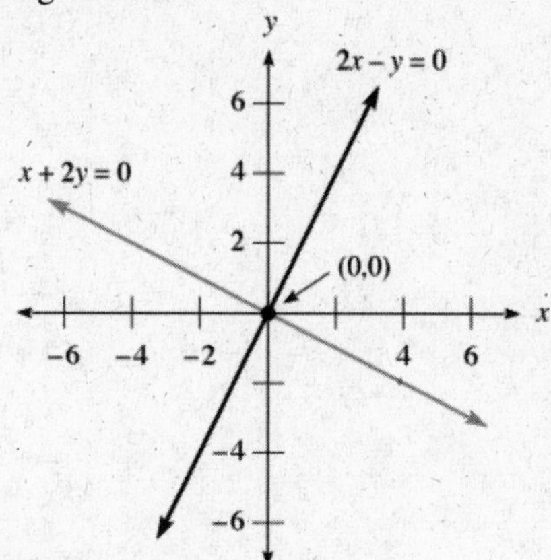

The intersection point is $(0,0)$.

15. Graphing both lines:

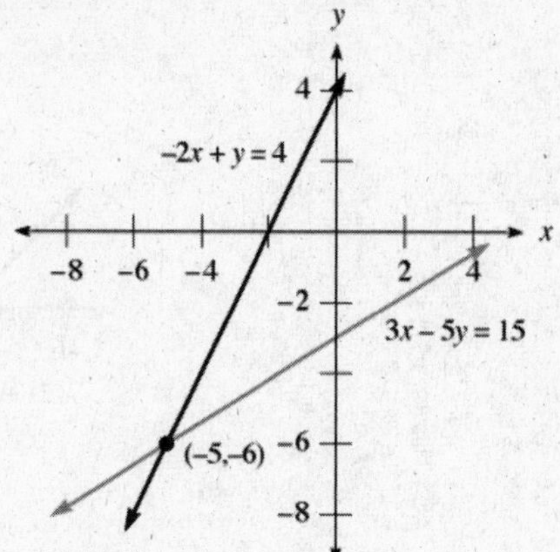

The intersection point is $(-5,-6)$.

17. Graphing both lines:

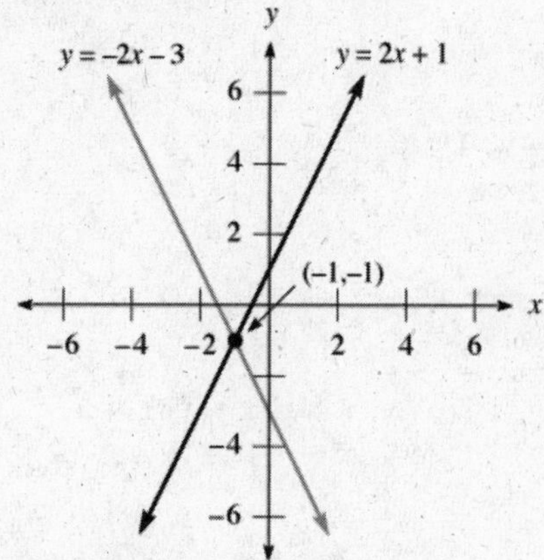

The intersection point is $(-1,-1)$.

19. Graphing both lines:

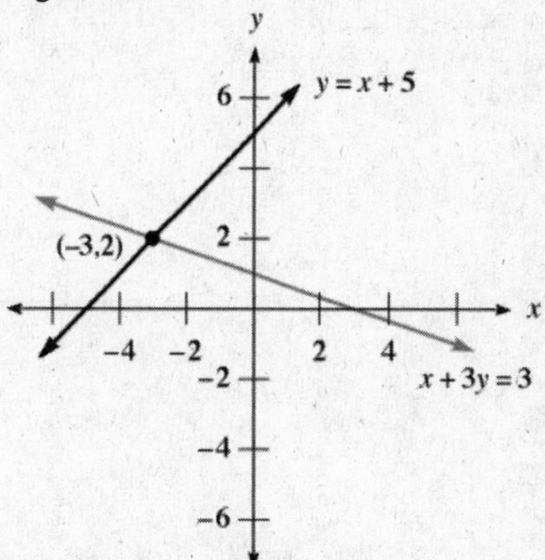

The intersection point is $(-3,2)$.

21. Graphing both lines:

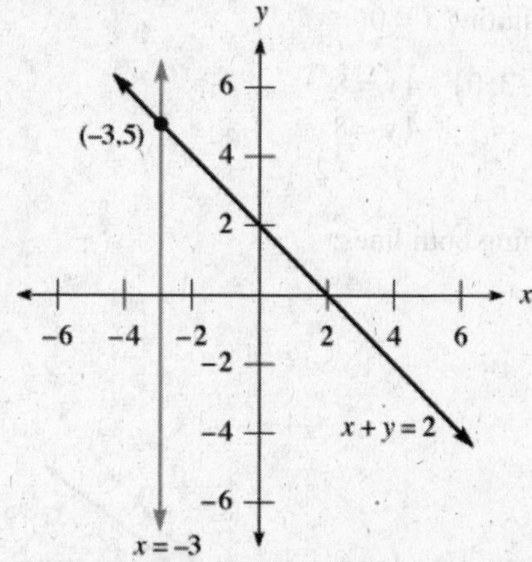

The intersection point is $(-3,5)$.

23. Graphing both lines:

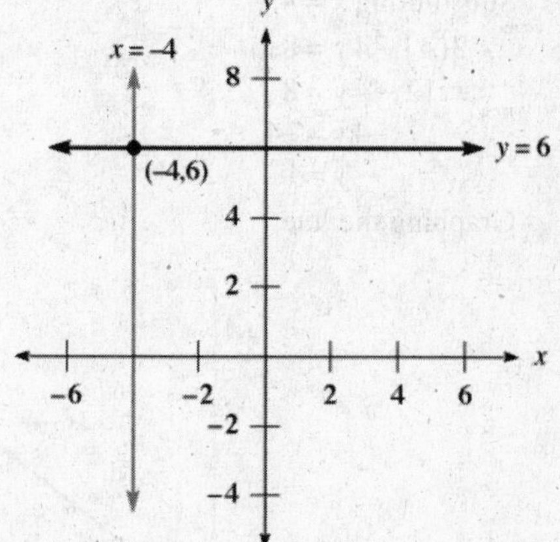

The intersection point is $(-4,6)$.

© 2012 Cengage Learning. All Rights Reserved. May not be scanned, copied or duplicated, or posted to a publicly accessible website, in whole or in part.

25. Graphing both lines:

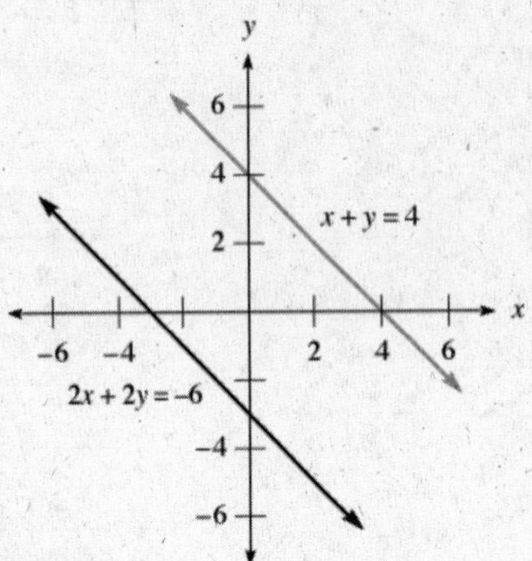

There is no intersection (the lines are parallel).

27. Graphing both lines:

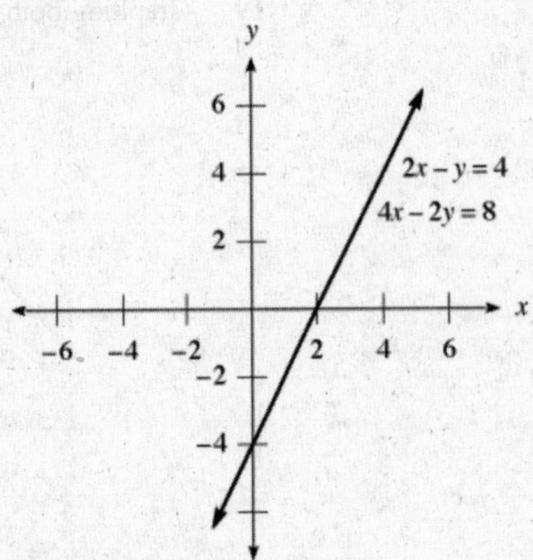

The system is dependent (both lines are the same, they coincide).

29. a. Simplifying: $(3x-4y)+(x-y)=4x-5y$

b. Substituting $x = 4$:

$$3(4)-4y=8$$
$$12-4y=8$$
$$-4y=-4$$
$$y=1$$

c. Substituting $x = 0$:

$$3(0)-4y=8$$
$$-4y=8$$
$$y=-2$$

d. Graphing the line:

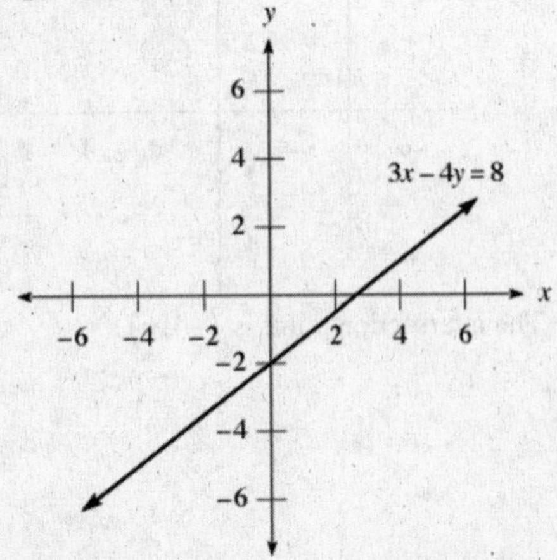

e. Graphing both lines:

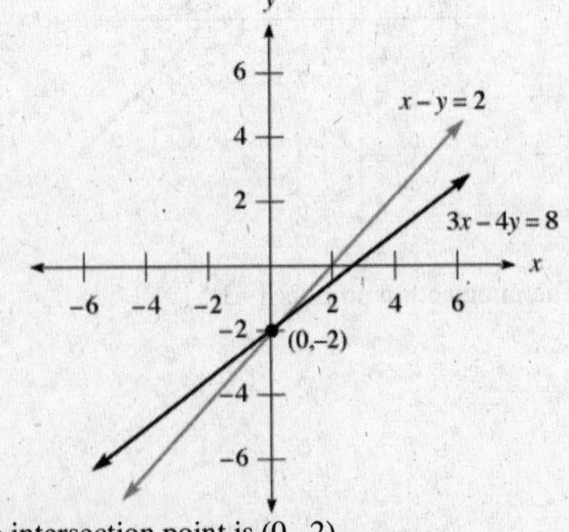

The intersection point is (0,–2).

© 2012 Cengage Learning. All Rights Reserved. May not be scanned, copied or duplicated, or posted to a publicly accessible website, in whole or in part.

31. Graphing both lines:

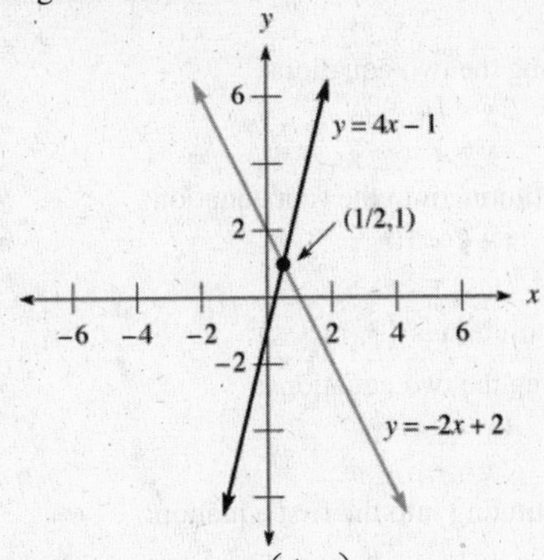

The intersection point is $\left(\frac{1}{2},1\right)$.

33. Graphing both lines:

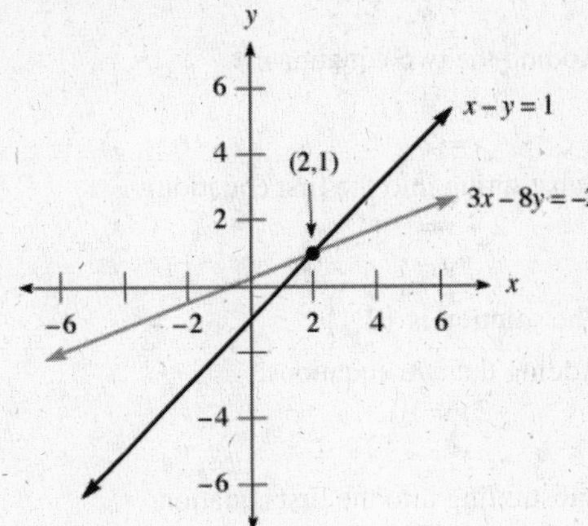

The intersection point is $(2,1)$.

35. No, not for positive values of light output (x) in the first quadrant.

37.
a. If Jane worked 25 hours, she would earn the same amount at each position.
b. If Jane worked less than 20 hours, she should choose Gigi's since she earns more in that position.
c. If Jane worked more than 30 hours, she should choose Marcy's since she earns more in that position.

39. Simplifying: $(x+y)+(x-y)=x+y+x-y=2x$

41. Simplifying: $(6x-3y)+(x+3y)=6x-3y+x+3y=7x$

43. Simplifying: $(-12x-20y)+(25x+20y)=-12x-20y+25x+20y=13x$

45. Simplifying: $-4(3x+5y)=-12x-20y$

47. Simplifying: $12\left(\frac{1}{4}x+\frac{2}{3}y\right)=12\bullet\frac{1}{4}x+12\bullet\frac{2}{3}y=3x+8y$

49. Simplifying: $-2(2x-y)=-4x+2y$

51. Substituting $x=3$:

$$\begin{aligned}3+y&=4\\y&=1\end{aligned}$$

53. Substituting $x=3$:

$$\begin{aligned}3+3y&=3\\3y&=0\\y&=0\end{aligned}$$

55. Substituting $x=6$:

$$\begin{aligned}3(6)+5y&=-7\\18+5y&=-7\\5y&=-25\\y&=-5\end{aligned}$$

57. Simplifying: $6x+100(0.04x+0.75)=6x+4x+75=10x+75$

59. Simplifying: $13x-1{,}000(0.002x+0.035)=13x-2x-35=11x-35$

61. Simplifying: $16x-10(1.7x-5.8)=16x-17x+58=-x+58$

63. Simplifying: $0.04x+0.06(100-x)=0.04x+6-0.06x=-0.02x+6$

65. Simplifying: $0.025x-0.028(1{,}000+x)=0.025x-28-0.028x=-0.003x-28$

67. Simplifying: $2.56x-1.25(100+x)=2.56x-125-1.25x=1.31x-125$

© 2012 Cengage Learning. All Rights Reserved. May not be scanned, copied or duplicated, or posted to a publicly accessible website, in whole or in part.

4.2 The Elimination Method

1. Adding the two equations:
$$2x = 4$$
$$x = 2$$
Substituting into the first equation:
$$2 + y = 3$$
$$y = 1$$
The solution is $(2,1)$.

3. Adding the two equations:
$$2y = 14$$
$$y = 7$$
Substituting into the first equation:
$$x + 7 = 10$$
$$x = 3$$
The solution is $(3,7)$.

5. Adding the two equations:
$$-2y = 10$$
$$y = -5$$
Substituting into the first equation:
$$x - (-5) = 7$$
$$x + 5 = 7$$
$$x = 2$$
The solution is $(2,-5)$.

7. Adding the two equations:
$$4x = -4$$
$$x = -1$$
Substituting into the first equation:
$$-1 + y = -1$$
$$y = 0$$
The solution is $(-1,0)$.

9. Adding the two equations: $0 = 0$
The lines coincide (the system is dependent).

11. Multiplying the first equation by 2:
$$6x - 2y = 8$$
$$2x + 2y = 24$$
Adding the two equations:
$$8x = 32$$
$$x = 4$$
Substituting into the first equation:
$$3(4) - y = 4$$
$$12 - y = 4$$
$$-y = -8$$
$$y = 8$$
The solution is $(4,8)$.

13. Multiplying the second equation by –3:
$$5x - 3y = -2$$
$$-30x + 3y = -3$$
Adding the two equations:
$$-25x = -5$$
$$x = \frac{1}{5}$$
Substituting into the first equation:
$$5\left(\frac{1}{5}\right) - 3y = -2$$
$$1 - 3y = -2$$
$$-3y = -3$$
$$y = 1$$
The solution is $\left(\frac{1}{5},1\right)$.

15. Multiplying the second equation by 4:
$$11x - 4y = 11$$
$$20x + 4y = 20$$
Adding the two equations:
$$31x = 31$$
$$x = 1$$
Substituting into the second equation:
$$5(1) + y = 5$$
$$5 + y = 5$$
$$y = 0$$
The solution is $(1,0)$.

17. Multiplying the second equation by 3:
$$3x - 5y = 7$$
$$-3x + 3y = -3$$
Adding the two equations:
$$-2y = 4$$
$$y = -2$$
Substituting into the second equation:
$$-x - 2 = -1$$
$$-x = 1$$
$$x = -1$$
The solution is $(-1,-2)$.

© 2012 Cengage Learning. All Rights Reserved. May not be scanned, copied or duplicated, or posted to a publicly accessible website, in whole or in part.

19. Multiplying the first equation by –2:
$$2x+16y=2$$
$$-2x+4y=13$$
Adding the two equations:
$$20y=15$$
$$y=\frac{3}{4}$$
Substituting into the first equation:
$$-x-8\left(\frac{3}{4}\right)=-1$$
$$-x-6=-1$$
$$-x=5$$
$$x=-5$$
The solution is $\left(-5,\frac{3}{4}\right)$.

21. Multiplying the first equation by 2:
$$-6x-2y=14$$
$$6x+7y=11$$
Adding the two equations:
$$5y=25$$
$$y=5$$
Substituting into the first equation:
$$-3x-5=7$$
$$-3x=12$$
$$x=-4$$
The solution is $(-4,5)$.

23. Adding the two equations:
$$8x=-24$$
$$x=-3$$
Substituting into the second equation:
$$2(-3)+y=-16$$
$$-6+y=-16$$
$$y=-10$$
The solution is $(-3,-10)$.

25. Multiplying the second equation by 3:
$$x+3y=9$$
$$6x-3y=12$$
Adding the two equations:
$$7x=21$$
$$x=3$$
Substituting into the first equation:
$$3+3y=9$$
$$3y=6$$
$$y=2$$
The solution is $(3,2)$.

27. Multiplying the second equation by 2:
$$x-6y=3$$
$$8x+6y=42$$
Adding the two equations:
$$9x=45$$
$$x=5$$
Substituting into the second equation:
$$4(5)+3y=21$$
$$20+3y=21$$
$$3y=1$$
$$y=\frac{1}{3}$$
The solution is $\left(5,\frac{1}{3}\right)$.

29. Multiplying the second equation by –3:
$$2x+9y=2$$
$$-15x-9y=24$$
Adding the two equations:
$$-13x=26$$
$$x=-2$$
Substituting into the first equation:
$$2(-2)+9y=2$$
$$-4+9y=2$$
$$9y=6$$
$$y=\frac{2}{3}$$
The solution is $\left(-2,\frac{2}{3}\right)$.

© 2012 Cengage Learning. All Rights Reserved. May not be scanned, copied or duplicated, or posted to a publicly accessible website, in whole or in part.

31. To clear each equation of fractions, multiply the first equation by 12 and the second equation by 6:

$$12\left(\frac{1}{3}x+\frac{1}{4}y\right)=12\left(\frac{7}{6}\right) \qquad 6\left(\frac{3}{2}x-\frac{1}{3}y\right)=6\left(\frac{7}{3}\right)$$
$$4x+3y=14 \qquad 9x-2y=14$$

The system of equations is:

$$4x+3y=14$$
$$9x-2y=14$$

Multiplying the first equation by 2 and the second equation by 3:

$$8x+6y=28$$
$$27x-6y=42$$

Adding the two equations:

$$35x=70$$
$$x=2$$

Substituting into $4x+3y=14$:

$$4(2)+3y=14$$
$$8+3y=14$$
$$3y=6$$
$$y=2$$

The solution is $(2,2)$.

33. Multiplying the first equation by –2:

$$-6x-4y=2$$
$$6x+4y=0$$

Adding the two equations:

$$0=2$$

Since this statement is false, the two lines are parallel, so the system has no solution.

35. Multiplying the first equation by 2 and the second equation by 3:

$$22x+12y=34$$
$$15x-12y=3$$

Adding the two equations:

$$37x=37$$
$$x=1$$

Substituting into the second equation:

$$5(1)-4y=1$$
$$5-4y=1$$
$$-4y=-4$$
$$y=1$$

The solution is $(1,1)$.

37. To clear each equation of fractions, multiply the first equation by 6 and the second equation by 6:

$$6\left(\frac{1}{2}x+\frac{1}{6}y\right)=6\left(\frac{1}{3}\right) \qquad 6\left(-x-\frac{1}{3}y\right)=6\left(-\frac{1}{6}\right)$$
$$3x+y=2 \qquad -6x-2y=-1$$

The system of equations is:

$$3x+y=2$$
$$-6x-2y=-1$$

Multiplying the first equation by 2:

$$6x+2y=4$$
$$-6x-2y=-1$$

Adding the two equations:

$$0=3$$

Since this statement is false, the two lines are parallel, so the system has no solution.

© 2012 Cengage Learning. All Rights Reserved. May not be scanned, copied or duplicated, or posted to a publicly accessible website, in whole or in part.

39. Multiplying the first equation by –5:

$$-5x-5y=-110$$
$$5x+10y=170$$

Adding the two equations:

$$5y=60$$
$$y=12$$

Substituting into the first equation:

$$x+12=22$$
$$x=10$$

The solution is $(10,12)$.

41. Multiplying the first equation by –5:

$$-5x-5y=-70$$
$$5x+25y=230$$

Adding the two equations:

$$20y=160$$
$$y=8$$

Substituting into the first equation:

$$x+8=14$$
$$x=6$$

The solution is $(6,8)$.

43. Multiplying the first equation by –6:

$$-6x-6y=-90,000$$
$$6x+7y=98,000$$

Adding the two equations:

$$y=8,000$$

Substituting into the first equation:

$$x+8,000=15,000$$
$$x=7,000$$

The solution is $(7000,8000)$.

45. Multiplying the first equation by –4:

$$-4x-4y=-44,000$$
$$4x+7y=68,000$$

Adding the two equations:

$$3y=24,000$$
$$y=8,000$$

Substituting into the first equation:

$$x+8,000=11,000$$
$$x=3,000$$

The solution is $(3000,8000)$.

47. Multiplying the first equation by –5:

$$-5x-5y=-115$$
$$5x+10y=175$$

Adding the two equations:

$$5y=60$$
$$y=12$$

Substituting into the first equation:

$$x+12=23$$
$$x=11$$

The solution is $(11,12)$.

49. Multiplying the second equation by 100 (to eliminate decimals):

$$x+y=22$$
$$5x+10y=170$$

Multiplying the first equation by –5:

$$-5x-5y=-110$$
$$5x+10y=170$$

Adding the two equations:

$$5y=60$$
$$y=12$$

Substituting into the first equation:

$$x+12=22$$
$$x=10$$

The solution is $(10,12)$.

51. Setting the non-camera phone sales equal to the total sales of phones results in the system:

$$60x-y=119,560$$
$$40x+y=80,560$$

Adding the two equations:

$$100x=200,120$$
$$x\approx 2001$$

The year was 2001.

© 2012 Cengage Learning. All Rights Reserved. May not be scanned, copied or duplicated, or posted to a publicly accessible website, in whole or in part.

53. Solving the equation:
$$\begin{aligned} x+(2x-1)&=2 \\ x+2x-1&=2 \\ 3x-1&=2 \\ 3x&=3 \\ x&=1 \end{aligned}$$

55. Solving the equation:
$$\begin{aligned} 2(3y-1)-3y&=4 \\ 6y-2-3y&=4 \\ 3y-2&=4 \\ 3y&=6 \\ y&=2 \end{aligned}$$

57. Solving the equation:
$$\begin{aligned} -2x+3(5x-1)&=10 \\ -2x+15x-3&=10 \\ 13x-3&=10 \\ 13x&=13 \\ x&=1 \end{aligned}$$

59. Solving for x:
$$\begin{aligned} x-3y&=-1 \\ x&=3y-1 \end{aligned}$$

61. Solving for y: $y=2(1)-1=2-1=1$

63. Solving for x: $x=3(2)-1=6-1=5$

65. Substituting $x=13$: $y=1.5(13)+15=19.5+15=34.5$

67. Substituting $x=12$: $y=0.75(12)+24.95=9+24.95=33.95$

69. Solving for y:
$$\begin{aligned} 3x-y&=3 \\ -y&=-3x+3 \\ y&=3x-3 \end{aligned}$$

The slope is 3 and the y-intercept is –3.

71. Solving for y:
$$\begin{aligned} 2x-5y&=25 \\ -5y&=-2x+25 \\ y&=\frac{2}{5}x-5 \end{aligned}$$

The slope is $\frac{2}{5}$ and the y-intercept is –5.

73. Finding the slope: $m=\dfrac{-5-3}{6-(-2)}=\dfrac{-8}{8}=-1$

75. Finding the slope: $m=\dfrac{-3-3}{2-5}=\dfrac{-6}{-3}=2$

77. Using the slope formula:
$$\begin{aligned} \frac{y-5}{-4-(-2)}&=-3 \\ \frac{y-5}{-2}&=-3 \\ y-5&=6 \\ y&=11 \end{aligned}$$

79. Using the slope formula:
$$\begin{aligned} \frac{y-(-6)}{6-3}&=5 \\ \frac{y+6}{3}&=5 \\ y+6&=15 \\ y&=9 \end{aligned}$$

81. Using the point-slope formula:
$$\begin{aligned} y-(-6)&=3(x-(-2)) \\ y+6&=3(x+2) \\ y+6&=3x+6 \\ y&=3x \end{aligned}$$

83. First find the slope: $m=\dfrac{1-(-5)}{3-(-3)}=\dfrac{1+5}{3+3}=\dfrac{6}{6}=1$

Using the point-slope formula:
$$\begin{aligned} y-1&=1(x-3) \\ y-1&=x-3 \\ y&=x-2 \end{aligned}$$

© 2012 Cengage Learning. All Rights Reserved. May not be scanned, copied or duplicated, or posted to a publicly accessible website, in whole or in part.

4.3 The Substitution Method

1. Substituting into the first equation:

$$\begin{aligned} x+(2x-1)&=11 \\ 3x-1&=11 \\ 3x&=12 \\ x&=4 \\ y&=2(4)-1=7 \end{aligned}$$

The solution is $(4,7)$.

3. Substituting into the first equation:

$$\begin{aligned} x+(5x+2)&=20 \\ 6x+2&=20 \\ 6x&=18 \\ x&=3 \\ y&=5(3)+2=17 \end{aligned}$$

The solution is $(3,17)$.

5. Substituting into the first equation:

$$\begin{aligned} -2x+(-4x+8)&=-1 \\ -6x+8&=-1 \\ -6x&=-9 \\ x&=\frac{3}{2} \\ y&=-4\left(\frac{3}{2}\right)+8=-6+8=2 \end{aligned}$$

The solution is $\left(\frac{3}{2},2\right)$.

7. Substituting into the first equation:

$$\begin{aligned} 3(-y+6)-2y&=-2 \\ -3y+18-2y&=-2 \\ -5y+18&=-2 \\ -5y&=-20 \\ y&=4 \\ x&=-4+6=2 \end{aligned}$$

The solution is $(2,4)$.

9. Substituting into the first equation:

$$\begin{aligned} 5x-4(4)&=-16 \\ 5x-16&=-16 \\ 5x&=0 \\ x&=0 \end{aligned}$$

The solution is $(0,4)$.

11. Substituting into the first equation:

$$\begin{aligned} 5x+4(-3x)&=7 \\ 5x-12x&=7 \\ -7x&=7 \\ x&=-1 \\ y&=-3(-1)=3 \end{aligned}$$

The solution is $(-1,3)$.

13. Solving the second equation for x:

$$\begin{aligned} x-2y&=-1 \\ x&=2y-1 \end{aligned}$$

Substituting into the first equation:

$$\begin{aligned} (2y-1)+3y&=4 \\ 5y-1&=4 \\ 5y&=5 \\ y&=1 \\ x&=2(1)-1=1 \end{aligned}$$

The solution is $(1,1)$.

15. Solving the second equation for x:

$$\begin{aligned} x-5y&=17 \\ x&=5y+17 \end{aligned}$$

Substituting into the first equation:

$$\begin{aligned} 2(5y+17)+y&=1 \\ 10y+34+y&=1 \\ 11y+34&=1 \\ 11y&=-33 \\ y&=-3 \\ x&=5(-3)+17=2 \end{aligned}$$

The solution is $(2,-3)$.

© 2012 Cengage Learning. All Rights Reserved. May not be scanned, copied or duplicated, or posted to a publicly accessible website, in whole or in part.

17. Solving the second equation for x:

$$x-5y=-5$$
$$x=5y-5$$

Substituting into the first equation:

$$3(5y-5)+5y=-3$$
$$15y-15+5y=-3$$
$$20y-15=-3$$
$$20y=12$$
$$y=\frac{3}{5}$$
$$x=5\left(\frac{3}{5}\right)-5=3-5=-2$$

The solution is $\left(-2,\frac{3}{5}\right)$.

19. Solving the second equation for x:

$$x-3y=-18$$
$$x=3y-18$$

Substituting into the first equation:

$$5(3y-18)+3y=0$$
$$15y-90+3y=0$$
$$18y-90=0$$
$$18y=90$$
$$y=5$$
$$x=3(5)-18=-3$$

The solution is $(-3,5)$.

21. Solving the second equation for x:

$$x+3y=12$$
$$x=-3y+12$$

Substituting into the first equation:

$$-3(-3y+12)-9y=7$$
$$9y-36-9y=7$$
$$-36=7$$

Since this statement is false, there is no solution to the system. The two lines are parallel.

23. Substituting into the first equation:

$$5x-8(2x-5)=7$$
$$5x-16x+40=7$$
$$-11x+40=7$$
$$-11x=-33$$
$$x=3$$
$$y=2(3)-5=1$$

The solution is $(3,1)$.

25. Substituting into the first equation:

$$7(2y-1)-6y=-1$$
$$14y-7-6y=-1$$
$$8y-7=-1$$
$$8y=6$$
$$y=\frac{3}{4}$$
$$x=2\left(\frac{3}{4}\right)-1=\frac{3}{2}-1=\frac{1}{2}$$

The solution is $\left(\frac{1}{2},\frac{3}{4}\right)$.

27. Substituting into the first equation:

$$-3x+2(3x)=6$$
$$-3x+6x=6$$
$$3x=6$$
$$x=2$$
$$y=3(2)=6$$

The solution is $(2,6)$.

29. Substituting into the first equation:

$$5(y)-6y=-4$$
$$-y=-4$$
$$y=4$$
$$x=4$$

The solution is $(4,4)$.

31. Substituting into the first equation:

$$3x+3(2x-12)=9$$
$$3x+6x-36=9$$
$$9x-36=9$$
$$9x=45$$
$$x=5$$
$$y=2(5)-12=-2$$

The solution is $(5,-2)$.

33. Substituting into the first equation:

$$7x-11(10)=16$$
$$7x-110=16$$
$$7x=126$$
$$x=18$$
$$y=10$$

The solution is $(18,10)$.

© 2012 Cengage Learning. All Rights Reserved. May not be scanned, copied or duplicated, or posted to a publicly accessible website, in whole or in part.

35. Substituting into the first equation:

$$\begin{aligned}-4x+4(x-2)&=-8\\-4x+4x-8&=-8\\-8&=-8\end{aligned}$$

Since this statement is true, the system is dependent. The two lines coincide.

37. Substituting into the first equation:

$$\begin{aligned}2x+5(12-x)&=36\\2x+60-5x&=36\\-3x+60&=36\\-3x&=-24\\x&=8\\y&=4\end{aligned}$$

The solution is $(8,4)$.

39. Substituting into the first equation:

$$\begin{aligned}5x+2(18-x)&=54\\5x+36-2x&=54\\3x+36&=54\\3x&=18\\x&=6\\y&=12\end{aligned}$$

The solution is $(6,12)$.

41. Substituting into the first equation:

$$\begin{aligned}2x+2(2x)&=96\\2x+4x&=96\\6x&=96\\x&=16\\y&=32\end{aligned}$$

The solution is $(16,32)$.

43. Substituting into the first equation:

$$\begin{aligned}0.05x+0.10(22-x)&=1.70\\0.05x+2.2-0.10x&=1.70\\-0.05x+2.2&=1.7\\-0.05x&=-0.5\\x&=10\\y&=22-10=12\end{aligned}$$

The solution is $(10,12)$.

45. a. Solving the equation:

$$\begin{aligned}4x-5&=20\\4x&=25\\x&=\frac{25}{4}\end{aligned}$$

b. Solving for *y*:

$$\begin{aligned}4x-5y&=20\\-5y&=-4x+20\\y&=\frac{4}{5}x-4\end{aligned}$$

c. Solving for *x*:

$$\begin{aligned}x-y&=5\\x&=y+5\end{aligned}$$

d. Substituting into the first equation:

$$\begin{aligned}4(y+5)-5y&=20\\4y+20-5y&=20\\-y+20&=20\\-y&=0\\y&=0\\x&=0+5=5\end{aligned}$$

The solution is $(5,0)$.

47. Setting the non-camera phone sales equal to the total sales of phones results in the system:

$$\begin{aligned}60x-y&=119,560\\40x+y&=80,560\end{aligned}$$

Solving the second equation for *y* yields $y=80,560-40x$. Substituting into the first equation:

$$\begin{aligned}60x-(80,560-40x)&=119,560\\60x-80,560+40x&=119,560\\100x-80,560&=119,560\\100x&=200,120\\x&\approx 2001\end{aligned}$$

The year was 2001.

© 2012 Cengage Learning. All Rights Reserved. May not be scanned, copied or duplicated, or posted to a publicly accessible website, in whole or in part.

49. **a.** Setting the two equations equal:

$$\frac{2.50}{20}x+150=\frac{2.50}{35}x+180$$
$$140\left(\frac{2.50}{20}x+150\right)=140\left(\frac{2.50}{35}x+180\right)$$
$$17.5x+21,000=10x+25,200$$
$$7.5x=4,200$$
$$x=560$$

At 560 miles the car and truck cost the same to operate.

b. If Daniel drives more than 800 miles, the car will be cheaper to operate.

c. If Daniel drives less than 400 miles, the truck will be cheaper to operate.

d. The graphs appear in the first quadrant only because all quantities are positive.

51. Let x and $5x + 8$ represent the two numbers. The equation is:

$$x+5x+8=26$$
$$6x+8=26$$
$$6x=18$$
$$x=3$$
$$5x+8=23$$

The two numbers are 3 and 23.

53. Let x represent the smaller number and $2x - 6$ represent the larger number. The equation is:

$$2x-6-x=9$$
$$x-6=9$$
$$x=15$$
$$2x-6=24$$

The two numbers are 15 and 24.

55. Let w represent the width and $3w + 5$ represent the length. Using the perimeter formula:

$$2(w)+2(3w+5)=58$$
$$2w+6w+10=58$$
$$8w+10=58$$
$$8w=48$$
$$w=6$$
$$3w+5=23$$

The width is 6 inches and the length is 23 inches.

57. Completing the table:

	Nickels	Dimes
Number	$x+4$	x
Value (cents)	$5(x+4)$	$10x$

The equation is:

$$5(x+4)+10x=170$$
$$5x+20+10x=170$$
$$15x+20=170$$
$$15x=150$$
$$x=10$$
$$x+4=14$$

John has 14 nickels and 10 dimes.

59. Substituting into the second equation:

$$x+(5x+2)=20$$
$$6x+2=20$$
$$6x=18$$
$$x=3$$
$$y=5(3)+2=17$$

The solution is (3,17).

© 2012 Cengage Learning. All Rights Reserved. May not be scanned, copied or duplicated, or posted to a publicly accessible website, in whole or in part.

61. Multiplying the second equation by 100 (to eliminate decimals):

$$x+y=15000$$
$$6x+7y=98000$$

Multiplying the first equation by –6:

$$-6x-6y=-90000$$
$$6x+7y=98000$$

Adding the two equations:

$$y=8000$$

Substituting into the first equation:

$$x+8000=15000$$
$$x=7000$$

The solution is $(7000,8000)$.

63. Simplifying: $6(3+4)+5=6(7)+5=42+5=47$

65. Simplifying: $1^2+2^2+3^2=1+4+9=14$

67. Simplifying: $5(6+3\cdot 2)+4+3\cdot 2=5(6+6)+4+3\cdot 2=5(12)+4+3\cdot 2=60+4+6=70$

69. Simplifying: $(1^3+2^3)+[(2\cdot 3)+(4\cdot 5)]=(1+8)+[6+20]=9+26=35$

71. Simplifying: $[2\cdot 3+4+5]\div 3=(6+4+5)\div 3=15\div 3=5$

73. Simplifying: $6\cdot 10^3+5\cdot 10^2+4\cdot 10^1=6,000+500+40=6,540$

75. Simplifying: $1\cdot 10^3+7\cdot 10^2+6\cdot 10^1+0=1,000+700+60+0=1,760$

77. Simplifying: $4\cdot 2-1+5\cdot 3-2=8-1+15-2=20$

79. Simplifying: $(2^3+3^2)\cdot 4-5=(8+9)\cdot 4-5=17\cdot 4-5=68-5=63$

81. Simplifying: $2(2^2+3^2)+3(3^2)=2(4+9)+3(9)=2(13)+3(9)=26+27=53$

4.4 Applications

1. Let x and y represent the two numbers. The system of equations is:

$$x+y=25$$
$$y=5+x$$

Substituting into the first equation:

$$x+(5+x)=25$$
$$2x+5=25$$
$$2x=20$$
$$x=10$$
$$y=5+10=15$$

The two numbers are 10 and 15.

3. Let x and y represent the two numbers. The system of equations is:

$$x+y=15$$
$$y=4x$$

Substituting into the first equation:

$$x+4x=15$$
$$5x=15$$
$$x=3$$
$$y=4(3)=12$$

The two numbers are 3 and 12.

© 2012 Cengage Learning. All Rights Reserved. May not be scanned, copied or duplicated, or posted to a publicly accessible website, in whole or in part.

5. Let x represent the larger number and y represent the smaller number. The system of equations is:

$$x - y = 5$$
$$x = 2y + 1$$

Substituting into the first equation:

$$2y + 1 - y = 5$$
$$y + 1 = 5$$
$$y = 4$$
$$x = 2(4) + 1 = 9$$

The two numbers are 4 and 9.

7. Let x and y represent the two numbers. The system of equations is:

$$y = 4x + 5$$
$$x + y = 35$$

Substituting into the second equation:

$$x + 4x + 5 = 35$$
$$5x + 5 = 35$$
$$5x = 30$$
$$x = 6$$
$$y = 4(6) + 5 = 29$$

The two numbers are 6 and 29.

9. Let x represent the amount invested at 6% and y represent the amount invested at 8%. The system of equations is:

$$x + y = 20000$$
$$0.06x + 0.08y = 1380$$

Multiplying the first equation by –0.06:

$$-0.06x - 0.06y = -1200$$
$$0.06x + 0.08y = 1380$$

Adding the two equations:

$$0.02y = 180$$
$$y = 9000$$

Substituting into the first equation:

$$x + 9000 = 20000$$
$$x = 11000$$

Mr. Wilson invested \$9,000 at 8% and \$11,000 at 6%.

11. Let x represent the amount invested at 5% and y represent the amount invested at 6%. The system of equations is:

$$x = 4y$$
$$0.05x + 0.06y = 520$$

Substituting into the second equation:

$$0.05(4y) + 0.06y = 520$$
$$0.20y + 0.06y = 520$$
$$0.26y = 520$$
$$y = 2000$$
$$x = 4(2000) = 8000$$

She invested \$8,000 at 5% and \$2,000 at 6%.

© 2012 Cengage Learning. All Rights Reserved. May not be scanned, copied or duplicated, or posted to a publicly accessible website, in whole or in part.

13. Let x represent the number of nickels and y represent the number of quarters. The system of equations is:

$$x+y=14$$
$$0.05x+0.25y=2.30$$

Multiplying the first equation by –0.05:

$$-0.05x-0.05y=-0.7$$
$$0.05x+0.25y=2.30$$

Adding the two equations:

$$0.20y=1.6$$
$$y=8$$

Substituting into the first equation:

$$x+8=14$$
$$x=6$$

Ron has 6 nickels and 8 quarters.

15. Let x represent the number of dimes and y represent the number of quarters. The system of equations is:

$$x+y=21$$
$$0.10x+0.25y=3.45$$

Multiplying the first equation by –0.10:

$$-0.10x-0.10y=-2.10$$
$$0.10x+0.25y=3.45$$

Adding the two equations:

$$0.15y=1.35$$
$$y=9$$

Substituting into the first equation:

$$x+9=21$$
$$x=12$$

Tom has 12 dimes and 9 quarters.

17. Let x represent the liters of 50% alcohol solution and y represent the liters of 20% alcohol solution. The system of equations is:

$$x+y=18$$
$$0.50x+0.20y=0.30(18)$$

Multiplying the first equation by –0.20:

$$-0.20x-0.20y=-3.6$$
$$0.50x+0.20y=5.4$$

Adding the two equations:

$$0.30x=1.8$$
$$x=6$$

Substituting into the first equation:

$$6+y=18$$
$$y=12$$

The mixture contains 6 liters of 50% alcohol solution and 12 liters of 20% alcohol solution.

© 2012 Cengage Learning. All Rights Reserved. May not be scanned, copied or duplicated, or posted to a publicly accessible website, in whole or in part.

19. Let x represent the gallons of 10% disinfectant and y represent the gallons of 7% disinfectant. The system of equations is:

$$x+y=30$$
$$0.10x+0.07y=0.08(30)$$

Multiplying the first equation by –0.07:

$$-0.07x-0.07y=-2.1$$
$$0.10x+0.07y=2.4$$

Adding the two equations:

$$0.03x=0.3$$
$$x=10$$

Substituting into the first equation:

$$10+y=30$$
$$y=20$$

The mixture contains 10 gallons of 10% disinfectant and 20 gallons of 7% disinfectant.

21. Let x represent the number of adult tickets and y represent the number of kids tickets. The system of equations is:

$$x+y=70$$
$$5.50x+4.00y=310$$

Multiplying the first equation by –4:

$$-4.00x-4.00y=-280$$
$$5.50x+4.00y=310$$

Adding the two equations:

$$1.5x=30$$
$$x=20$$

Substituting into the first equation:

$$20+y=70$$
$$y=50$$

The matinee had 20 adult tickets sold and 50 kids tickets sold.

23. Let x represent the width and y represent the length. The system of equations is:

$$2x+2y=96$$
$$y=2x$$

Substituting into the first equation:

$$2x+2(2x)=96$$
$$2x+4x=96$$
$$6x=96$$
$$x=16$$
$$y=2(16)=32$$

The width is 16 feet and the length is 32 feet.

25. Let x represent the number of \$5 chips and y represent the number of \$25 chips. The system of equations is:

$$x+y=45$$
$$5x+25y=465$$

Multiplying the first equation by –5:

$$-5x-5y=-225$$
$$5x+25y=465$$

Adding the two equations:

$$20y=240$$
$$y=12$$

Substituting into the first equation:

$$x+12=45$$
$$x=33$$

The gambler has 33 \$5 chips and 12 \$25 chips.

© 2012 Cengage Learning. All Rights Reserved. May not be scanned, copied or duplicated, or posted to a publicly accessible website, in whole or in part.

27. Let x represent the number of shares of \$11 stock and y represent the number of shares of \$20 stock. The system of equations is:

$$x+y=150$$
$$11x+20y=2550$$

Multiplying the first equation by –11:

$$-11x-11y=-1650$$
$$11x+20y=2550$$

Adding the two equations:

$$9y=900$$
$$y=100$$

Substituting into the first equation:

$$x+100=150$$
$$x=50$$

She bought 50 shares at \$11 and 100 shares at \$20.

29. Let x represent the number of modules and y represent the number of fixed racks. The system of equations is:

$$x+y=8$$
$$6200x+1570y=21,820$$

Multiplying the first equation by –1570:

$$-1570x-1570y=-12,560$$
$$6200x+1570y=21,820$$

Adding the two equations:

$$4630x=9260$$
$$x=2$$
$$y=8-2=6$$

They bought 2 modules and 6 fixed racks.

31. Substituting $x=-2$, $x=0$, and $x=2$:

$$y=\frac{1}{2}(-2)+3=-1+3=2 \qquad y=\frac{1}{2}(0)+3=0+3=3 \qquad y=\frac{1}{2}(2)+3=1+3=4$$

The ordered pairs are $(-2,2)$, $(0,3)$, and $(2,4)$.

33. Graphing the line:

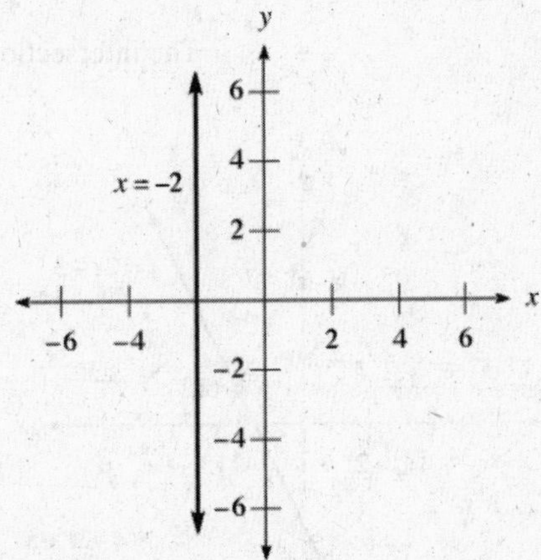

35. Computing the slope: $m=\dfrac{1-5}{0-2}=\dfrac{-4}{-2}=2$

© 2012 Cengage Learning. All Rights Reserved. May not be scanned, copied or duplicated, or posted to a publicly accessible website, in whole or in part.

37. Using the point-slope formula:

$$y-1=\frac{1}{2}(x-(-2))$$
$$y-1=\frac{1}{2}(x+2)$$
$$y-1=\frac{1}{2}x+1$$
$$y=\frac{1}{2}x+2$$

39. Computing the slope: $m=\frac{1-5}{0-2}=\frac{-4}{-2}=2$. Using the point-slope formula:

$$y-1=2(x-0)$$
$$y-1=2x$$
$$y=2x+1$$

Chapter 4 Review

1. Graphing both lines:

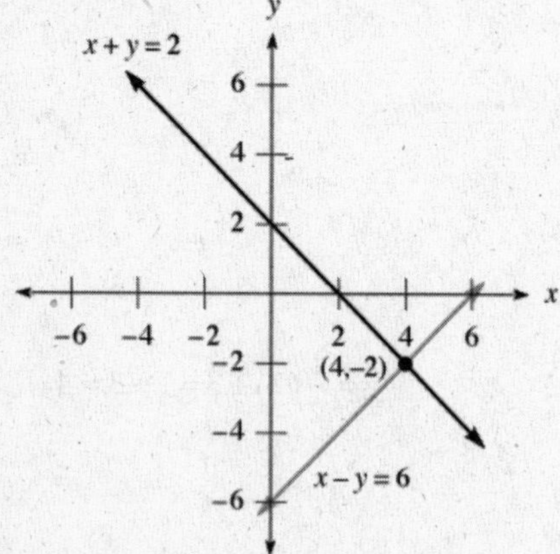

The intersection point is $(4,-2)$.

3. Graphing both lines:

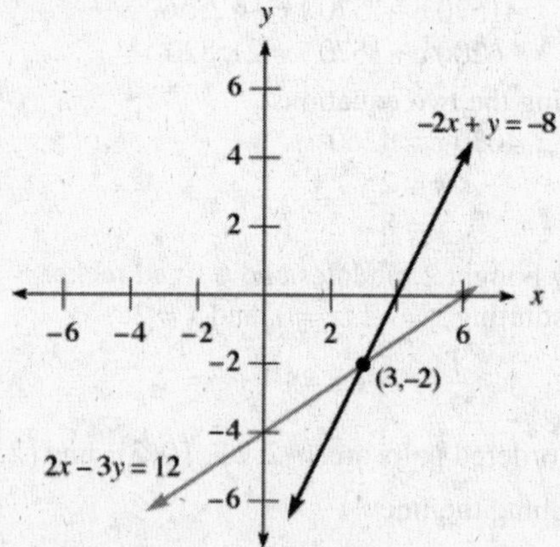

The intersection point is $(3,-2)$.

5. Graphing both lines:

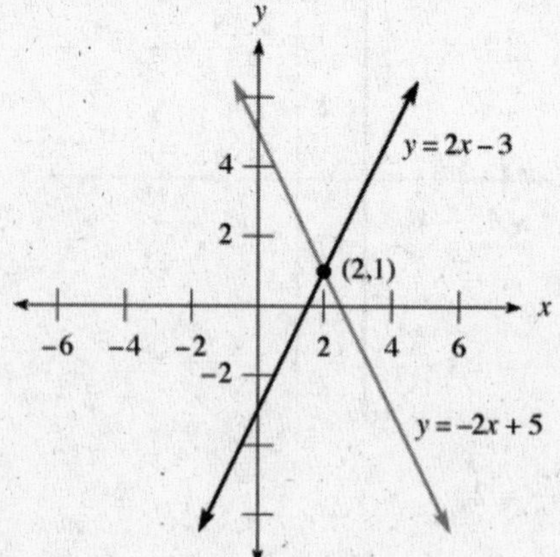

The intersection point is $(2,1)$.

© 2012 Cengage Learning. All Rights Reserved. May not be scanned, copied or duplicated, or posted to a publicly accessible website, in whole or in part.

7. Adding the two equations:

$$2x = 2$$
$$x = 1$$

Substituting into the second equation:

$$1 + y = -2$$
$$y = -3$$

The solution is $(1,-3)$.

9. Multiplying the first equation by 2:

$$10x - 6y = 4$$
$$-10x + 6y = -4$$

Adding the two equations:

$$0 = 0$$

Since this statement is true, the system is dependent. The two lines coincide.

11. Multiplying the second equation by –4:

$$-3x + 4y = 1$$
$$16x - 4y = 12$$

Adding the two equations:

$$13x = 13$$
$$x = 1$$

Substituting into the second equation:

$$-4(1) + y = -3$$
$$-4 + y = -3$$
$$y = 1$$

The solution is $(1,1)$.

13. Multiplying the first equation by 3 and the second equation by 5:

$$-6x + 15y = -33$$
$$35x - 15y = -25$$

Adding the two equations:

$$29x = -58$$
$$x = -2$$

Substituting into the first equation:

$$-2(-2) + 5y = -11$$
$$4 + 5y = -11$$
$$5y = -15$$
$$y = -3$$

The solution is $(-2,-3)$.

15. Substituting into the first equation:

$$x + (-3x + 1) = 5$$
$$-2x + 1 = 5$$
$$-2x = 4$$
$$x = -2$$

Substituting into the second equation: $y = -3(-2) + 1 = 6 + 1 = 7$. The solution is $(-2,7)$.

17. Substituting into the first equation:

$$4x - 3(3x + 7) = -16$$
$$4x - 9x - 21 = -16$$
$$-5x - 21 = -16$$
$$-5x = 5$$
$$x = -1$$

Substituting into the second equation: $y = 3(-1) + 7 = -3 + 7 = 4$. The solution is $(-1,4)$.

© 2012 Cengage Learning. All Rights Reserved. May not be scanned, copied or duplicated, or posted to a publicly accessible website, in whole or in part.

19. Solving the first equation for x:

$$x-4y=2$$
$$x=4y+2$$

Substituting into the second equation:

$$-3(4y+2)+12y=-8$$
$$-12y-6+12y=-8$$
$$-6=-8$$

Since this statement is false, there is no solution to the system. The two lines are parallel.

21. Solving the second equation for x:

$$x+6y=-11$$
$$x=-6y-11$$

Substituting into the first equation:

$$10(-6y-11)-5y=20$$
$$-60y-110-5y=20$$
$$-65y-110=20$$
$$-65y=130$$
$$y=-2$$

Substituting into $x=-6y-11$: $x=-6(-2)-11=12-11=1$. The solution is $(1,-2)$.

23. Let x represent the smaller number and y represent the larger number. The system of equations is:

$$x+y=18$$
$$2x=6+y$$

Solving the first equation for y:

$$x+y=18$$
$$y=-x+18$$

Substituting into the second equation:

$$2x=6+(-x+18)$$
$$2x=-x+24$$
$$3x=24$$
$$x=8$$

Substituting into the first equation:

$$8+y=18$$
$$y=10$$

The two numbers are 8 and 10.

25. Let x represent the amount invested at 4% and y represent the amount invested at 5%. The system of equations is:

$$x+y=12000$$
$$0.04x+0.05y=560$$

Multiplying the first equation by –0.04:

$$-0.04x-0.04y=-480$$
$$0.04x+0.05y=560$$

Adding the two equations:

$$0.01y=80$$
$$y=8000$$

Substituting into the first equation:

$$x+8000=12000$$
$$x=4000$$

So \$4,000 was invested at 4% and \$8,000 was invested at 5%.

© 2012 Cengage Learning. All Rights Reserved. May not be scanned, copied or duplicated, or posted to a publicly accessible website, in whole or in part.

27. Let x represent the number of dimes and y represent the number of nickels. The system of equations is:

$$x+y=17$$
$$0.10x+0.05y=1.35$$

Multiplying the first equation by –0.05:

$$-0.05x-0.05y=-0.85$$
$$0.10x+0.05y=1.35$$

Adding the two equations:

$$0.05x=0.50$$
$$x=10$$

Substituting into the first equation:

$$10+y=17$$
$$y=7$$

Barbara has 10 dimes and 7 nickels.

29. Let x represent the liters of 20% alcohol solution and y represent the liters of 10% alcohol solution. The system of equations is:

$$x+y=50$$
$$0.20x+0.10y=0.12(50)$$

Multiplying the first equation by –0.10:

$$-0.10x-0.10y=-5$$
$$0.20x+0.10y=6$$

Adding the two equations:

$$0.10x=1$$
$$x=10$$

Substituting into the first equation:

$$10+y=50$$
$$y=40$$

The solution contains 40 liters of 10% alcohol solution and 10 liters of 20% alcohol solution.

Chapter 4 Cumulative Review

1. Simplifying using order of operations: $3\bullet 4+5=12+5=17$

3. Simplifying using order of operations: $4\bullet 3^2+4(6-3)=4\bullet 9+4\bullet 3=36+12=48$

5. Simplifying using order of operations: $\frac{12-3}{8-8}=\frac{9}{0}$, which is undefined

7. Simplifying: $\frac{11}{60}-\frac{13}{84}=\frac{11\bullet 7}{60\bullet 7}-\frac{13\bullet 5}{84\bullet 5}=\frac{77}{420}-\frac{65}{420}=\frac{12}{420}=\frac{1}{35}$

9. Simplifying the expression: $2(x-5)+8=2x-10+8=2x-2$

11. Solving the equation:

$$-5-6=-y-3+2y$$
$$-11=y-3$$
$$-11+3=y-3+3$$
$$y=-8$$

13. Solving the equation:

$$3(x-4)=9$$
$$3x-12=9$$
$$3x-12+12=9+12$$
$$3x=21$$
$$\frac{1}{3}(3x)=\frac{1}{3}(21)$$
$$x=7$$

© 2012 Cengage Learning. All Rights Reserved. May not be scanned, copied or duplicated, or posted to a publicly accessible website, in whole or in part.

15. Solving the inequality:

$$0.3x + 0.7 \le -2$$
$$0.3x + 0.7 - 0.7 \le -2 - 0.7$$
$$0.3x \le -2.7$$
$$\frac{0.3x}{0.3} \le \frac{-2.7}{0.3}$$
$$x \le -9$$

Graphing the solution set:

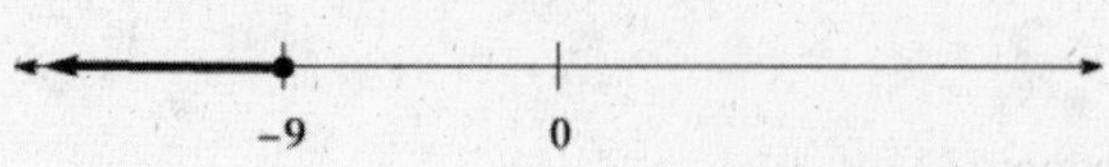

17. Graphing the line:

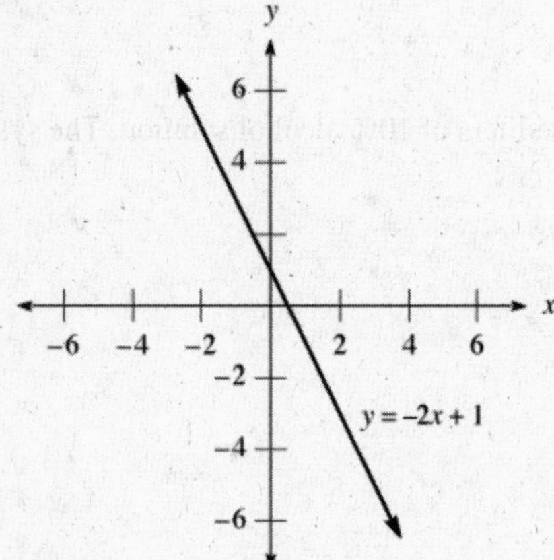

19. Graphing the two equations:

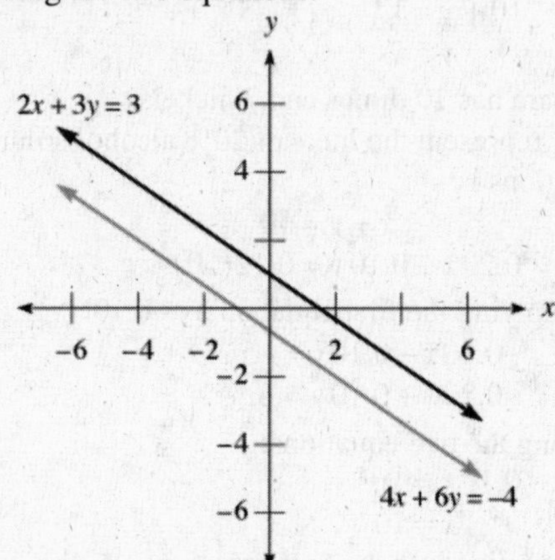

The two lines are parallel.

21. Multiplying the first equation by –2:

$$-2x - 2y = -14$$
$$2x + 2y = 14$$

Adding the two equations:

$$0 = 0$$

Since this statement is true, the system is dependent. The two lines coincide.

23. Multiplying the second equation by 3:

$$2x + 3y = 13$$
$$3x - 3y = -3$$

Adding the two equations:

$$5x = 10$$
$$x = 2$$

Substituting into the first equation:

$$2(2) + 3y = 13$$
$$4 + 3y = 13$$
$$3y = 9$$
$$y = 3$$

The solution is $(2,3)$.

25. Multiplying the second equation by –2:

$$2x + 5y = 33$$
$$-2x + 6y = 0$$

Adding the two equations:

$$11y = 33$$
$$y = 3$$

Substituting into the second equation:

$$x - 3(3) = 0$$
$$x - 9 = 0$$
$$x = 9$$

The solution is $(9,3)$.

© 2012 Cengage Learning. All Rights Reserved. May not be scanned, copied or duplicated, or posted to a publicly accessible website, in whole or in part.

27. Multiplying the second equation by 7:

$$3x-7y=12$$
$$14x+7y=56$$

Adding the two equations:

$$17x=68$$
$$x=4$$

Substituting into the second equation:

$$2(4)+y=8$$
$$8+y=8$$
$$y=0$$

The solution is $(4,0)$.

29. Substituting into the first equation:

$$2x-3(5x+2)=7$$
$$2x-15x-6=7$$
$$-13x-6=7$$
$$-13x=13$$
$$x=-1$$

Substituting into the second equation: $y=5(-1)+2=-5+2=-3$. The solution is $(-1,-3)$.

31. The pattern is to add –3, so the next number is: $-5+(-3)=-8$

33. The quotient is: $\frac{-30}{6}=-5$

35. Factoring into primes: $180=10\bullet 18=(2\bullet 5)\bullet(2\bullet 9)=(2\bullet 5)\bullet(2\bullet 3\bullet 3)=2^2\bullet 3^2\bullet 5$

37. Simplifying, then evaluating when $x=3$: $-3x+7+5x=2x+7=2(3)+7=6+7=13$

39. Substituting $x=-2$:

$$4(-2)-5y=12$$
$$-8-5y=12$$
$$-5y=20$$
$$y=-4$$

The ordered pair is $(-2,-4)$.

41. To find the x-intercept, let $y=0$:

$$3x-4(0)=12$$
$$3x=12$$
$$x=4$$

To find the y-intercept, let $x=0$:

$$3(0)-4y=12$$
$$-4y=12$$
$$y=-3$$

The x-intercept is 4 and the y-intercept is –3.

43. Computing the slope: $m=\frac{-4-1}{-5-(-1)}=\frac{-5}{-4}=\frac{5}{4}$

45. The slope-intercept form is $y=\frac{2}{3}x+3$.

47. First compute the slope: $m=\frac{6-3}{6-4}=\frac{3}{2}$. Using the point-slope formula:

$$y-3=\frac{3}{2}(x-4)$$
$$y-3=\frac{3}{2}x-6$$
$$y=\frac{3}{2}x-3$$

© 2012 Cengage Learning. All Rights Reserved. May not be scanned, copied or duplicated, or posted to a publicly accessible website, in whole or in part.

49. Let x represent the number of nickels and y represent the number of dimes. The system of equations is:

$$x+y=15$$
$$0.05x+0.10y=1.10$$

Multiplying the first equation by –0.05:

$$-0.05x-0.05y=-0.75$$
$$0.05x+0.10y=1.10$$

Adding the two equations:

$$0.05y=0.35$$
$$y=7$$

Substituting into the first equation:

$$x+7=15$$
$$x=8$$

Joy has 8 nickels and 7 dimes.

51. Finding the amount: $0.78(200)=156$ people

Chapter 4 Test

1. The solution is (–4,2).
2. Graphing the two equations:

The intersection point is $(1,2)$.

3. Graphing the two equations:

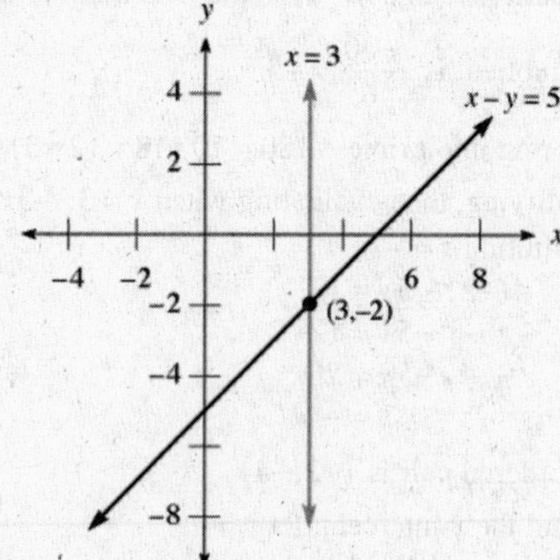

The intersection point is $(3,-2)$.

4. Graphing both lines:

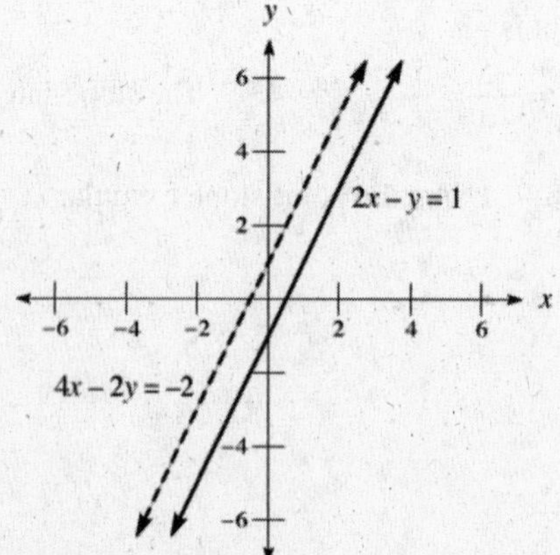

The lines are parallel, so there is no intersection (empty set).

© 2012 Cengage Learning. All Rights Reserved. May not be scanned, copied or duplicated, or posted to a publicly accessible website, in whole or in part.

5. Adding the two equations:

$$\begin{aligned} 3x &= -9 \\ x &= -3 \end{aligned}$$

Substituting into the first equation:

$$\begin{aligned} -3 - y &= 1 \\ -y &= 4 \\ y &= -4 \end{aligned}$$

The solution is $(-3,-4)$.

6. Multiplying the first equation by –1:

$$\begin{aligned} -2x - y &= -7 \\ 3x + y &= 12 \end{aligned}$$

Adding the two equations:

$$x = 5$$

Substituting into the first equation:

$$\begin{aligned} 2(5) + y &= 7 \\ 10 + y &= 7 \\ y &= -3 \end{aligned}$$

The solution is $(5,-3)$.

7. Multiplying the second equation by 4:

$$\begin{aligned} 7x + 8y &= -2 \\ 12x - 8y &= 40 \end{aligned}$$

Adding the two equations:

$$\begin{aligned} 19x &= 38 \\ x &= 2 \end{aligned}$$

Substituting into the first equation:

$$\begin{aligned} 7(2) + 8y &= -2 \\ 14 + 8y &= -2 \\ 8y &= -16 \\ y &= -2 \end{aligned}$$

The solution is $(2,-2)$.

8. Multiplying the first equation by –3 and the second equation by 2:

$$\begin{aligned} -18x + 30y &= -18 \\ 18x - 30y &= 18 \end{aligned}$$

Adding the two equations: $0 = 0$. Since this equation is true, the system is dependent. The two lines coincide.

9. Substituting into the first equation:

$$\begin{aligned} 3x + 2(2x+3) &= 20 \\ 3x + 4x + 6 &= 20 \\ 7x + 6 &= 20 \\ 7x &= 14 \\ x &= 2 \end{aligned}$$

Substituting into the second equation: $y = 2(2) + 3 = 4 + 3 = 7$. The solution is $(2,7)$.

10. Substituting into the first equation:

$$\begin{aligned} 3(y+1) - 6y &= -6 \\ 3y + 3 - 6y &= -6 \\ -3y + 3 &= -6 \\ -3y &= -9 \\ y &= 3 \end{aligned}$$

Substituting into the second equation: $x = 3 + 1 = 4$. The solution is $(4,3)$.

11. Solving the second equation for *y*:

$$\begin{aligned} -3x + y &= 3 \\ y &= 3x + 3 \end{aligned}$$

Substituting into the first equation:

$$\begin{aligned} 7x - 2(3x+3) &= -4 \\ 7x - 6x - 6 &= -4 \\ x - 6 &= -4 \\ x &= 2 \end{aligned}$$

Substituting into $y = 3x + 3$: $y = 3(2) + 3 = 6 + 3 = 9$. The solution is $(2,9)$.

© 2012 Cengage Learning. All Rights Reserved. May not be scanned, copied or duplicated, or posted to a publicly accessible website, in whole or in part.

12. Solving the second equation for x:

$$x + 3y = -8$$
$$x = -3y - 8$$

Substituting into the first equation:

$$2(-3y - 8) - 3y = -7$$
$$-6y - 16 - 3y = -7$$
$$-9y - 16 = -7$$
$$-9y = 9$$
$$y = -1$$

Substituting into $x = -3y - 8$: $x = -3(-1) - 8 = 3 - 8 = -5$. The solution is $(-5, -1)$.

13. Let x and y represent the two numbers. The system of equations is:

$$x + y = 12$$
$$x - y = 2$$

Adding the two equations:

$$2x = 14$$
$$x = 7$$

Substituting into the first equation:

$$7 + y = 12$$
$$y = 5$$

The two numbers are 5 and 7.

14. Let x and y represent the two numbers. The system of equations is:

$$x + y = 15$$
$$y = 6 + 2x$$

Substituting into the first equation:

$$x + 6 + 2x = 15$$
$$3x + 6 = 15$$
$$3x = 9$$
$$x = 3$$

Substituting into the second equation: $y = 6 + 2(3) = 6 + 6 = 12$. The two numbers are 3 and 12.

15. Let x represent the amount invested at 9% and y represent the amount invested at 11%. The system of equations is:

$$x + y = 10000$$
$$0.09x + 0.11y = 980$$

Multiplying the first equation by –0.09:

$$-0.09x - 0.09y = -900$$
$$0.09x + 0.11y = 980$$

Adding the two equations:

$$0.02y = 80$$
$$y = 4000$$

Substituting into the first equation:

$$x + 4000 = 10000$$
$$x = 6000$$

Dr. Stork should invest \$6,000 at 9% and \$4,000 at 11%.

© 2012 Cengage Learning. All Rights Reserved. May not be scanned, copied or duplicated, or posted to a publicly accessible website, in whole or in part.

16. Let x represent the number of nickels and y represent the number of quarters. The system of equations is:

$$x+y=12$$
$$0.05x+0.25y=1.60$$

Multiplying the first equation by –0.05:

$$-0.05x-0.05y=-0.60$$
$$0.05x+0.25y=1.60$$

Adding the two equations:

$$0.20y=1.00$$
$$y=5$$

Substituting into the first equation:

$$x+5=12$$
$$x=7$$

Diane has 7 nickels and 5 quarters.

17. Let w represent the width and l represent the length. The system of equations is:

$$l=2w+10$$
$$2w+2l=170$$

Substituting into the second equation:

$$2w+2(2w+10)=170$$
$$2w+4w+20=170$$
$$6w+20=170$$
$$6w=150$$
$$w=25$$
$$l=2(25)+10=60$$

The width is 25 feet and the length is 60 feet.

18. **a.** They charge the same for approximately 150 miles (where the lines intersect).

b. Company 1 would charge less if the truck is driven less than 100 miles.

c. Company 2 will charge less if $x > 150$ miles.

© 2012 Cengage Learning. All Rights Reserved. May not be scanned, copied or duplicated, or posted to a publicly accessible website, in whole or in part.

Chapter 5
Exponents and Polynomials

Getting Ready for Chapter 5

1. Simplifying: $\frac{1}{2} \cdot \frac{5}{7} = \frac{5}{14}$
2. Simplifying: $\frac{9.6}{3} = 3.2$
3. Simplifying: $9 - 20 = -11$
4. Simplifying: $1 - 8 = -7$
5. Simplifying: $-8 - 2(3) = -8 - 6 = -14$
6. Simplifying: $2(3) - 4 - 3(-4) = 6 - 4 - (-12) = 2 + 12 = 14$
7. Simplifying: $2(3) + 4(5) - 5(2) = 6 + 20 - 10 = 26 - 10 = 16$
8. Simplifying: $3(-2)^2 - 5(-2) + 4 = 3(4) - 5(-2) + 4 = 12 + 10 + 4 = 26$
9. Simplifying: $-(4x^2 - 2x - 6) = -4x^2 + 2x + 6$
10. Simplifying: $3x \cdot 3x = 9x^2$
11. Simplifying: $2(2x)(-3) = -12x$
12. Simplifying: $(3x)(2x) = 6x^2$
13. Simplifying: $-3x - 10x = -13x$
14. Simplifying: $x - 2x = -x$
15. Simplifying: $(2x^3 + 0x^2) - (2x^3 - 10x^2) = 2x^3 + 0x^2 - 2x^3 + 10x^2 = 10x^2$
16. Simplifying: $(4x - 14) - (4x - 10) = 4x - 14 - 4x + 10 = -4$
17. Simplifying: $-4x(x + 5) = -4x^2 - 20x$
18. Simplifying: $-2x(2x + 7) = -4x^2 - 14x$
19. Simplifying: $2x^2(x - 5) = 2x^3 - 10x^2$
20. Simplifying: $10x(x - 5) = 10x^2 - 50x$

5.1 Multiplication with Exponents

1. The base is 4 and the exponent is 2. Evaluating the expression: $4^2 = 4 \cdot 4 = 16$
3. The base is 0.3 and the exponent is 2. Evaluating the expression: $0.3^2 = 0.3 \cdot 0.3 = 0.09$
5. The base is 4 and the exponent is 3. Evaluating the expression: $4^3 = 4 \cdot 4 \cdot 4 = 64$
7. The base is –5 and the exponent is 2. Evaluating the expression: $(-5)^2 = (-5) \cdot (-5) = 25$
9. The base is 2 and the exponent is 3. Evaluating the expression: $-2^3 = -2 \cdot 2 \cdot 2 = -8$
11. The base is 3 and the exponent is 4. Evaluating the expression: $3^4 = 3 \cdot 3 \cdot 3 \cdot 3 = 81$
13. The base is $\frac{2}{3}$ and the exponent is 2. Evaluating the expression: $\left(\frac{2}{3}\right)^2 = \left(\frac{2}{3}\right) \cdot \left(\frac{2}{3}\right) = \frac{4}{9}$

© 2012 Cengage Learning. All Rights Reserved. May not be scanned, copied or duplicated, or posted to a publicly accessible website, in whole or in part.

15. The base is $\frac{1}{2}$ and the exponent is 4. Evaluating the expression: $\left(\frac{1}{2}\right)^4=\left(\frac{1}{2}\right)\bullet\left(\frac{1}{2}\right)\bullet\left(\frac{1}{2}\right)\bullet\left(\frac{1}{2}\right)=\frac{1}{16}$

17. **a.** Completing the table:

Number (x)	1	2	3	4	5	6	7
Square (x^2)	1	4	9	16	25	36	49

b. For numbers larger than 1, the square of the number is larger than the number.

19. Simplifying the expression: $x^4\bullet x^5=x^{4+5}=x^9$

21. Simplifying the expression: $y^{10}\bullet y^{20}=y^{10+20}=y^{30}$

23. Simplifying the expression: $2^5\bullet 2^4\bullet 2^3=2^{5+4+3}=2^{12}$

25. Simplifying the expression: $x^4\bullet x^6\bullet x^8\bullet x^{10}=x^{4+6+8+10}=x^{28}$

27. Simplifying the expression: $\left(x^2\right)^5=x^{2\bullet 5}=x^{10}$

29. Simplifying the expression: $\left(5^4\right)^3=5^{4\bullet 3}=5^{12}$

31. Simplifying the expression: $\left(y^3\right)^3=y^{3\bullet 3}=y^9$

33. Simplifying the expression: $\left(2^5\right)^{10}=2^{5\bullet 10}=2^{50}$

35. Simplifying the expression: $\left(a^3\right)^x=a^{3x}$

37. Simplifying the expression: $\left(b^x\right)^y=b^{xy}$

39. Simplifying the expression: $(4x)^2=4^2\bullet x^2=16x^2$

41. Simplifying the expression: $(2y)^5=2^5\bullet y^5=32y^5$

43. Simplifying the expression: $(-3x)^4=(-3)^4\bullet x^4=81x^4$

45. Simplifying the expression: $(0.5ab)^2=(0.5)^2\bullet a^2b^2=0.25a^2b^2$

47. Simplifying the expression: $(4xyz)^3=4^3\bullet x^3y^3z^3=64x^3y^3z^3$

49. Simplifying using properties of exponents: $\left(2x^4\right)^3=2^3\left(x^4\right)^3=8x^{12}$

51. Simplifying using properties of exponents: $\left(4a^3\right)^2=4^2\left(a^3\right)^2=16a^6$

53. Simplifying using properties of exponents: $\left(x^2\right)^3\left(x^4\right)^2=x^6\bullet x^8=x^{14}$

55. Simplifying using properties of exponents: $\left(a^3\right)^1\left(a^2\right)^4=a^3\bullet a^8=a^{11}$

57. Simplifying using properties of exponents: $(2x)^3(2x)^4=(2x)^7=2^7x^7=128x^7$

59. Simplifying using properties of exponents: $\left(3x^2\right)^3(2x)^4=3^3x^6\bullet 2^4x^4=27x^6\bullet 16x^4=432x^{10}$

61. Simplifying using properties of exponents: $\left(4x^2y^3\right)^2=4^2x^4y^6=16x^4y^6$

63. Simplifying using properties of exponents: $\left(\frac{2}{3}a^4b^5\right)^3=\left(\frac{2}{3}\right)^3a^{12}b^{15}=\frac{8}{27}a^{12}b^{15}$

65. Writing as a perfect square: $x^4=\left(x^2\right)^2$

67. Writing as a perfect square: $16x^2=(4x)^2$

69. Writing as a perfect cube: $8=(2)^3$

71. Writing as a perfect cube: $64x^3=(4x)^3$

73. **a.** Substituting $x=2$: $x^3x^2=(2)^3(2)^2=(8)(4)=32$

b. Substituting $x=2$: $\left(x^3\right)^2=\left(2^3\right)^2=(8)^2=64$

c. Substituting $x=2$: $x^5=(2)^5=32$

d. Substituting $x=2$: $x^6=(2)^6=64$

© 2012 Cengage Learning. All Rights Reserved. May not be scanned, copied or duplicated, or posted to a publicly accessible website, in whole or in part.

75. **a.** Completing the table:

Number (x)	Square (x^2)
–3	9
–2	4
–1	1
0	0
1	1
2	4
3	9

b. Constructing a line graph:

77. Completing the table:

Number (x)	Square (x^2)
–2.5	6.25
–1.5	2.25
–0.5	0.25
0	0
0.5	0.25
1.5	2.25
2.5	6.25

79. Writing in scientific notation: $43,200 = 4.32 \times 10^4$

81. Writing in scientific notation: $570 = 5.7 \times 10^2$

83. Writing in scientific notation: $238,000 = 2.38 \times 10^5$

85. Writing in expanded form: $2.49 \times 10^3 = 2,490$

87. Writing in expanded form: $3.52 \times 10^2 = 352$

89. Writing in expanded form: $2.8 \times 10^4 = 28,000$

91. The area of the base is: $A = (525 \text{ ft})^2 = 275,625 \text{ feet}^2$

93. The volume is given by: $V = (3 \text{ in.})^3 = 27 \text{ inches}^3$

95. The volume is given by: $V = (2.5 \text{ in.})^3 \approx 15.6 \text{ inches}^3$

97. The volume is given by: $V = (8 \text{ in.})(4.5 \text{ in.})(1 \text{ in.}) = 36 \text{ inches}^3$

99. Writing in scientific notation: $650,000,000 \text{ seconds} = 6.5 \times 10^8 \text{ seconds}$

101. Writing in expanded form: $7.4 \times 10^5 \text{ dollars} = \$740,000$

103. Writing in expanded form: $1.8 \times 10^5 \text{ dollars} = \$180,000$

105. Substitute $c = 8$, $b = 3.35$, and $s = 3.11$: $d = \pi \bullet 3.11 \bullet 8 \bullet \left(\frac{1}{2} \bullet 3.35\right)^2 \approx 219 \text{ inches}^3$

107. Substitute $c = 6$, $b = 3.59$, and $s = 2.99$: $d = \pi \bullet 2.99 \bullet 6 \bullet \left(\frac{1}{2} \bullet 3.59\right)^2 \approx 182 \text{ inches}^3$

109. Subtracting: $4 - 7 = 4 + (-7) = -3$

111. Subtracting: $4 - (-7) = 4 + 7 = 11$

113. Subtracting: $15 - 20 = 15 + (-20) = -5$

115. Subtracting: $-15 - (-20) = -15 + 20 = 5$

117. Simplifying: $2(3) - 4 = 6 - 4 = 2$

119. Simplifying: $4(3) - 3(2) = 12 - 6 = 6$

121. Simplifying: $2(5 - 3) = 2(2) = 4$

123. Simplifying: $5 + 4(-2) - 2(-3) = 5 - 8 + 6 = 3$

© 2012 Cengage Learning. All Rights Reserved. May not be scanned, copied or duplicated, or posted to a publicly accessible website, in whole or in part.

125. Factoring into primes: $128 = 4 \cdot 32 = (2 \cdot 2) \cdot (4 \cdot 8) = (2 \cdot 2) \cdot (2 \cdot 2 \cdot 2 \cdot 2 \cdot 2) = 2^7$

127. Factoring into primes: $250 = 10 \cdot 25 = (2 \cdot 5) \cdot (5 \cdot 5) = 2 \cdot 5^3$

129. Factoring into primes: $720 = 10 \cdot 72 = (2 \cdot 5) \cdot (8 \cdot 9) = (2 \cdot 5) \cdot (2 \cdot 2 \cdot 2 \cdot 3 \cdot 3) = 2^4 \cdot 3^2 \cdot 5$

131. Factoring into primes: $820 = 10 \cdot 82 = (2 \cdot 5) \cdot (2 \cdot 41) = 2^2 \cdot 5 \cdot 41$

133. Factoring into primes: $6^3 = (2 \cdot 3)^3 = 2^3 \cdot 3^3$

135. Factoring into primes: $30^3 = (2 \cdot 3 \cdot 5)^3 = 2^3 \cdot 3^3 \cdot 5^3$

137. Factoring into primes: $25^3 = (5 \cdot 5)^3 = 5^3 \cdot 5^3 = 5^6$

139. Factoring into primes: $12^3 = (2 \cdot 2 \cdot 3)^3 = 2^3 \cdot 2^3 \cdot 3^3 = 2^6 \cdot 3^3$

5.2 Division with Exponents

1. Writing with positive exponents: $3^{-2} = \frac{1}{3^2} = \frac{1}{9}$

3. Writing with positive exponents: $6^{-2} = \frac{1}{6^2} = \frac{1}{36}$

5. Writing with positive exponents: $8^{-2} = \frac{1}{8^2} = \frac{1}{64}$

7. Writing with positive exponents: $5^{-3} = \frac{1}{5^3} = \frac{1}{125}$

9. Writing with positive exponents: $2x^{-3} = 2 \cdot \frac{1}{x^3} = \frac{2}{x^3}$

11. Writing with positive exponents: $(2x)^{-3} = \frac{1}{(2x)^3} = \frac{1}{8x^3}$

13. Writing with positive exponents: $(5y)^{-2} = \frac{1}{(5y)^2} = \frac{1}{25y^2}$

15. Writing with positive exponents: $10^{-2} = \frac{1}{10^2} = \frac{1}{100}$

17. Completing the table:

Number (x)	Square (x^2)	Power of 2 (2^x)
–3	9	$\frac{1}{8}$
–2	4	$\frac{1}{4}$
–1	1	$\frac{1}{2}$
0	0	1
1	1	2
2	4	4
3	9	8

19. Simplifying: $\frac{5^1}{5^3} = 5^{1-3} = 5^{-2} = \frac{1}{5^2} = \frac{1}{25}$

21. Simplifying: $\frac{x^{10}}{x^4} = x^{10-4} = x^6$

23. Simplifying: $\frac{4^3}{4^0} = 4^{3-0} = 4^3 = 64$

25. Simplifying: $\frac{(2x)^7}{(2x)^4} = (2x)^{7-4} = (2x)^3 = 2^3 x^3 = 8x^3$

27. Simplifying: $\frac{6^{11}}{6} = \frac{6^{11}}{6^1} = 6^{11-1} = 6^{10} \quad (= 60{,}466{,}176)$

29. Simplifying: $\frac{6}{6^{11}} = \frac{6^1}{6^{11}} = 6^{1-11} = 6^{-10} = \frac{1}{6^{10}} \quad \left(= \frac{1}{60{,}466{,}176}\right)$

31. Simplifying: $\frac{2^{-5}}{2^3} = 2^{-5-3} = 2^{-8} = \frac{1}{2^8} = \frac{1}{256}$

33. Simplifying: $\frac{2^5}{2^{-3}} = 2^{5-(-3)} = 2^{5+3} = 2^8 = 256$

© 2012 Cengage Learning. All Rights Reserved. May not be scanned, copied or duplicated, or posted to a publicly accessible website, in whole or in part.

35. Simplifying: $\dfrac{(3x)^{-5}}{(3x)^{-8}}=(3x)^{-5-(-8)}=(3x)^{-5+8}=(3x)^3=3^3x^3=27x^3$

37. Simplifying: $(3xy)^4=3^4x^4y^4=81x^4y^4$

39. Simplifying: $10^0=1$

41. Simplifying: $\left(2a^2b\right)^1=2a^2b$

43. Simplifying: $\left(7y^3\right)^{-2}=\dfrac{1}{\left(7y^3\right)^2}=\dfrac{1}{49y^6}$

45. Simplifying: $x^{-3}\bullet x^{-5}=x^{-3-5}=x^{-8}=\dfrac{1}{x^8}$

47. Simplifying: $y^7\bullet y^{-10}=y^{7-10}=y^{-3}=\dfrac{1}{y^3}$

49. Simplifying: $\dfrac{\left(x^2\right)^3}{x^4}=\dfrac{x^6}{x^4}=x^{6-4}=x^2$

51. Simplifying: $\dfrac{\left(a^4\right)^3}{\left(a^3\right)^2}=\dfrac{a^{12}}{a^6}=a^{12-6}=a^6$

53. Simplifying: $\dfrac{y^7}{\left(y^2\right)^8}=\dfrac{y^7}{y^{16}}=y^{7-16}=y^{-9}=\dfrac{1}{y^9}$

55. Simplifying: $\left(\dfrac{y^7}{y^2}\right)^8=\left(y^{7-2}\right)^8=\left(y^5\right)^8=y^{40}$

57. Simplifying: $\dfrac{\left(x^{-2}\right)^3}{x^{-5}}=\dfrac{x^{-6}}{x^{-5}}=x^{-6-(-5)}=x^{-6+5}=x^{-1}=\dfrac{1}{x}$

59. Simplifying: $\left(\dfrac{x^{-2}}{x^{-5}}\right)^3=\left(x^{-2+5}\right)^3=\left(x^3\right)^3=x^9$

61. Simplifying: $\dfrac{\left(a^3\right)^2\left(a^4\right)^5}{\left(a^5\right)^2}=\dfrac{a^6\bullet a^{20}}{a^{10}}=\dfrac{a^{26}}{a^{10}}=a^{26-10}=a^{16}$

63. Simplifying: $\dfrac{\left(a^{-2}\right)^3\left(a^4\right)^2}{\left(a^{-3}\right)^{-2}}=\dfrac{a^{-6}\bullet a^8}{a^6}=\dfrac{a^2}{a^6}=a^{2-6}=a^{-4}=\dfrac{1}{a^4}$

65. **a.** Substituting $x=2$: $\dfrac{x^7}{x^2}=\dfrac{2^7}{2^2}=\dfrac{128}{4}=32$

b. Substituting $x=2$: $x^5=2^5=32$

c. Substituting $x=2$: $\dfrac{x^2}{x^7}=\dfrac{2^2}{2^7}=\dfrac{4}{128}=\dfrac{1}{32}$

d. Substituting $x=2$: $x^{-5}=2^{-5}=\dfrac{1}{2^5}=\dfrac{1}{32}$

67. **a.** Writing as a perfect square: $\dfrac{1}{25}=\left(\dfrac{1}{5}\right)^2$

b. Writing as a perfect square: $\dfrac{1}{64}=\left(\dfrac{1}{8}\right)^2$

c. Writing as a perfect square: $\dfrac{1}{x^2}=\left(\dfrac{1}{x}\right)^2$

d. Writing as a perfect square: $\dfrac{1}{x^4}=\left(\dfrac{1}{x^2}\right)^2$

© 2012 Cengage Learning. All Rights Reserved. May not be scanned, copied or duplicated, or posted to a publicly accessible website, in whole or in part.

69. Completing the table:

Number (x)	Power of 2 (2^x)
−3	$\frac{1}{8}$
−2	$\frac{1}{4}$
−1	$\frac{1}{2}$
0	1
1	2
2	4
3	8

Constructing the line graph:

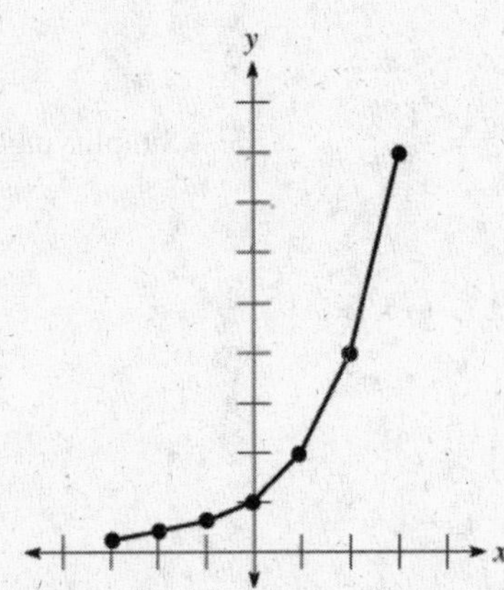

71. Writing in scientific notation: $0.0048 = 4.8 \times 10^{-3}$

73. Writing in scientific notation: $25 = 2.5 \times 10^{1}$

75. Writing in scientific notation: $0.000009 = 9 \times 10^{-6}$

77. Completing the table:

Expanded Form	Scientific Notation $(n \times 10^r)$
0.000357	3.57×10^{-4}
0.00357	3.57×10^{-3}
0.0357	3.57×10^{-2}
0.357	3.57×10^{-1}
3.57	3.57×10^{0}
35.7	3.57×10^{1}
357	3.57×10^{2}
3,570	3.57×10^{3}
35,700	3.57×10^{4}

79. Writing in expanded form: $4.23 \times 10^{-3} = 0.00423$

81. Writing in expanded form: $8 \times 10^{-5} = 0.00008$

83. Writing in expanded form: $4.2 \times 10^{0} = 4.2$

85. Writing in scientific notation: \$52 million = $\$5.2 \times 10^{7}$ million

87. Writing in expanded form: 2×10^{-3} seconds = 0.002 seconds

89. Writing in scientific notation: $25 \times 10^{3} = 2.5 \times 10^{4}$

91. Writing in scientific notation: $23.5 \times 10^{4} = 2.35 \times 10^{5}$

93. Writing in scientific notation: $0.82 \times 10^{-3} = 8.2 \times 10^{-4}$

© 2012 Cengage Learning. All Rights Reserved. May not be scanned, copied or duplicated, or posted to a publicly accessible website, in whole or in part.

95. The area of the smaller square is $(10 \text{ in.})^2 = 100 \text{ inches}^2$, while the area of the larger square is $(20 \text{ in.})^2 = 400 \text{ inches}^2$. It would take 4 smaller squares to cover the larger square.

97. The area of the smaller square is x^2, while the area of the larger square is $(2x)^2 = 4x^2$. It would take 4 smaller squares to cover the larger square.

99. The volume of the smaller box is $(6 \text{ in.})^3 = 216 \text{ inches}^3$, while the volume of the larger box is $(12 \text{ in.})^3 = 1,728 \text{ inches}^3$. Thus 8 smaller boxes will fit inside the larger box ($8 \bullet 216 = 1,728$).

101. The volume of the smaller box is x^3, while the volume of the larger box is $(2x)^3 = 8x^3$. Thus 8 smaller boxes will fit inside the larger box.

103. Simplifying: $3(4.5) = 13.5$

105. Simplifying: $\frac{4}{5}(10) = \frac{40}{5} = 8$

107. Simplifying: $6.8(3.9) = 26.52$

109. Simplifying: $-3+15 = 12$

111. Simplifying: $x^5 \bullet x^3 = x^{5+3} = x^8$

113. Simplifying: $\frac{x^3}{x^2} = x^{3-2} = x$

115. Simplifying: $\frac{y^3}{y^5} = y^{3-5} = y^{-2} = \frac{1}{y^2}$

117. Writing in expanded form: $3.4 \times 10^2 = 340$

119. Simplifying: $4x + 3x = (4+3)x = 7x$

121. Simplifying: $5a - 3a = (5-3)a = 2a$

123. Simplifying: $4y + 5y + y = (4+5+1)y = 10y$

5.3 Operations with Monomials

1. Multiplying the monomials: $(3x^4)(4x^3) = 12x^{4+3} = 12x^7$

3. Multiplying the monomials: $(-2y^4)(8y^7) = -16y^{4+7} = -16y^{11}$

5. Multiplying the monomials: $(8x)(4x) = 32x^{1+1} = 32x^2$

7. Multiplying the monomials: $(10a^3)(10a)(2a^2) = 200a^{3+1+2} = 200a^6$

9. Multiplying the monomials: $(6ab^2)(-4a^2b) = -24a^{1+2}b^{2+1} = -24a^3b^3$

11. Multiplying the monomials: $(4x^2y)(3x^3y^3)(2xy^4) = 24x^{2+3+1}y^{1+3+4} = 24x^6y^8$

13. Dividing the monomials: $\frac{15x^3}{5x^2} = \frac{15}{5} \bullet \frac{x^3}{x^2} = 3x$

15. Dividing the monomials: $\frac{18y^9}{3y^{12}} = \frac{18}{3} \bullet \frac{y^9}{y^{12}} = 6 \bullet \frac{1}{y^3} = \frac{6}{y^3}$

17. Dividing the monomials: $\frac{32a^3}{64a^4} = \frac{32}{64} \bullet \frac{a^3}{a^4} = \frac{1}{2} \bullet \frac{1}{a} = \frac{1}{2a}$

19. Dividing the monomials: $\frac{21a^2b^3}{-7ab^5} = \frac{21}{-7} \bullet \frac{a^2}{a} \bullet \frac{b^3}{b^5} = -3 \bullet a \bullet \frac{1}{b^2} = -\frac{3a}{b^2}$

21. Dividing the monomials: $\frac{3x^3y^2z}{27xy^2z^3} = \frac{3}{27} \bullet \frac{x^3}{x} \bullet \frac{y^2}{y^2} \bullet \frac{z}{z^3} = \frac{1}{9} \bullet x^2 \bullet \frac{1}{z^2} = \frac{x^2}{9z^2}$

© 2012 Cengage Learning. All Rights Reserved. May not be scanned, copied or duplicated, or posted to a publicly accessible website, in whole or in part.

23. Completing the table:

a	b	ab	$\frac{a}{b}$	$\frac{b}{a}$
10	$5x$	$50x$	$\frac{2}{x}$	$\frac{x}{2}$
$20x^3$	$6x^2$	$120x^5$	$\frac{10x}{3}$	$\frac{3}{10x}$
$25x^5$	$5x^4$	$125x^9$	$5x$	$\frac{1}{5x}$
$3x^{-2}$	$3x^2$	9	$\frac{1}{x^4}$	x^4
$-2y^4$	$8y^7$	$-16y^{11}$	$-\frac{1}{4y^3}$	$-4y^3$

25. Finding the product: $(3\times10^3)(2\times10^5)=6\times10^8$

27. Finding the product: $(3.5\times10^4)(5\times10^{-6})=17.5\times10^{-2}=1.75\times10^{-1}$

29. Finding the product: $(5.5\times10^{-3})(2.2\times10^{-4})=12.1\times10^{-7}=1.21\times10^{-6}$

31. Finding the quotient: $\frac{8.4\times10^5}{2\times10^2}=4.2\times10^3$

33. Finding the quotient: $\frac{6\times10^8}{2\times10^{-2}}=3\times10^{10}$

35. Finding the quotient: $\frac{2.5\times10^{-6}}{5\times10^{-4}}=0.5\times10^{-2}=5\times10^{-3}$

37. Combining the monomials: $3x^2+5x^2=(3+5)x^2=8x^2$

39. Combining the monomials: $8x^5-19x^5=(8-19)x^5=-11x^5$

41. Combining the monomials: $2a+a-3a=(2+1-3)a=0a=0$

43. Combining the monomials: $10x^3-8x^3+2x^3=(10-8+2)x^3=4x^3$

45. Combining the monomials: $20ab^2-19ab^2+30ab^2=(20-19+30)ab^2=31ab^2$

47. Completing the table:

a	b	ab	$a+b$
$5x$	$3x$	$15x^2$	$8x$
$4x^2$	$2x^2$	$8x^4$	$6x^2$
$3x^3$	$6x^3$	$18x^6$	$9x^3$
$2x^4$	$-3x^4$	$-6x^8$	$-x^4$
x^5	$7x^5$	$7x^{10}$	$8x^5$

49. Simplifying the expression: $\frac{(3x^2)(8x^5)}{6x^4}=\frac{24x^7}{6x^4}=\frac{24}{6}\bullet\frac{x^7}{x^4}=4x^3$

51. Simplifying the expression: $\frac{(9a^2b)(2a^3b^4)}{18a^5b^7}=\frac{18a^5b^5}{18a^5b^7}=\frac{18}{18}\bullet\frac{a^5}{a^5}\bullet\frac{b^5}{b^7}=1\bullet\frac{1}{b^2}=\frac{1}{b^2}$

53. Simplifying the expression: $\frac{(4x^3y^2)(9x^4y^{10})}{(3x^5y)(2x^6y)}=\frac{36x^7y^{12}}{6x^{11}y^2}=\frac{36}{6}\bullet\frac{x^7}{x^{11}}\bullet\frac{y^{12}}{y^2}=6\bullet\frac{1}{x^4}\bullet y^{10}=\frac{6y^{10}}{x^4}$

55. Simplifying the expression: $\frac{(6\times10^8)(3\times10^5)}{9\times10^7}=\frac{18\times10^{13}}{9\times10^7}=2\times10^6$

© 2012 Cengage Learning. All Rights Reserved. May not be scanned, copied or duplicated, or posted to a publicly accessible website, in whole or in part.

57. Simplifying the expression: $\dfrac{\left(5\times10^{3}\right)\left(4\times10^{-5}\right)}{2\times10^{-2}}=\dfrac{20\times10^{-2}}{2\times10^{-2}}=10=1\times10^{1}$

59. Simplifying the expression: $\dfrac{\left(2.8\times10^{-7}\right)\left(3.6\times10^{4}\right)}{2.4\times10^{3}}=\dfrac{10.08\times10^{-3}}{2.4\times10^{3}}=4.2\times10^{-6}$

61. Simplifying the expression: $\dfrac{18x^4}{3x}+\dfrac{21x^7}{7x^4}=6x^3+3x^3=9x^3$

63. Simplifying the expression: $\dfrac{45a^6}{9a^4}-\dfrac{50a^8}{2a^6}=5a^2-25a^2=-20a^2$

65. Simplifying the expression: $\dfrac{6x^7y^4}{3x^2y^2}+\dfrac{8x^5y^8}{2y^6}=2x^5y^2+4x^5y^2=6x^5y^2$

67. Applying the distributive property: $xy\left(x+\dfrac{1}{y}\right)=xy\bullet x+xy\bullet\dfrac{1}{y}=x^2y+x$

69. Applying the distributive property: $xy\left(\dfrac{1}{y}+\dfrac{1}{x}\right)=xy\bullet\dfrac{1}{y}+xy\bullet\dfrac{1}{x}=x+y$

71. Applying the distributive property: $x^2\left(1-\dfrac{4}{x^2}\right)=x^2\bullet1-x^2\bullet\dfrac{4}{x^2}=x^2-4$

73. Applying the distributive property: $x^2\left(1-\dfrac{1}{x}-\dfrac{6}{x^2}\right)=x^2\bullet1-x^2\bullet\dfrac{1}{x}-x^2\bullet\dfrac{6}{x^2}=x^2-x-6$

75. Applying the distributive property: $x^2\left(1-\dfrac{5}{x}\right)=x^2\bullet1-x^2\bullet\dfrac{5}{x}=x^2-5x$

77. Applying the distributive property: $x^2\left(1-\dfrac{8}{x}\right)=x^2\bullet1-x^2\bullet\dfrac{8}{x}=x^2-8x$

79. **a.** Dividing by $5a^2$: $\dfrac{10a^2}{5a^2}=2$

b. Dividing by $5a^2$: $\dfrac{-15a^2b}{5a^2}=-3b$

c. Dividing by $5a^2$: $\dfrac{25a^2b^2}{5a^2}=5b^2$

81. **a.** Dividing by $8x^2y$: $\dfrac{24x^3y^2}{8x^2y}=3xy$

b. Dividing by $8x^2y$: $\dfrac{16x^2y^2}{8x^2y}=2y$

c. Dividing by $8x^2y$: $\dfrac{-4x^2y^3}{8x^2y}=-\dfrac{y^2}{2}$

83. Simplifying: $3-8=-5$

85. Simplifying: $-1+7=6$

87. Simplifying: $3(5)^2+1=3(25)+1=75+1=76$

89. Simplifying: $2x^2+4x^2=6x^2$

91. Simplifying: $-5x+7x=2x$

93. Simplifying: $-(2x+9)=-2x-9$

95. Substituting $x=4$: $2x+3=2(4)+3=8+3=11$

97. Evaluating when $x=-2$: $4x=4(-2)=-8$

99. Evaluating when $x=-2$: $-2x+5=-2(-2)+5=4+5=9$

101. Evaluating when $x=-2$: $x^2+5x+6=(-2)^2+5(-2)+6=4-10+6=0$

© 2012 Cengage Learning. All Rights Reserved. May not be scanned, copied or duplicated, or posted to a publicly accessible website, in whole or in part.

103. The ordered pairs are $(-2,-2)$, $(0,2)$, and $(2,6)$:

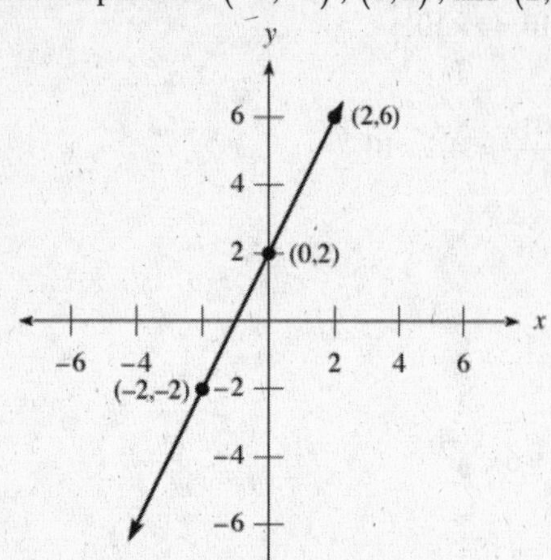

105. The ordered pairs are $(-3,0)$, $(0,1)$, and $(3,2)$:

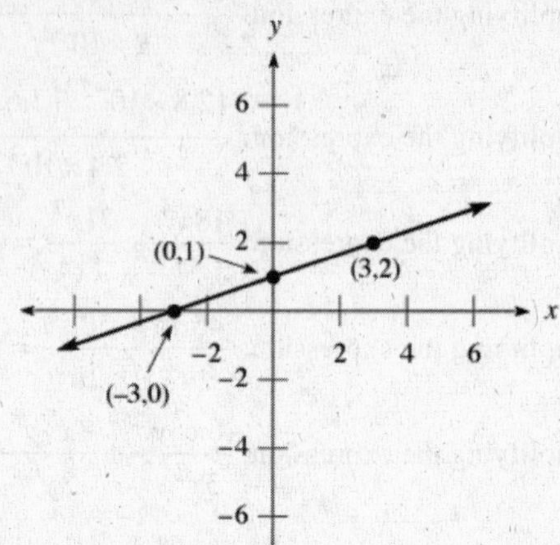

5.4 Addition and Subtraction of Polynomials

1. This is a trinomial of degree 3.

3. This is a trinomial of degree 3.

5. This is a binomial of degree 1.

7. This is a binomial of degree 2.

9. This is a monomial of degree 2.

11. This is a monomial of degree 0.

13. Combining the polynomials: $(2x^2+3x+4)+(3x^2+2x+5)=(2x^2+3x^2)+(3x+2x)+(4+5)=5x^2+5x+9$

15. Combining the polynomials: $(3a^2-4a+1)+(2a^2-5a+6)=(3a^2+2a^2)+(-4a-5a)+(1+6)=5a^2-9a+7$

17. Combining the polynomials: $x^2+4x+2x+8=x^2+(4x+2x)+8=x^2+6x+8$

19. Combining the polynomials: $6x^2-3x-10x+5=6x^2+(-3x-10x)+5=6x^2-13x+5$

21. Combining the polynomials: $x^2-3x+3x-9=x^2+(-3x+3x)-9=x^2-9$

23. Combining the polynomials: $3y^2-5y-6y+10=3y^2+(-5y-6y)+10=3y^2-11y+10$

25. Combining the polynomials:

$$(6x^3-4x^2+2x)+(9x^2-6x+3)=6x^3+(-4x^2+9x^2)+(2x-6x)+3=6x^3+5x^2-4x+3$$

27. Combining the polynomials:

$$\left(\frac{2}{3}x^2-\frac{1}{5}x-\frac{3}{4}\right)+\left(\frac{4}{3}x^2-\frac{4}{5}x+\frac{7}{4}\right)=\left(\frac{2}{3}x^2+\frac{4}{3}x^2\right)+\left(-\frac{1}{5}x-\frac{4}{5}x\right)+\left(-\frac{3}{4}+\frac{7}{4}\right)=2x^2-x+1$$

29. Combining the polynomials: $(a^2-a-1)-(-a^2+a+1)=a^2-a-1+a^2-a-1=2a^2-2a-2$

31. Combining the polynomials:

$$\left(\frac{5}{9}x^3+\frac{1}{3}x^2-2x+1\right)-\left(\frac{2}{3}x^3+x^2+\frac{1}{2}x-\frac{3}{4}\right)=\frac{5}{9}x^3+\frac{1}{3}x^2-2x+1-\frac{2}{3}x^3-x^2-\frac{1}{2}x+\frac{3}{4}$$
$$=-\frac{1}{9}x^3-\frac{2}{3}x^2-\frac{5}{2}x+\frac{7}{4}$$

33. Combining the polynomials:

$$(4y^2-3y+2)+(5y^2+12y-4)-(13y^2-6y+20)=4y^2-3y+2+5y^2+12y-4-13y^2+6y-20$$
$$=(4y^2+5y^2-13y^2)+(-3y+12y+6y)+(2-4-20)$$
$$=-4y^2+15y-22$$

35. Simplifying: $(x^2-5x)-(x^2-3x)=x^2-5x-x^2+3x=-2x$

37. Simplifying: $(6x^2-11x)-(6x^2-15x)=6x^2-11x-6x^2+15x=4x$

© 2012 Cengage Learning. All Rights Reserved. May not be scanned, copied or duplicated, or posted to a publicly accessible website, in whole or in part.

39. Simplifying: $\left(x^3+3x^2+9x\right)-\left(3x^2+9x+27\right)=x^3+3x^2+9x-3x^2-9x-27=x^3-27$

41. Simplifying: $\left(x^3+4x^2+4x\right)+\left(2x^2+8x+8\right)=x^3+4x^2+4x+2x^2+8x+8=x^3+6x^2+12x+8$

43. Simplifying: $\left(x^2-4\right)-\left(x^2-4x+4\right)=x^2-4-x^2+4x-4=4x-8$

45. Performing the subtraction:

$$\begin{aligned}\left(11x^2-10x+13\right)-\left(10x^2+23x-50\right)&=11x^2-10x+13-10x^2-23x+50\\&=\left(11x^2-10x^2\right)+\left(-10x-23x\right)+\left(13+50\right)\\&=x^2-33x+63\end{aligned}$$

47. Performing the subtraction:

$$\begin{aligned}\left(11y^2+11y+11\right)-\left(3y^2+7y-15\right)&=11y^2+11y+11-3y^2-7y+15\\&=\left(11y^2-3y^2\right)+\left(11y-7y\right)+\left(11+15\right)\\&=8y^2+4y+26\end{aligned}$$

49. Performing the addition:

$$\left(25x^2-50x+75\right)+\left(50x^2-100x-150\right)=\left(25x^2+50x^2\right)+\left(-50x-100x\right)+\left(75-150\right)=75x^2-150x-75$$

51. Performing the operations:

$$(3x-2)+(11x+5)-(2x+1)=3x-2+11x+5-2x-1=(3x+11x-2x)+(-2+5-1)=12x+2$$

53. Evaluating when $x=3$: $x^2-2x+1=(3)^2-2(3)+1=9-6+1=4$

55. a. Substituting $p=5$: $100p^2-1,300p+4,000=100(5)^2-1,300(5)+4,000=2,500-6,500+4,000=0$

b. Substituting $p=8$: $100p^2-1,300p+4,000=100(8)^2-1,300(8)+4,000=6,400-10,400+4,000=0$

57. a. Substituting $x=8$: $600+1,000x-100x^2=600+1,000(8)-100(8)^2=600+8,000-6,400=2,200$

b. Substituting $x=-2$: $600+1,000x-100x^2=600+1,000(-2)-100(-2)^2=600-2,000-400=-1,800$

59. Finding the area: $A=\left(1.2\times10^6\text{ in.}\right)^2=1.44\times10^{12}\text{ in.}^2$

61. Finding the volume of the cylinder and sphere:

$$V_{\text{cylinder}}=\pi\left(3^2\right)(6)=54\pi \qquad V_{\text{sphere}}=\tfrac{4}{3}\pi\left(3^3\right)=36\pi$$

Subtracting to find the amount of space to pack: $V=54\pi-36\pi=18\pi$ inches^3

63. For the first year: $53,160x+63,935y$

For the second year: $53,160(2x)+63,935y=106,320x+63,935y$

Total for both years: $(53,160x+63,935y)+(106,320x+63,935y)=159,480x+127,870y$

65. Simplifying: $(-5)(-1)=5$

67. Simplifying: $(-1)(6)=-6$

69. Simplifying: $(5x)(-4x)=-20x^2$

71. Simplifying: $3x(-7)=-21x$

73. Simplifying: $5x+(-3x)=2x$

75. Multiplying: $3(2x-6)=6x-18$

77. Multiplying: $3x(-5x)=-15x^2$

79. Multiplying: $2x\left(3x^2\right)=6x^{1+2}=6x^3$

81. Multiplying: $3x^2\left(2x^2\right)=6x^{2+2}=6x^4$

© 2012 Cengage Learning. All Rights Reserved. May not be scanned, copied or duplicated, or posted to a publicly accessible website, in whole or in part.

5.5 Multiplication of Polynomials

1. Using the distributive property: $2x(3x+1)=2x(3x)+2x(1)=6x^2+2x$

3. Using the distributive property: $2x^2(3x^2-2x+1)=2x^2(3x^2)-2x^2(2x)+2x^2(1)=6x^4-4x^3+2x^2$

5. Using the distributive property: $2ab(a^2-ab+1)=2ab(a^2)-2ab(ab)+2ab(1)=2a^3b-2a^2b^2+2ab$

7. Using the distributive property: $y^2(3y^2+9y+12)=y^2(3y^2)+y^2(9y)+y^2(12)=3y^4+9y^3+12y^2$

9. Using the distributive property:
$$4x^2y(2x^3y+3x^2y^2+8y^3)=4x^2y(2x^3y)+4x^2y(3x^2y^2)+4x^2y(8y^3)=8x^5y^2+12x^4y^3+32x^2y^4$$

11. Multiplying using the FOIL method: $(x+3)(x+4)=x^2+3x+4x+12=x^2+7x+12$

13. Multiplying using the FOIL method: $(x+6)(x+1)=x^2+6x+1x+6=x^2+7x+6$

15. Multiplying using the FOIL method: $\left(x+\frac{1}{2}\right)\left(x+\frac{3}{2}\right)=x^2+\frac{1}{2}x+\frac{3}{2}x+\frac{3}{4}=x^2+2x+\frac{3}{4}$

17. Multiplying using the FOIL method: $(a+5)(a-3)=a^2+5a-3a-15=a^2+2a-15$

19. Multiplying using the FOIL method: $(x-a)(y+b)=xy-ay+bx-ab$

21. Multiplying using the FOIL method: $(x+6)(x-6)=x^2+6x-6x-36=x^2-36$

23. Multiplying using the FOIL method: $\left(y+\frac{5}{6}\right)\left(y-\frac{5}{6}\right)=y^2+\frac{5}{6}y-\frac{5}{6}y-\frac{25}{36}=y^2-\frac{25}{36}$

25. Multiplying using the FOIL method: $(2x-3)(x-4)=2x^2-3x-8x+12=2x^2-11x+12$

27. Multiplying using the FOIL method: $(a+2)(2a-1)=2a^2+4a-a-2=2a^2+3a-2$

29. Multiplying using the FOIL method: $(2x-5)(3x-2)=6x^2-15x-4x+10=6x^2-19x+10$

31. Multiplying using the FOIL method: $(2x+3)(a+4)=2ax+3a+8x+12$

33. Multiplying using the FOIL method: $(5x-4)(5x+4)=25x^2-20x+20x-16=25x^2-16$

35. Multiplying using the FOIL method: $\left(2x-\frac{1}{2}\right)\left(x+\frac{3}{2}\right)=2x^2-\frac{1}{2}x+3x-\frac{3}{4}=2x^2+\frac{5}{2}x-\frac{3}{4}$

37. Multiplying using the FOIL method: $(1-2a)(3-4a)=3-6a-4a+8a^2=3-10a+8a^2$

39. The product is $(x+2)(x+3)=x^2+5x+6$:

	x	3
x	x^2	$3x$
2	$2x$	6

41. The product is $(x+1)(2x+2)=2x^2+4x+2$:

	x	x	2
x	x^2	x^2	$2x$
1	x	x	2

43. Multiplying using the column method:
$$\begin{array}{r} a^2-3a+2 \\ a-3 \\ \hline a^3-3a^2+2a \\ -3a^2+9a-6 \\ \hline a^3-6a^2+11a-6 \end{array}$$

45. Multiplying using the column method:
$$\begin{array}{r} x^2-2x+4 \\ x+2 \\ \hline x^3-2x^2+4x \\ 2x^2-4x+8 \\ \hline x^3+8 \end{array}$$

© 2012 Cengage Learning. All Rights Reserved. May not be scanned, copied or duplicated, or posted to a publicly accessible website, in whole or in part.

47. Multiplying using the column method:

$$\begin{array}{r} x^2 \;+8x \;+9 \\ 2x \;+1 \\ \hline 2x^3 +16x^2 +18x \qquad \\ x^2 +8x+9 \\ \hline 2x^3 +17x^2 +26x+9 \end{array}$$

49. Multiplying using the column method:

$$\begin{array}{r} 5x^2 +2x+1 \\ x^2 -3x+5 \\ \hline 5x^4 +2x^3 + x^2 \qquad\qquad \\ -15x^3 -6x^2 -3x \quad \\ 25x^2 +10x+5 \\ \hline 5x^4 -13x^3 +20x^2 +7x+5 \end{array}$$

51. Multiplying using the FOIL method: $\left(x^2+3\right)\left(2x^2-5\right)=2x^4-5x^2+6x^2-15=2x^4+x^2-15$

53. Multiplying using the FOIL method: $\left(3a^4+2\right)\left(2a^2+5\right)=6a^6+15a^4+4a^2+10$

55. First multiply two polynomials using the FOIL method: $(x+3)(x+4)=x^2+3x+4x+12=x^2+7x+12$

Now using the column method:

$$\begin{array}{r} x^2 +7x \;+12 \\ x \;+5 \\ \hline x^3 +7x^2 +12x \qquad \\ 5x^2 +35x+60 \\ \hline x^3 +12x^2 +47x+60 \end{array}$$

57. Simplifying: $(x-3)(x-2)+2=x^2-3x-2x+6+2=x^2-5x+8$

59. Simplifying: $(2x-3)(4x+3)+4=8x^2+6x-12x-9+4=8x^2-6x-5$

61. Simplifying: $(x+4)(x-5)+(-5)(2)=x^2-5x+4x-20-10=x^2-x-30$

63. Simplifying: $2(x-3)+x(x+2)=2x-6+x^2+2x=x^2+4x-6$

65. Simplifying: $3x(x+1)-2x(x-5)=3x^2+3x-2x^2+10x=x^2+13x$

67. Simplifying: $x(x+2)-3=x^2+2x-3$

69. Simplifying: $a(a-3)+6=a^2-3a+6$

71.
a. Finding the product: $(x+1)(x-1)=x^2-x+x-1=x^2-1$

b. Finding the product: $(x+1)(x+1)=x^2+x+x+1=x^2+2x+1$

c. Finding the product: $(x+1)\left(x^2+2x+1\right)=x^3+2x^2+x+x^2+2x+1=x^3+3x^2+3x+1$

d. Finding the product:

$$(x+1)\left(x^3+3x^2+3x+1\right)=x^4+3x^3+3x^2+x+x^3+3x^2+3x+1=x^4+4x^3+6x^2+4x+1$$

73.
a. Finding the product: $(x+1)(x-1)=x^2-x+x-1=x^2-1$

b. Finding the product: $(x+1)(x-2)=x^2-2x+x-2=x^2-x-2$

c. Finding the product: $(x+1)(x-3)=x^2-3x+x-3=x^2-2x-3$

d. Finding the product: $(x+1)(x-4)=x^2-4x+x-4=x^2-3x-4$

75. The expression is equal to 0 if $x=0$.

77. The expression is equal to 0 if $x=-5$.

79. The expression is equal to 0 if $x=3$ or $x=-2$.

81. The expression is never equal to 0.

83. The expression is: $x(6,200-25x)=6,200x-25x^2$

Substituting $x=3$: $6,200(3)-25(3)^2=18,600-225=\$18,375$

85. Let x represent the width and $2x+5$ represent the length. The area is given by: $A=x(2x+5)=2x^2+5x$

87. Let x and $x+1$ represent the width and length, respectively. The area is given by: $A=x(x+1)=x^2+x$

89. The revenue is: $R=xp=(1,200-100p)p=1,200p-100p^2$

91. Simplifying: $13\cdot 13=169$

93. Simplifying: $2(x)(-5)=-10x$

© 2012 Cengage Learning. All Rights Reserved. May not be scanned, copied or duplicated, or posted to a publicly accessible website, in whole or in part.

95. Simplifying: $6x+(-6x)=0$

97. Simplifying: $(2x)(-3)+(2x)(3)=-6x+6x=0$

99. Multiplying: $-4(3x-4)=-12x+16$

101. Multiplying: $(x-1)(x+2)=x^2+2x-x-2=x^2+x-2$

103. Multiplying: $(x+3)(x+3)=x^2+3x+3x+9=x^2+6x+9$

105. Graphing each line:

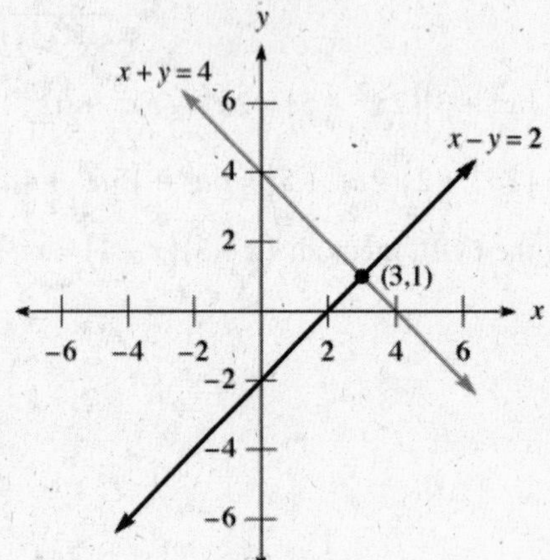

The intersection point is $(3,1)$.

107. Multiplying the first equation by 3 and the second equation by –2:

$$6x+9y=-3$$
$$-6x-10y=4$$

Adding the two equations:

$$-y=1$$
$$y=-1$$

Substituting into the first equation:

$$2x+3(-1)=-1$$
$$2x-3=-1$$
$$2x=2$$
$$x=1$$

The solution is $(1,-1)$.

109. Substituting into the first equation:

$$2x-6(3x+1)=2$$
$$2x-18x-6=2$$
$$-16x-6=2$$
$$-16x=8$$
$$x=-\frac{1}{2}$$

Substituting into the second equation: $y=3\left(-\frac{1}{2}\right)+1=-\frac{3}{2}+1=-\frac{1}{2}$. The solution is $\left(-\frac{1}{2},-\frac{1}{2}\right)$.

© 2012 Cengage Learning. All Rights Reserved. May not be scanned, copied or duplicated, or posted to a publicly accessible website, in whole or in part.

111. Let x represent the number of dimes and y represent the number of quarters. The system of equations is:

$$x+y=11$$
$$0.10x+0.25y=1.85$$

Multiplying the first equation by –0.10:

$$-0.10x-0.10y=-1.10$$
$$0.10x+0.25y=1.85$$

Adding the two equations:

$$0.15y=0.75$$
$$y=5$$

Substituting into the first equation:

$$x+5=11$$
$$x=6$$

Amy has 6 dimes and 5 quarters.

5.6 Binomial Squares and Other Special Products

1. Multiplying using the FOIL method: $(x-2)^2=(x-2)(x-2)=x^2-2x-2x+4=x^2-4x+4$

3. Multiplying using the FOIL method: $(a+3)^2=(a+3)(a+3)=a^2+3a+3a+9=a^2+6a+9$

5. Multiplying using the FOIL method: $(x-5)^2=(x-5)(x-5)=x^2-5x-5x+25=x^2-10x+25$

7. Multiplying using the FOIL method: $\left(a-\frac{1}{2}\right)^2=\left(a-\frac{1}{2}\right)\left(a-\frac{1}{2}\right)=a^2-\frac{1}{2}a-\frac{1}{2}a+\frac{1}{4}=a^2-a+\frac{1}{4}$

9. Multiplying using the FOIL method: $(x+10)^2=(x+10)(x+10)=x^2+10x+10x+100=x^2+20x+100$

11. Multiplying using the square of binomial formula: $(a+0.8)^2=a^2+2(a)(0.8)+(0.8)^2=a^2+1.6a+0.64$

13. Multiplying using the square of binomial formula: $(2x-1)^2=(2x)^2-2(2x)(1)+(1)^2=4x^2-4x+1$

15. Multiplying using the square of binomial formula: $(4a+5)^2=(4a)^2+2(4a)(5)+(5)^2=16a^2+40a+25$

17. Multiplying using the square of binomial formula: $(3x-2)^2=(3x)^2-2(3x)(2)+(2)^2=9x^2-12x+4$

19. Multiplying using the square of binomial formula: $(3a+5b)^2=(3a)^2+2(3a)(5b)+(5b)^2=9a^2+30ab+25b^2$

21. Multiplying using the square of binomial formula: $(4x-5y)^2=(4x)^2-2(4x)(5y)+(5y)^2=16x^2-40xy+25y^2$

23. Multiplying using the square of binomial formula: $(7m+2n)^2=(7m)^2+2(7m)(2n)+(2n)^2=49m^2+28mn+4n^2$

25. Multiplying using the square of binomial formula:

$$(6x-10y)^2=(6x)^2-2(6x)(10y)+(10y)^2=36x^2-120xy+100y^2$$

27. Multiplying using the square of binomial formula: $\left(x^2+5\right)^2=\left(x^2\right)^2+2\left(x^2\right)(5)+(5)^2=x^4+10x^2+25$

29. Multiplying using the square of binomial formula: $\left(a^2+1\right)^2=\left(a^2\right)^2+2\left(a^2\right)(1)+(1)^2=a^4+2a^2+1$

31. Multiplying using the square of binomial formula: $\left(y+\frac{3}{2}\right)^2=(y)^2+2(y)\left(\frac{3}{2}\right)+\left(\frac{3}{2}\right)^2=y^2+3y+\frac{9}{4}$

33. Multiplying using the square of binomial formula: $\left(a+\frac{1}{2}\right)^2=(a)^2+2(a)\left(\frac{1}{2}\right)+\left(\frac{1}{2}\right)^2=a^2+a+\frac{1}{4}$

35. Multiplying using the square of binomial formula: $\left(x+\frac{3}{4}\right)^2=(x)^2+2(x)\left(\frac{3}{4}\right)+\left(\frac{3}{4}\right)^2=x^2+\frac{3}{2}x+\frac{9}{16}$

© 2012 Cengage Learning. All Rights Reserved. May not be scanned, copied or duplicated, or posted to a publicly accessible website, in whole or in part.

37. Multiplying using the square of binomial formula: $\left(t+\frac{1}{5}\right)^2=(t)^2+2(t)\left(\frac{1}{5}\right)+\left(\frac{1}{5}\right)^2=t^2+\frac{2}{5}t+\frac{1}{25}$

39. Completing the table:

x	$(x+3)^2$	x^2+9	x^2+6x+9
1	16	10	16
2	25	13	25
3	36	18	36
4	49	25	49

41. Completing the table:

a	b	$(a+b)^2$	a^2+b^2	a^2+ab+b^2	$a^2+2ab+b^2$
1	1	4	2	3	4
3	5	64	34	49	64
3	4	49	25	37	49
4	5	81	41	61	81

43. Multiplying using the FOIL method: $(a+5)(a-5)=a^2+5a-5a-25=a^2-25$

45. Multiplying using the FOIL method: $(y-1)(y+1)=y^2-y+y-1=y^2-1$

47. Multiplying using the difference of squares formula: $(9+x)(9-x)=(9)^2-(x)^2=81-x^2$

49. Multiplying using the difference of squares formula: $(2x+5)(2x-5)=(2x)^2-(5)^2=4x^2-25$

51. Multiplying using the difference of squares formula: $\left(4x+\frac{1}{3}\right)\left(4x-\frac{1}{3}\right)=(4x)^2-\left(\frac{1}{3}\right)^2=16x^2-\frac{1}{9}$

53. Multiplying using the difference of squares formula: $(2a+7)(2a-7)=(2a)^2-(7)^2=4a^2-49$

55. Multiplying using the difference of squares formula: $(6-7x)(6+7x)=(6)^2-(7x)^2=36-49x^2$

57. Multiplying using the difference of squares formula: $\left(x^2+3\right)\left(x^2-3\right)=\left(x^2\right)^2-(3)^2=x^4-9$

59. Multiplying using the difference of squares formula: $\left(a^2+4\right)\left(a^2-4\right)=\left(a^2\right)^2-(4)^2=a^4-16$

61. Multiplying using the difference of squares formula: $\left(5y^4-8\right)\left(5y^4+8\right)=\left(5y^4\right)^2-(8)^2=25y^8-64$

63. Multiplying and simplifying: $(x+3)(x-3)+(x+5)(x-5)=\left(x^2-9\right)+\left(x^2-25\right)=2x^2-34$

65. Multiplying and simplifying:

$$(2x+3)^2-(4x-1)^2=\left(4x^2+12x+9\right)-\left(16x^2-8x+1\right)=4x^2+12x+9-16x^2+8x-1=-12x^2+20x+8$$

67. Multiplying and simplifying:

$$\begin{aligned}(a+1)^2-(a+2)^2+(a+3)^2&=\left(a^2+2a+1\right)-\left(a^2+4a+4\right)+\left(a^2+6a+9\right)\\&=a^2+2a+1-a^2-4a-4+a^2+6a+9\\&=a^2+4a+6\end{aligned}$$

69. Multiplying and simplifying:

$$\begin{aligned}(2x+3)^3&=(2x+3)(2x+3)^2\\&=(2x+3)\left(4x^2+12x+9\right)\\&=8x^3+24x^2+18x+12x^2+36x+27\\&=8x^3+36x^2+54x+27\end{aligned}$$

© 2012 Cengage Learning. All Rights Reserved. May not be scanned, copied or duplicated, or posted to a publicly accessible website, in whole or in part.

71. **a.** Evaluating when $x = 6$: $x^2 - 25 = (6)^2 - 25 = 36 - 25 = 11$

b. Evaluating when $x = 6$: $(x-5)^2 = (6-5)^2 = (1)^2 = 1$

c. Evaluating when $x = 6$: $(x+5)(x-5) = (6+5)(6-5) = (11)(1) = 11$

73. **a.** Evaluating when $x = -2$: $(x+3)^2 = (-2+3)^2 = (1)^2 = 1$

b. Evaluating when $x = -2$: $x^2 + 9 = (-2)^2 + 9 = 4 + 9 = 13$

c. Evaluating when $x = -2$: $x^2 + 6x + 9 = (-2)^2 + 6(-2) + 9 = 4 - 12 + 9 = 1$

75. Evaluating when $x = 2$:

$$(x+3)^2 = (2+3)^2 = (5)^2 = 25$$
$$x^2 + 6x + 9 = (2)^2 + 6(2) + 9 = 4 + 12 + 9 = 25$$

Both expressions are equal to 25.

77. Let x and $x + 1$ represent the two integers. The expression can be written as:

$$(x)^2 + (x+1)^2 = x^2 + \left(x^2 + 2x + 1\right) = 2x^2 + 2x + 1$$

79. Let x, $x + 1$, and $x + 2$ represent the three integers. The expression can be written as:

$$(x)^2 + (x+1)^2 + (x+2)^2 = x^2 + \left(x^2 + 2x + 1\right) + \left(x^2 + 4x + 4\right) = 3x^2 + 6x + 5$$

81. Verifying the areas: $(a+b)^2 = a^2 + ab + ab + b^2 = a^2 + 2ab + b^2$

83. Simplifying: $\dfrac{10x^3}{5x} = 2x^{3-1} = 2x^2$

85. Simplifying: $\dfrac{3x^2}{3} = x^2$

87. Simplifying: $\dfrac{9x^2}{3x} = 3x^{2-1} = 3x$

89. Simplifying: $\dfrac{24x^3y^2}{8x^2y} = 3x^{3-2}y^{2-1} = 3xy$

91. Simplifying: $\dfrac{15x^2y}{3xy} = \dfrac{15}{3} \bullet \dfrac{x^2}{x} \bullet \dfrac{y}{y} = 5x$

93. Simplifying: $\dfrac{35a^6b^8}{70a^2b^{10}} = \dfrac{35}{70} \bullet \dfrac{a^6}{a^2} \bullet \dfrac{b^8}{b^{10}} = \dfrac{1}{2} \bullet a^4 \bullet \dfrac{1}{b^2} = \dfrac{a^4}{2b^2}$

95. Writing the ratio and simplifying: $\dfrac{12}{20} = \dfrac{4 \bullet 3}{4 \bullet 5} = \dfrac{3}{5}$

97. Graphing both lines:

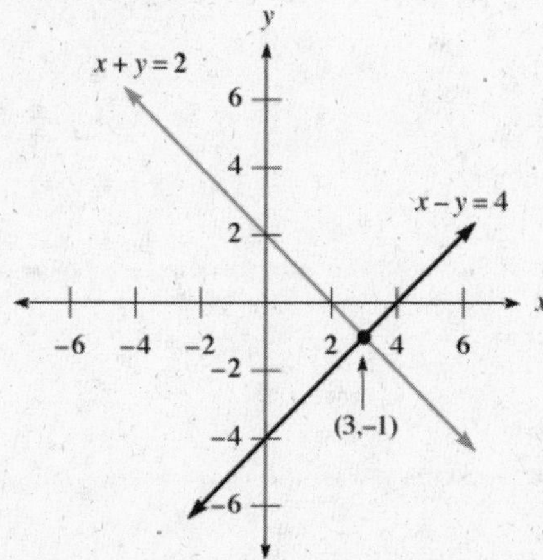

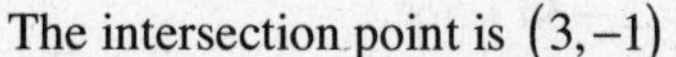
The intersection point is $(3,-1)$.

99. Graphing both lines:

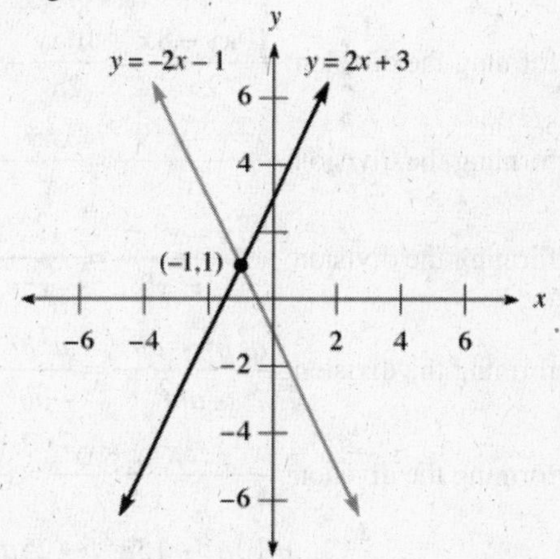

The intersection point is $(-1,1)$.

© 2012 Cengage Learning. All Rights Reserved. May not be scanned, copied or duplicated, or posted to a publicly accessible website, in whole or in part.

5.7 Dividing a Polynomial by a Monomial

1. Performing the division: $\dfrac{5x^2-10x}{5x}=\dfrac{5x^2}{5x}-\dfrac{10x}{5x}=x-2$

3. Performing the division: $\dfrac{15x-10x^3}{5x}=\dfrac{15x}{5x}-\dfrac{10x^3}{5x}=3-2x^2$

5. Performing the division: $\dfrac{25x^2y-10xy}{5x}=\dfrac{25x^2y}{5x}-\dfrac{10xy}{5x}=5xy-2y$

7. Performing the division: $\dfrac{35x^5-30x^4+25x^3}{5x}=\dfrac{35x^5}{5x}-\dfrac{30x^4}{5x}+\dfrac{25x^3}{5x}=7x^4-6x^3+5x^2$

9. Performing the division: $\dfrac{50x^5-25x^3+5x}{5x}=\dfrac{50x^5}{5x}-\dfrac{25x^3}{5x}+\dfrac{5x}{5x}=10x^4-5x^2+1$

11. Performing the division: $\dfrac{8a^2-4a}{-2a}=\dfrac{8a^2}{-2a}+\dfrac{-4a}{-2a}=-4a+2$

13. Performing the division: $\dfrac{16a^5+24a^4}{-2a}=\dfrac{16a^5}{-2a}+\dfrac{24a^4}{-2a}=-8a^4-12a^3$

15. Performing the division: $\dfrac{8ab+10a^2}{-2a}=\dfrac{8ab}{-2a}+\dfrac{10a^2}{-2a}=-4b-5a$

17. Performing the division: $\dfrac{12a^3b-6a^2b^2+14ab^3}{-2a}=\dfrac{12a^3b}{-2a}+\dfrac{-6a^2b^2}{-2a}+\dfrac{14ab^3}{-2a}=-6a^2b+3ab^2-7b^3$

19. Performing the division: $\dfrac{a^2+2ab+b^2}{-2a}=\dfrac{a^2}{-2a}+\dfrac{2ab}{-2a}+\dfrac{b^2}{-2a}=-\dfrac{a}{2}-b-\dfrac{b^2}{2a}$

21. Performing the division: $\dfrac{6x+8y}{2}=\dfrac{6x}{2}+\dfrac{8y}{2}=3x+4y$

23. Performing the division: $\dfrac{7y-21}{-7}=\dfrac{7y}{-7}+\dfrac{-21}{-7}=-y+3$

25. Performing the division: $\dfrac{2x^2+16x-18}{2}=\dfrac{2x^2}{2}+\dfrac{16x}{2}-\dfrac{18}{2}=x^2+8x-9$

27. Performing the division: $\dfrac{3y^2-9y+3}{3}=\dfrac{3y^2}{3}-\dfrac{9y}{3}+\dfrac{3}{3}=y^2-3y+1$

29. Performing the division: $\dfrac{10xy-8x}{2x}=\dfrac{10xy}{2x}-\dfrac{8x}{2x}=5y-4$

31. Performing the division: $\dfrac{x^2y-x^3y^2}{x}=\dfrac{x^2y}{x}-\dfrac{x^3y^2}{x}=xy-x^2y^2$

33. Performing the division: $\dfrac{x^2y-x^3y^2}{-x^2y}=\dfrac{x^2y}{-x^2y}+\dfrac{-x^3y^2}{-x^2y}=-1+xy$

35. Performing the division: $\dfrac{a^2b^2-ab^2}{-ab^2}=\dfrac{a^2b^2}{-ab^2}+\dfrac{-ab^2}{-ab^2}=-a+1$

37. Performing the division: $\dfrac{x^3-3x^2y+xy^2}{x}=\dfrac{x^3}{x}-\dfrac{3x^2y}{x}+\dfrac{xy^2}{x}=x^2-3xy+y^2$

39. Performing the division: $\dfrac{10a^2-15a^2b+25a^2b^2}{5a^2}=\dfrac{10a^2}{5a^2}-\dfrac{15a^2b}{5a^2}+\dfrac{25a^2b^2}{5a^2}=2-3b+5b^2$

41. Performing the division: $\dfrac{26x^2y^2-13xy}{-13xy}=\dfrac{26x^2y^2}{-13xy}+\dfrac{-13xy}{-13xy}=-2xy+1$

© 2012 Cengage Learning. All Rights Reserved. May not be scanned, copied or duplicated, or posted to a publicly accessible website, in whole or in part.

43. Performing the division: $\dfrac{4x^2y^2-2xy}{4xy}=\dfrac{4x^2y^2}{4xy}-\dfrac{2xy}{4xy}=xy-\dfrac{1}{2}$

45. Performing the division: $\dfrac{5a^2x-10ax^2+15a^2x^2}{20a^2x^2}=\dfrac{5a^2x}{20a^2x^2}-\dfrac{10ax^2}{20a^2x^2}+\dfrac{15a^2x^2}{20a^2x^2}=\dfrac{1}{4x}-\dfrac{1}{2a}+\dfrac{3}{4}$

47. Performing the division: $\dfrac{16x^5+8x^2+12x}{12x^3}=\dfrac{16x^5}{12x^3}+\dfrac{8x^2}{12x^3}+\dfrac{12x}{12x^3}=\dfrac{4x^2}{3}+\dfrac{2}{3x}+\dfrac{1}{x^2}$

49. Performing the division: $\dfrac{9a^{5m}-27a^{3m}}{3a^{2m}}=\dfrac{9a^{5m}}{3a^{2m}}-\dfrac{27a^{3m}}{3a^{2m}}=3a^{5m-2m}-9a^{3m-2m}=3a^{3m}-9a^m$

51. Performing the division:

$$\frac{10x^{5m}-25x^{3m}+35x^m}{5x^m}=\frac{10x^{5m}}{5x^m}-\frac{25x^{3m}}{5x^m}+\frac{35x^m}{5x^m}=2x^{5m-m}-5x^{3m-m}+7x^{m-m}=2x^{4m}-5x^{2m}+7$$

53. Simplifying and then dividing:

$$\begin{aligned}\frac{2x^3(3x+2)-3x^2(2x-4)}{2x^2}&=\frac{6x^4+4x^3-6x^3+12x^2}{2x^2}\\&=\frac{6x^4-2x^3+12x^2}{2x^2}\\&=\frac{6x^4}{2x^2}-\frac{2x^3}{2x^2}+\frac{12x^2}{2x^2}\\&=3x^2-x+6\end{aligned}$$

55. Simplifying and then dividing:

$$\frac{(x+2)^2-(x-2)^2}{2x}=\frac{\left(x^2+4x+4\right)-\left(x^2-4x+4\right)}{2x}=\frac{x^2+4x+4-x^2+4x-4}{2x}=\frac{8x}{2x}=4$$

57. Simplifying and then dividing:

$$\frac{(x+5)^2+(x+5)(x-5)}{2x}=\frac{\left(x^2+10x+25\right)+\left(x^2-25\right)}{2x}=\frac{2x^2+10x}{2x}=\frac{2x^2}{2x}+\frac{10x}{2x}=x+5$$

59. **a.** Evaluating when $x=2$: $2x+3=2(2)+3=4+3=7$

b. Evaluating when $x=2$: $\dfrac{10x+15}{5}=\dfrac{10(2)+15}{5}=\dfrac{20+15}{5}=\dfrac{35}{5}=7$

c. Evaluating when $x=2$: $10x+3=10(2)+3=20+3=23$

61. **a.** Evaluating when $x=10$: $\dfrac{3x+8}{2}=\dfrac{3(10)+8}{2}=\dfrac{30+8}{2}=\dfrac{38}{2}=19$

b. Evaluating when $x=10$: $3x+4=3(10)+4=30+4=34$

c. Evaluating when $x=10$: $\frac{3}{2}x+4=\frac{3}{2}(10)+4=15+4=19$

63. **a.** Evaluating when $x=2$: $2x^2-3x=2(2)^2-3(2)=2(4)-3(2)=8-6=2$

b. Evaluating when $x=2$: $\dfrac{10x^3-15x^2}{5x}=\dfrac{10(2)^3-15(2)^2}{5(2)}=\dfrac{10(8)-15(4)}{5(2)}=\dfrac{80-60}{10}=\dfrac{20}{10}=2$

65. Evaluating each expression when $x=2$:

$\dfrac{10x+15}{5}=\dfrac{10(2)+15}{5}=\dfrac{20+15}{5}=\dfrac{35}{5}=7$ $\qquad 2x+3=2(2)+3=4+3=7$

Both expressions equal 7.

© 2012 Cengage Learning. All Rights Reserved. May not be scanned, copied or duplicated, or posted to a publicly accessible website, in whole or in part.

67. Evaluating each expression when $x = 10$:

$$\frac{3x+8}{2}=\frac{3(10)+8}{2}=\frac{30+8}{2}=\frac{38}{2}=19 \qquad 3x+4=3(10)+4=30+4=34$$

Thus $\frac{3x+8}{2} \neq 3x+4$.

69. Dividing:

$$\begin{array}{r} 146 \\ 27\overline{)3962} \\ \underline{27} \\ 126 \\ \underline{108} \\ 182 \\ \underline{162} \\ 20 \end{array}$$

The quotient is $146\frac{20}{27}$.

71. Dividing: $\frac{2x^2+5x}{x}=\frac{2x^2}{x}+\frac{5x}{x}=2x+5$

73. Multiplying: $(x-3)x=x^2-3x$

75. Multiplying: $2x^2(x-5)=2x^3-10x^2$

77. Subtracting: $(x^2-5x)-(x^2-3x)=x^2-5x-x^2+3x=-2x$

79. Subtracting: $(-2x+8)-(-2x+6)=-2x+8+2x-6=2$

81. **a.** Let x represent the total number of DVDs rented. The equation is:

$$0.168x = 365$$
$$x=\frac{365}{0.168}$$
$$x \approx 2{,}173$$

There were 2,173 million DVDs rented in 2009.

b. Finding the percent: $0.247(2{,}173) \approx 536.7$

Blockbuster rented approximately 536.7 million DVDs in 2009.

83. Adding the two equations:

$$2x = 14$$
$$x = 7$$

Substituting into the first equation:

$$7+y=6$$
$$y=-1$$

The solution is $(7,-1)$.

85. Multiplying the second equation by 3:

$$2x-3y=-5$$
$$3x+3y=15$$

Adding the two equations:

$$5x=10$$
$$x=2$$

Substituting into the second equation:

$$2+y=5$$
$$y=3$$

The solution is $(2,3)$.

© 2012 Cengage Learning. All Rights Reserved. May not be scanned, copied or duplicated, or posted to a publicly accessible website, in whole or in part.

87. Substituting into the first equation:

$$\begin{aligned} x+2x-1&=2 \\ 3x-1&=2 \\ 3x&=3 \\ x&=1 \end{aligned}$$

Substituting into the second equation: $y=2(1)-1=2-1=1$. The solution is $(1,1)$.

89. Substituting into the first equation:

$$\begin{aligned} 4x+2(-2x+4)&=8 \\ 4x-4x+8&=8 \\ 8&=8 \end{aligned}$$

Since this statement is true, the system is dependent. The two lines coincide.

5.8 Dividing a Polynomial by a Polynomial

1. Using long division:

$$\begin{array}{r} x-2 \\ x-3\overline{)x^2-5x+6} \\ \underline{x^2-3x} \\ -2x+6 \\ \underline{-2x+6} \\ 0 \end{array}$$

The quotient is $x-2$.

3. Using long division:

$$\begin{array}{r} a+4 \\ a+5\overline{)a^2+9a+20} \\ \underline{a^2+5a} \\ 4a+20 \\ \underline{4a+20} \\ 0 \end{array}$$

The quotient is $a+4$.

5. Using long division:

$$\begin{array}{r} x-3 \\ x-3\overline{)x^2-6x+9} \\ \underline{x^2-3x} \\ -3x+9 \\ \underline{-3x+9} \\ 0 \end{array}$$

The quotient is $x-3$.

7. Using long division:

$$\begin{array}{r} x+3 \\ 2x-1\overline{)2x^2+5x-3} \\ \underline{2x^2-x} \\ 6x-3 \\ \underline{6x-3} \\ 0 \end{array}$$

The quotient is $x+3$.

9. Using long division:

$$\begin{array}{r} a-5 \\ 2a+1\overline{)2a^2-9a-5} \\ \underline{2a^2+a} \\ -10a-5 \\ \underline{-10a-5} \\ 0 \end{array}$$

The quotient is $a-5$.

11. Using long division:

$$\begin{array}{r} x+2 \\ x+3\overline{)x^2+5x+8} \\ \underline{x^2+3x} \\ 2x+8 \\ \underline{2x+6} \\ 2 \end{array}$$

The quotient is $x+2+\dfrac{2}{x+3}$.

13. Using long division:

$$\begin{array}{r} a-2 \\ a+5\overline{)a^2+3a+2} \\ \underline{a^2+5a} \\ -2a+2 \\ \underline{-2a-10} \\ 12 \end{array}$$

The quotient is $a-2+\dfrac{12}{a+5}$.

15. Using long division:

$$\begin{array}{r} x+4 \\ x-2\overline{)x^2+2x+1} \\ \underline{x^2-2x} \\ 4x+1 \\ \underline{4x-8} \\ 9 \end{array}$$

The quotient is $x+4+\dfrac{9}{x-2}$.

© 2012 Cengage Learning. All Rights Reserved. May not be scanned, copied or duplicated, or posted to a publicly accessible website, in whole or in part.

17. Using long division:

$$\begin{array}{r} x+4 \\ x+1\overline{)x^2+5x-6} \\ \underline{x^2+x} \\ 4x-6 \\ \underline{4x+4} \\ -10 \end{array}$$

The quotient is $x+4+\dfrac{-10}{x+1}$.

19. Using long division:

$$\begin{array}{r} a+1 \\ a+2\overline{)a^2+3a+1} \\ \underline{a^2+2a} \\ a+1 \\ \underline{a+2} \\ -1 \end{array}$$

The quotient is $a+1+\dfrac{-1}{a+2}$.

21. Using long division:

$$\begin{array}{r} x-3 \\ 2x+4\overline{)2x^2-2x+5} \\ \underline{2x^2+4x} \\ -6x+5 \\ \underline{-6x-12} \\ 17 \end{array}$$

The quotient is $x-3+\dfrac{17}{2x+4}$.

23. Using long division:

$$\begin{array}{r} 3a-2 \\ 2a+3\overline{)6a^2+5a+1} \\ \underline{6a^2+9a} \\ -4a+1 \\ \underline{-4a-6} \\ 7 \end{array}$$

The quotient is $3a-2+\dfrac{7}{2a+3}$.

25. Using long division:

$$\begin{array}{r} 2a^2-a-3 \\ 3a-5\overline{)6a^3-13a^2-4a+15} \\ \underline{6a^3-10a^2} \\ -3a^2-4a \\ \underline{-3a^2+5a} \\ -9a+15 \\ \underline{-9a+15} \\ 0 \end{array}$$

The quotient is $2a^2-a-3$.

27. Using long division:

$$\begin{array}{r} x^2-x+5 \\ x+1\overline{)x^3+0x^2+4x+5} \\ \underline{x^3+x^2} \\ -x^2+4x \\ \underline{-x^2-x} \\ 5x+5 \\ \underline{5x+5} \\ 0 \end{array}$$

The quotient is x^2-x+5.

29. Using long division:

$$\begin{array}{r} x^2+x+1 \\ x-1\overline{)x^3+0x^2+0x-1} \\ \underline{x^3-x^2} \\ x^2+0x \\ \underline{x^2-x} \\ x-1 \\ \underline{x-1} \\ 0 \end{array}$$

The quotient is x^2+x+1.

31. Using long division:

$$\begin{array}{r} x^2+2x+4 \\ x-2\overline{)x^3+0x^2+0x-8} \\ \underline{x^3-2x^2} \\ 2x^2+0x \\ \underline{2x^2-4x} \\ 4x-8 \\ \underline{4x-8} \\ 0 \end{array}$$

The quotient is x^2+2x+4.

33. **a.** Evaluating when $x=3$: $x^2+2x+4=(3)^2+2(3)+4=9+6+4=19$

b. Evaluating when $x=3$: $\dfrac{x^3-8}{x-2}=\dfrac{(3)^3-8}{3-2}=\dfrac{27-8}{3-2}=\dfrac{19}{1}=19$

c. Evaluating when $x=3$: $x^2-4=(3)^2-4=9-4=5$

© 2012 Cengage Learning. All Rights Reserved. May not be scanned, copied or duplicated, or posted to a publicly accessible website, in whole or in part.

35. **a.** Evaluating when $x=4$: $x+3=4+3=7$

b. Evaluating when $x=4$: $\frac{x^2+9}{x+3}=\frac{(4)^2+9}{4+3}=\frac{16+9}{4+3}=\frac{25}{7}$

c. Evaluating when $x=4$: $x-3+\frac{18}{x+3}=4-3+\frac{18}{4+3}=1+\frac{18}{7}=\frac{7}{7}+\frac{18}{7}=\frac{25}{7}$

37. Finding the average: $\frac{7,264}{18}\approx 404$ yards per hole

39. Let x and y represent the two numbers. The system of equations is:

$$\begin{aligned} x+y&=25 \\ y&=4x \end{aligned}$$

Substituting into the first equation:

$$\begin{aligned} x+4x&=25 \\ 5x&=25 \\ x&=5 \\ y&=4(5)=20 \end{aligned}$$

The two numbers are 5 and 20.

41. Let x represent the amount invested at 8% and y represent the amount invested at 9%. The system of equations is:

$$\begin{aligned} x+y&=1200 \\ 0.08x+0.09y&=100 \end{aligned}$$

Multiplying the first equation by –0.08:

$$\begin{aligned} -0.08x-0.08y&=-96 \\ 0.08x+0.09y&=100 \end{aligned}$$

Adding the two equations:

$$\begin{aligned} 0.01y&=4 \\ y&=400 \end{aligned}$$

Substituting into the first equation:

$$\begin{aligned} x+400&=1200 \\ x&=800 \end{aligned}$$

You have \$800 invested at 8% and \$400 invested at 9%.

43. Let x represent the number of \$5 bills and $x + 4$ represent the number of \$10 bills. The equation is:

$$\begin{aligned} 5(x)+10(x+4)&=160 \\ 5x+10x+40&=160 \\ 15x+40&=160 \\ 15x&=120 \\ x&=8 \\ x+4&=12 \end{aligned}$$

You have 8 \$5 bills and 12 \$10 bills.

45. Let x represent the gallons of 20% antifreeze and y represent the gallons of 60% antifreeze. The system of equations is:

$$\begin{aligned} x+y&=16 \\ 0.20x+0.60y&=0.35(16) \end{aligned}$$

Multiplying the first equation by –0.20:

$$\begin{aligned} -0.20x-0.20y&=-3.2 \\ 0.20x+0.60y&=5.6 \end{aligned}$$

Adding the two equations:

$$\begin{aligned} 0.40y&=2.4 \\ y&=6 \end{aligned}$$

Substituting into the first equation:

$$\begin{aligned} x+6&=16 \\ x&=10 \end{aligned}$$

The mixture contains 10 gallons of 20% antifreeze and 6 gallons of 60% antifreeze.

© 2012 Cengage Learning. All Rights Reserved. May not be scanned, copied or duplicated, or posted to a publicly accessible website, in whole or in part.

Chapter 5 Review

1. Simplifying the expression: $(-1)^3=(-1)(-1)(-1)=-1$

3. Simplifying the expression: $\left(\frac{3}{7}\right)^2=\left(\frac{3}{7}\right)\bullet\left(\frac{3}{7}\right)=\frac{9}{49}$

5. Simplifying the expression: $x^{15}\bullet x^7\bullet x^5\bullet x^3=x^{15+7+5+3}=x^{30}$

7. Simplifying the expression: $\left(2^6\right)^4=2^{6\bullet 4}=2^{24}$

9. Simplifying the expression: $(-2xyz)^3=(-2)^3\bullet x^3y^3z^3=-8x^3y^3z^3$

11. Writing with positive exponents: $4x^{-5}=4\bullet\frac{1}{x^5}=\frac{4}{x^5}$

13. Simplifying the expression: $\frac{a^9}{a^3}=a^{9-3}=a^6$

15. Simplifying the expression: $\frac{x^9}{x^{-6}}=x^{9-(-6)}=x^{9+6}=x^{15}$

17. Simplifying the expression: $(-3xy)^0=1$

19. Simplifying the expression: $\left(3x^3y^2\right)^2=3^2x^6y^4=9x^6y^4$

21. Simplifying the expression: $\left(-3xy^2\right)^{-3}=\frac{1}{\left(-3xy^2\right)^3}=\frac{1}{(-3)^3x^3y^6}=-\frac{1}{27x^3y^6}$

23. Simplifying the expression: $\frac{\left(x^{-3}\right)^3\left(x^6\right)^{-1}}{\left(x^{-5}\right)^{-4}}=\frac{x^{-9}\bullet x^{-6}}{x^{20}}=\frac{x^{-15}}{x^{20}}=x^{-15-20}=x^{-35}=\frac{1}{x^{35}}$

25. Simplifying the expression: $\frac{\left(10x^3y^5\right)\left(21x^2y^6\right)}{\left(7xy^3\right)\left(5x^9y\right)}=\frac{210x^5y^{11}}{35x^{10}y^4}=\frac{210}{35}\bullet\frac{x^5}{x^{10}}\bullet\frac{y^{11}}{y^4}=6\bullet\frac{1}{x^5}\bullet y^7=\frac{6y^7}{x^5}$

27. Simplifying the expression: $\frac{8x^8y^3}{2x^3y}-\frac{10x^6y^9}{5xy^7}=4x^5y^2-2x^5y^2=2x^5y^2$

29. Finding the quotient: $\frac{4.6\times 10^5}{2\times 10^{-3}}=2.3\times 10^8$

31. Performing the operations: $\left(3a^2-5a+5\right)+\left(5a^2-7a-8\right)=\left(3a^2+5a^2\right)+(-5a-7a)+(5-8)=8a^2-12a-3$

33. Performing the operations:

$$\begin{aligned}\left(4x^2-3x-2\right)-\left(8x^2+3x-2\right)&=4x^2-3x-2-8x^2-3x+2\\&=\left(4x^2-8x^2\right)+(-3x-3x)+(-2+2)\\&=-4x^2-6x\end{aligned}$$

35. Multiplying: $3x(4x-7)=3x(4x)-3x(7)=12x^2-21x$

37. Multiplying using the column method:

$$\begin{array}{r}a^2+5a-4\\ a+1\\ \hline a^3+5a^2-4a\\ a^2+5a-4\\ \hline a^3+6a^2+a-4\end{array}$$

© 2012 Cengage Learning. All Rights Reserved. May not be scanned, copied or duplicated, or posted to a publicly accessible website, in whole or in part.

39. Multiplying using the FOIL method: $(3x-7)(2x-5)=6x^2-14x-15x+35=6x^2-29x+35$

41. Multiplying using the difference of squares formula: $(a^2-3)(a^2+3)=(a^2)^2-(3)^2=a^4-9$

43. Multiplying using the square of binomial formula: $(3x+4)^2=(3x)^2+2(3x)(4)+(4)^2=9x^2+24x+16$

45. Performing the division: $\frac{10ab+20a^2}{-5a}=\frac{10ab}{-5a}+\frac{20a^2}{-5a}=-2b-4a$

47. Using long division:

$$\begin{array}{r} x+9 \\ x+6\overline{)x^2+15x+54} \\ \underline{x^2+6x} \\ 9x+54 \\ \underline{9x+54} \\ 0 \end{array}$$

The quotient is $x+9$.

49. Using long division:

$$\begin{array}{r} x^2-4x+16 \\ x+4\overline{)x^3+0x^2+0x+64} \\ \underline{x^3+4x^2} \\ -4x^2+0x \\ \underline{-4x^2-16x} \\ 16x+64 \\ \underline{16x+64} \\ 0 \end{array}$$

The quotient is $x^2-4x+16$.

51. Using long division:

$$\begin{array}{r} x^2-4x+5 \\ 2x+1\overline{)2x^3-7x^2+6x+10} \\ \underline{2x^3+x^2} \\ -8x^2+6x \\ \underline{-8x^2-4x} \\ 10x+10 \\ \underline{10x+5} \\ 5 \end{array}$$

The quotient is $x^2-4x+5+\frac{5}{2x+1}$.

Chapter 5 Cumulative Review

1. Simplifying the expression: $-\left(-\frac{3}{4}\right)=\frac{3}{4}$

3. Simplifying the expression: $2\bullet7+10=14+10=24$

5. Simplifying the expression: $10+(-15)=10-15=-5$

7. Simplifying the expression: $-7\left(-\frac{1}{7}\right)=1$

9. Simplifying the expression: $6\left(-\frac{1}{2}x+\frac{1}{6}y\right)=6\left(-\frac{1}{2}x\right)+6\left(\frac{1}{6}y\right)=-3x+y$

11. Simplifying the expression: $y^{10}\bullet y^6=y^{10+6}=y^{16}$

13. Simplifying the expression: $(3a^2b^5)^3(2a^6b^7)^2=27a^6b^{15}\bullet4a^{12}b^{14}=108a^{18}b^{29}$

15. Multiplying the values: $(3.5\times10^3)(8\times10^{-6})=28\times10^{-3}=2.8\times10^{-2}$

17. Simplifying the expression by combining like terms: $10a^2b-5a^2b+a^2b=(10-5+1)a^2b=6a^2b$

19. Multiplying using the square of binomial formula: $(5x-1)^2=(5x)^2-2(5x)(1)+(1)^2=25x^2-10x+1$

© 2012 Cengage Learning. All Rights Reserved. May not be scanned, copied or duplicated, or posted to a publicly accessible website, in whole or in part.

21. Solving the equation:

$$\begin{aligned} 8-2y+3y&=12\\ 8+y&=12\\ y&=4 \end{aligned}$$

23. Solving the equation:

$$\begin{aligned} 18&=3(2x-2)\\ 18&=6x-6\\ 24&=6x\\ x&=4 \end{aligned}$$

25. Solving the inequality:

$$\begin{aligned} -\frac{a}{6}&>4\\ -6\left(-\frac{a}{6}\right)&<-6(4)\\ a&<-24 \end{aligned}$$

Graphing the solution set:

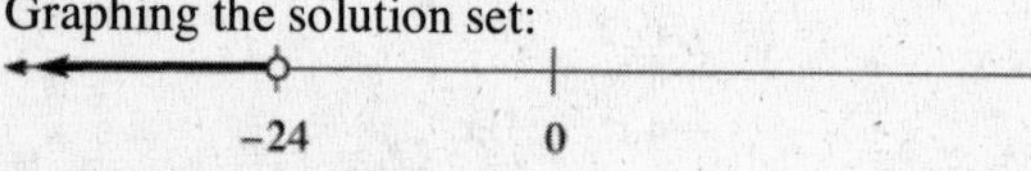

27. Solving the compound inequality:

$$\begin{aligned} -3<2x+3<5\\ -6<2x<2\\ -3<x<1 \end{aligned}$$

Graphing the solution set:

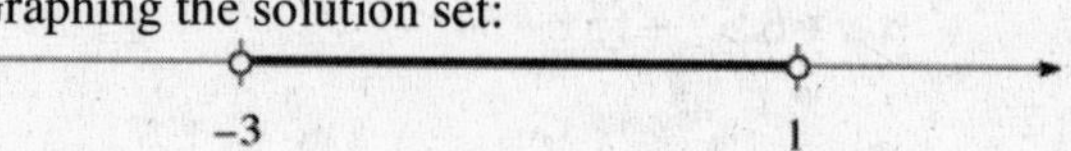

29. Using long division:

$$\begin{array}{r} 2x+7\\ 2x-3\overline{)4x^2+8x-10}\\ \underline{4x^2-6x}\\ 14x-10\\ \underline{14x-21}\\ 11 \end{array}$$

The quotient is $2x+7+\dfrac{11}{2x-3}$.

31. Graphing the equation:

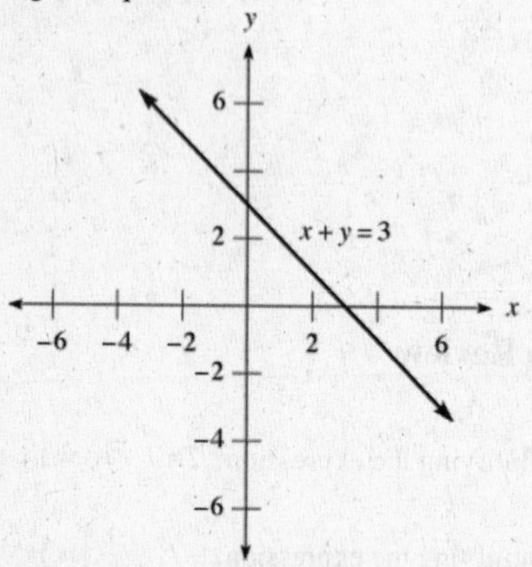

33. Graphing the two lines:

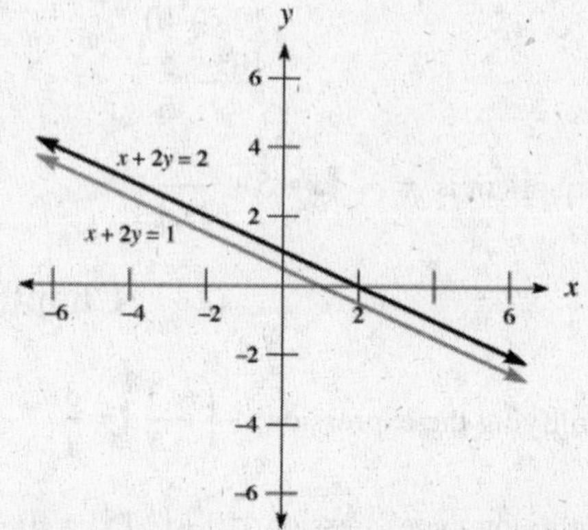

There is no solution to the system. The lines are parallel.

35. Multiplying the second equation by –1:

$$\begin{aligned} x+2y&=5\\ -x+y&=-2 \end{aligned}$$

Adding the two equations:

$$\begin{aligned} 3y&=3\\ y&=1 \end{aligned}$$

Substituting into the second equation:

$$\begin{aligned} x-1&=2\\ x&=3 \end{aligned}$$

The solution is $(3,1)$.

© 2012 Cengage Learning. All Rights Reserved. May not be scanned, copied or duplicated, or posted to a publicly accessible website, in whole or in part.

37. Multiplying the second equation by –2:

$$\begin{aligned} 2x+y&=-3 \\ -2x+6y&=10 \end{aligned}$$

Adding the two equations:

$$\begin{aligned} 7y&=7 \\ y&=1 \end{aligned}$$

Substituting into the first equation:

$$\begin{aligned} 2x+1&=-3 \\ 2x&=-4 \\ x&=-2 \end{aligned}$$

The solution is $(-2,1)$.

39. Substituting into the first equation:

$$\begin{aligned} x+(x+1)&=9 \\ 2x+1&=9 \\ 2x&=8 \\ x&=4 \\ y&=4+1=5 \end{aligned}$$

The solution is $(4,5)$.

41. Evaluating when $x=3$: $8x-3=8(3)-3=24-3=21$

43. Solving for h:

$$\begin{aligned} \frac{1}{2}bh&=A \\ bh&=2A \\ h&=\frac{2A}{b} \end{aligned}$$

45. The only whole number is 0.

47. This is the additive inverse property.

49. To find the x-intercept, let $y=0$:

$$\begin{aligned} 0&=2x+4 \\ -4&=2x \\ x&=-2 \end{aligned}$$

To find the y-intercept, let $x=0$: $y=2(0)+4=0+4=4$

The x-intercept is –2 and the y-intercept is 4.

51. Finding the difference: $\$63,935-\$53,160=\$10,775$

© 2012 Cengage Learning. All Rights Reserved. May not be scanned, copied or duplicated, or posted to a publicly accessible website, in whole or in part.

Chapter 5 Test

1. Simplifying the expression: $(-3)^4 = (-3)(-3)(-3)(-3) = 81$
2. Simplifying the expression: $\left(\frac{3}{4}\right)^2 = \left(\frac{3}{4}\right)\left(\frac{3}{4}\right) = \frac{9}{16}$
3. Simplifying the expression: $(3x^3)^2(2x^4)^3 = 9x^6 \bullet 8x^{12} = 72x^{18}$
4. Simplifying the expression: $3^{-2} = \frac{1}{3^2} = \frac{1}{9}$
5. Simplifying the expression: $(3a^4b^2)^0 = 1$
6. Simplifying the expression: $\frac{a^{-3}}{a^{-5}} = a^{-3-(-5)} = a^{-3+5} = a^2$
7. Simplifying the expression: $\frac{(x^{-2})^3(x^{-3})^{-5}}{(x^{-4})^{-2}} = \frac{x^{-6}x^{15}}{x^8} = \frac{x^9}{x^8} = x^{9-8} = x$
8. Writing in scientific notation: $0.0278 = 2.78 \times 10^{-2}$
9. Writing in expanded form: $2.43 \times 10^5 = 243{,}000$
10. Simplifying the expression: $\frac{35x^2y^4z}{70x^6y^2z} = \frac{35}{70} \bullet \frac{x^2}{x^6} \bullet \frac{y^4}{y^2} \bullet \frac{z}{z} = \frac{1}{2} \bullet \frac{1}{x^4} \bullet y^2 = \frac{y^2}{2x^4}$
11. Simplifying the expression: $\frac{(6a^2b)(9a^3b^2)}{18a^4b^3} = \frac{54a^5b^3}{18a^4b^3} = 3a$
12. Simplifying the expression: $\frac{24x^7}{3x^2} + \frac{14x^9}{7x^4} = 8x^5 + 2x^5 = 10x^5$
13. Simplifying the expression: $\frac{(2.4 \times 10^5)(4.5 \times 10^{-2})}{1.2 \times 10^{-6}} = \frac{10.8 \times 10^3}{1.2 \times 10^{-6}} = 9.0 \times 10^9$
14. Performing the operations: $8x^2 - 4x + 6x + 2 = 8x^2 + 2x + 2$
15. Performing the operations: $(5x^2 - 3x + 4) - (2x^2 - 7x - 2) = 5x^2 - 3x + 4 - 2x^2 + 7x + 2 = 3x^2 + 4x + 6$
16. Performing the operations: $(6x - 8) - (3x - 4) = 6x - 8 - 3x + 4 = 3x - 4$
17. Evaluating when $y = -2$: $2y^2 - 3y - 4 = 2(-2)^2 - 3(-2) - 4 = 2(4) + 6 - 4 = 8 + 6 - 4 = 10$
18. Multiplying using the distributive property:
$$2a^2(3a^2 - 5a + 4) = 2a^2(3a^2) - 2a^2(5a) + 2a^2(4) = 6a^4 - 10a^3 + 8a^2$$
19. Multiplying using the FOIL method: $\left(x + \frac{1}{2}\right)\left(x + \frac{1}{3}\right) = x^2 + \frac{1}{2}x + \frac{1}{3}x + \frac{1}{6} = x^2 + \frac{5}{6}x + \frac{1}{6}$
20. Multiplying using the FOIL method: $(4x - 5)(2x + 3) = 8x^2 - 10x + 12x - 15 = 8x^2 + 2x - 15$
21. Multiplying using the column method:
$$\begin{array}{l} x^2 + 3x + 9 \\ \underline{x - 3} \\ x^3 + 3x^2 + 9x \\ \underline{ - 3x^2 - 9x - 27} \\ x^3 - 27 \end{array}$$

© 2012 Cengage Learning. All Rights Reserved. May not be scanned, copied or duplicated, or posted to a publicly accessible website, in whole or in part.

22. Multiplying using the square of binomial formula: $(x+5)^2 = x^2 + 2(x)(5) + (5)^2 = x^2 + 10x + 25$

23. Multiplying using the square of binomial formula: $(3a-2b)^2 = (3a)^2 - 2(3a)(2b) + (2b)^2 = 9a^2 - 12ab + 4b^2$

24. Multiplying using the difference of squares formula: $(3x-4y)(3x+4y) = (3x)^2 - (4y)^2 = 9x^2 - 16y^2$

25. Multiplying using the difference of squares formula: $(a^2-3)(a^2+3) = (a^2)^2 - (3)^2 = a^4 - 9$

26. Performing the division: $\dfrac{10x^3 + 15x^2 - 5x}{5x} = \dfrac{10x^3}{5x} + \dfrac{15x^2}{5x} - \dfrac{5x}{5x} = 2x^2 + 3x - 1$

27. Using long division:

$$\begin{array}{r} 4x+3 \\ 2x-3\overline{)8x^2 - 6x - 5} \\ \underline{8x^2 - 12x} \\ 6x - 5 \\ \underline{6x - 9} \\ 4 \end{array}$$

The quotient is $4x + 3 + \dfrac{4}{2x-3}$.

28. Using long division:

$$\begin{array}{r} 3x^2 + 9x + 25 \\ x-3\overline{)3x^3 + 0x^2 - 2x + 1} \\ \underline{3x^3 - 9x^2} \\ 9x^2 - 2x \\ \underline{9x^2 - 27x} \\ 25x + 1 \\ \underline{25x - 75} \\ 76 \end{array}$$

The quotient is $3x^2 + 9x + 25 + \dfrac{76}{x-3}$.

29. Using the volume formula: $V = (2.5 \text{ cm})^3 = 15.625 \text{ cm}^3$

30. Let w represent the width, $5w$ represent the length, and $\frac{1}{5}w$ represent the height. The volume is given by:

$$V = (w)(5w)\left(\frac{1}{5}w\right) = w^3$$

31. Writing in scientific notation: $63.7 \times 10^6 = 6.37 \times 10^7$ dollars

32. Writing in scientific notation: $51.1 \times 10^6 = 5.11 \times 10^7$ dollars

33. Writing in scientific notation: $0.005 = 5 \times 10^{-3}$ seconds

34. Writing in scientific notation: $0.002 = 2 \times 10^{-3}$ seconds

© 2012 Cengage Learning. All Rights Reserved. May not be scanned, copied or duplicated, or posted to a publicly accessible website, in whole or in part.

Chapter 6
Factoring

Getting Ready for Chapter 6

1. Dividing: $\dfrac{y^3-16y^2+64y}{y}=\dfrac{y^3}{y}-\dfrac{16y^2}{y}+\dfrac{64y}{y}=y^2-16y+64$

2. Dividing: $\dfrac{5x^3+35x^2+60x}{5x}=\dfrac{5x^3}{5x}+\dfrac{35x^2}{5x}+\dfrac{60x}{5x}=x^2+7x+12$

3. Dividing: $\dfrac{-12x^4+48x^3+144x^2}{-12x^2}=\dfrac{-12x^4}{-12x^2}+\dfrac{48x^3}{-12x^2}+\dfrac{144x^2}{-12x^2}=x^2-4x-12$

4. Dividing: $\dfrac{16x^5+20x^4+60x^3}{4x^3}=\dfrac{16x^5}{4x^3}+\dfrac{20x^4}{4x^3}+\dfrac{60x^3}{4x^3}=4x^2+5x+15$

5. Multiplying: $(x-4)(x+4)=(x)^2-(4)^2=x^2-16$

6. Multiplying: $(x-6)(x+6)=(x)^2-(6)^2=x^2-36$

7. Multiplying: $(x-8)(x+8)=(x)^2-(8)^2=x^2-64$

8. Multiplying: $(x^2-4)(x^2+4)=(x^2)^2-(4)^2=x^4-16$

9. Multiplying: $x(x^2+2x+4)=x^3+2x^2+4x$

10. Multiplying: $2x(4x^2-10x+25)=8x^3-20x^2+50x$

11. Multiplying: $-2(x^2+2x+4)=-2x^2-4x-8$

12. Multiplying: $5(4x^2-10x+25)=20x^2-50x+125$

13. Multiplying: $(x-2)(x^2+2x+4)=(x)^3-(2)^3=x^3-8$

14. Multiplying: $(2x+5)(4x^2-10x+25)=(2x)^3+(5)^3=8x^3+125$

15. Multiplying: $3x^2(x+3)(x-3)=3x^2(x^2-9)=3x^4-27x^2$

16. Multiplying: $2x^3(x+5)^2=2x^3(x^2+10x+25)=2x^5+20x^4+50x^3$

17. Multiplying: $y^3(y^2+36)=y^5+36y^3$

18. Multiplying: $(3a-4)(2a-1)=6a^2-8a-3a+4=6a^2-11a+4$

19. Multiplying: $6x(x-4)(x+2)=6x(x^2-2x-8)=6x^3-12x^2-48x$

20. Multiplying: $2a^3b(a^2+3a+1)=2a^5b+6a^4b+2a^3b$

© 2012 Cengage Learning. All Rights Reserved. May not be scanned, copied or duplicated, or posted to a publicly accessible website, in whole or in part.

6.1 The Greatest Common Factor and Factoring by Grouping

1. Factoring out the greatest common factor: $15x+25=5(3x+5)$
3. Factoring out the greatest common factor: $6a+9=3(2a+3)$
5. Factoring out the greatest common factor: $4x-8y=4(x-2y)$
7. Factoring out the greatest common factor: $3x^2-6x-9=3(x^2-2x-3)$
9. Factoring out the greatest common factor: $3a^2-3a-60=3(a^2-a-20)$
11. Factoring out the greatest common factor: $24y^2-52y+24=4(6y^2-13y+6)$
13. Factoring out the greatest common factor: $9x^2-8x^3=x^2(9-8x)$
15. Factoring out the greatest common factor: $13a^2-26a^3=13a^2(1-2a)$
17. Factoring out the greatest common factor: $21x^2y-28xy^2=7xy(3x-4y)$
19. Factoring out the greatest common factor: $22a^2b^2-11ab^2=11ab^2(2a-1)$
21. Factoring out the greatest common factor: $7x^3+21x^2-28x=7x(x^2+3x-4)$
23. Factoring out the greatest common factor: $121y^4-11x^4=11(11y^4-x^4)$
25. Factoring out the greatest common factor: $100x^4-50x^3+25x^2=25x^2(4x^2-2x+1)$
27. Factoring out the greatest common factor: $8a^2+16b^2+32c^2=8(a^2+2b^2+4c^2)$
29. Factoring out the greatest common factor: $4a^2b-16ab^2+32a^2b^2=4ab(a-4b+8ab)$
31. Factoring out the greatest common factor: $121a^3b^2-22a^2b^3+33a^3b^3=11a^2b^2(11a-2b+3ab)$
33. Factoring out the greatest common factor: $12x^2y^3-72x^5y^3-36x^4y^4=12x^2y^3(1-6x^3-3x^2y)$
35. Factoring by grouping: $xy+5x+3y+15=x(y+5)+3(y+5)=(y+5)(x+3)$
37. Factoring by grouping: $xy+6x+2y+12=x(y+6)+2(y+6)=(y+6)(x+2)$
39. Factoring by grouping: $ab+7a-3b-21=a(b+7)-3(b+7)=(b+7)(a-3)$
41. Factoring by grouping: $ax-bx+ay-by=x(a-b)+y(a-b)=(a-b)(x+y)$
43. Factoring by grouping: $2ax+6x-5a-15=2x(a+3)-5(a+3)=(a+3)(2x-5)$
45. Factoring by grouping: $3xb-4b-6x+8=b(3x-4)-2(3x-4)=(3x-4)(b-2)$
47. Factoring by grouping: $x^2+ax+2x+2a=x(x+a)+2(x+a)=(x+a)(x+2)$
49. Factoring by grouping: $x^2-ax-bx+ab=x(x-a)-b(x-a)=(x-a)(x-b)$
51. Factoring by grouping: $ax+ay+bx+by+cx+cy=a(x+y)+b(x+y)+c(x+y)=(x+y)(a+b+c)$
53. Factoring by grouping: $6x^2+9x+4x+6=3x(2x+3)+2(2x+3)=(2x+3)(3x+2)$
55. Factoring by grouping: $20x^2-2x+50x-5=2x(10x-1)+5(10x-1)=(10x-1)(2x+5)$
57. Factoring by grouping: $20x^2+4x+25x+5=4x(5x+1)+5(5x+1)=(5x+1)(4x+5)$
59. Factoring by grouping: $x^3+2x^2+3x+6=x^2(x+2)+3(x+2)=(x+2)(x^2+3)$
61. Factoring by grouping: $6x^3-4x^2+15x-10=2x^2(3x-2)+5(3x-2)=(3x-2)(2x^2+5)$
63. Its greatest common factor is $3\cdot 2=6$.
65. The correct factoring is: $12x^2+6x+3=3(4x^2+2x+1)$

© 2012 Cengage Learning. All Rights Reserved. May not be scanned, copied or duplicated, or posted to a publicly accessible website, in whole or in part.

67. Multiplying using the FOIL method: $(x-7)(x+2)=x^2-7x+2x-14=x^2-5x-14$

69. Multiplying using the FOIL method: $(x-3)(x+2)=x^2-3x+2x-6=x^2-x-6$

71. Multiplying using the column method:

$$\begin{array}{r} x^2-3x+9 \\ x+3 \\ \hline x^3-3x^2+9x \\ 3x^2-9x+27 \\ \hline x^3+27 \end{array}$$

73. Multiplying using the column method:

$$\begin{array}{r} x^2+4x-3 \\ 2x+1 \\ \hline 2x^3+8x^2-6x \\ x^2+4x-3 \\ \hline 2x^3+9x^2-2x-3 \end{array}$$

75. Multiplying: $3x^4\left(6x^3-4x^2+2x\right)=3x^4\bullet 6x^3-3x^4\bullet 4x^2+3x^4\bullet 2x=18x^7-12x^6+6x^5$

77. Multiplying: $\left(x+\frac{1}{3}\right)\left(x+\frac{2}{3}\right)=x^2+\frac{2}{3}x+\frac{1}{3}x+\frac{2}{9}=x^2+x+\frac{2}{9}$

79. Multiplying: $(6x+4y)(2x-3y)=12x^2-18xy+8xy-12y^2=12x^2-10xy-12y^2$

81. Multiplying: $(9a+1)(9a-1)=81a^2-9a+9a-1=81a^2-1$

83. Multiplying: $(x-9)(x-9)=x^2-9x-9x+81=x^2-18x+81$

85. Multiplying: $(x+2)\left(x^2-2x+4\right)=x\left(x^2-2x+4\right)+2\left(x^2-2x+4\right)=x^3-2x^2+4x+2x^2-4x+8=x^3+8$

87. Dividing: $\frac{y^3-16y^2+64y}{y}=\frac{y^3}{y}-\frac{16y^2}{y}+\frac{64y}{y}=y^2-16y+64$

89. Dividing: $\frac{-12x^4+48x^3+144x^2}{-12x^2}=\frac{-12x^4}{-12x^2}+\frac{48x^3}{-12x^2}+\frac{144x^2}{-12x^2}=x^2-4x-12$

91. Dividing: $\frac{-18y^5+63y^4-108y^3}{-9y^3}=\frac{-18y^5}{-9y^3}+\frac{63y^4}{-9y^3}-\frac{108y^3}{-9y^3}=2y^2-7y+12$

93. Subtracting the polynomials: $\left(5x^2+5x-4\right)-\left(3x^2-2x+7\right)=5x^2+5x-4-3x^2+2x-7=2x^2+7x-11$

95. Subtracting the polynomials: $(7x+3)-(4x-5)=7x+3-4x+5=3x+8$

97. Subtracting the polynomials: $\left(5x^2-5\right)-\left(2x^2-4x\right)=5x^2-5-2x^2+4x=3x^2+4x-5$

6.2 Factoring Trinomials

1. Factoring the trinomial: $x^2+7x+12=(x+3)(x+4)$
3. Factoring the trinomial: $x^2+3x+2=(x+2)(x+1)$
5. Factoring the trinomial: $a^2+10a+21=(a+7)(a+3)$
7. Factoring the trinomial: $x^2-7x+10=(x-5)(x-2)$
9. Factoring the trinomial: $y^2-10y+21=(y-7)(y-3)$
11. Factoring the trinomial: $x^2-x-12=(x-4)(x+3)$
13. Factoring the trinomial: $y^2+y-12=(y+4)(y-3)$
15. Factoring the trinomial: $x^2+5x-14=(x+7)(x-2)$
17. Factoring the trinomial: $r^2-8r-9=(r-9)(r+1)$
19. Factoring the trinomial: $x^2-x-30=(x-6)(x+5)$
21. Factoring the trinomial: $a^2+15a+56=(a+7)(a+8)$
23. Factoring the trinomial: $y^2-y-42=(y-7)(y+6)$
25. Factoring the trinomial: $x^2+13x+42=(x+7)(x+6)$
27. Factoring the trinomial: $2x^2+6x+4=2\left(x^2+3x+2\right)=2(x+2)(x+1)$
29. Factoring the trinomial: $3a^2-3a-60=3\left(a^2-a-20\right)=3(a-5)(a+4)$
31. Factoring the trinomial: $100x^2-500x+600=100\left(x^2-5x+6\right)=100(x-3)(x-2)$

© 2012 Cengage Learning. All Rights Reserved. May not be scanned, copied or duplicated, or posted to a publicly accessible website, in whole or in part.

33. Factoring the trinomial: $100p^2-1300p+4000=100\left(p^2-13p+40\right)=100(p-8)(p-5)$

35. Factoring the trinomial: $x^4-x^3-12x^2=x^2\left(x^2-x-12\right)=x^2(x-4)(x+3)$

37. Factoring the trinomial: $2r^3+4r^2-30r=2r\left(r^2+2r-15\right)=2r(r+5)(r-3)$

39. Factoring the trinomial: $2y^4-6y^3-8y^2=2y^2\left(y^2-3y-4\right)=2y^2(y-4)(y+1)$

41. Factoring the trinomial: $x^5+4x^4+4x^3=x^3\left(x^2+4x+4\right)=x^3(x+2)(x+2)=x^3(x+2)^2$

43. Factoring the trinomial: $3y^4-12y^3-15y^2=3y^2\left(y^2-4y-5\right)=3y^2(y-5)(y+1)$

45. Factoring the trinomial: $4x^4-52x^3+144x^2=4x^2\left(x^2-13x+36\right)=4x^2(x-9)(x-4)$

47. Factoring the trinomial: $x^2+5xy+6y^2=(x+2y)(x+3y)$

49. Factoring the trinomial: $x^2-9xy+20y^2=(x-4y)(x-5y)$

51. Factoring the trinomial: $a^2+2ab-8b^2=(a+4b)(a-2b)$

53. Factoring the trinomial: $a^2-10ab+25b^2=(a-5b)(a-5b)=(a-5b)^2$

55. Factoring the trinomial: $a^2+10ab+25b^2=(a+5b)(a+5b)=(a+5b)^2$

57. Factoring the trinomial: $x^2+2xa-48a^2=(x+8a)(x-6a)$

59. Factoring the trinomial: $x^2-5xb-36b^2=(x-9b)(x+4b)$

61. Factoring the trinomial: $x^4-5x^2+6=\left(x^2-2\right)\left(x^2-3\right)$

63. Factoring the trinomial: $x^2-80x-2000=(x-100)(x+20)$

65. Factoring the trinomial: $x^2-x+\frac{1}{4}=\left(x-\frac{1}{2}\right)\left(x-\frac{1}{2}\right)=\left(x-\frac{1}{2}\right)^2$

67. Factoring the trinomial: $x^2+0.6x+0.08=(x+0.4)(x+0.2)$

69. We can use long division to find the other factor:

$$\begin{array}{r}
x+16 \\
x+8\overline{)x^2+24x+128} \\
\underline{x^2+8x} \\
16x+128 \\
\underline{16x+128} \\
0
\end{array}$$

The other factor is $x+16$.

71. Using FOIL to multiply out the factors: $(4x+3)(x-1)=4x^2+3x-4x-3=4x^2-x-3$

73. Multiplying using the FOIL method: $(6a+1)(a+2)=6a^2+a+12a+2=6a^2+13a+2$

75. Multiplying using the FOIL method: $(3a+2)(2a+1)=6a^2+4a+3a+2=6a^2+7a+2$

77. Multiplying using the FOIL method: $(6a+2)(a+1)=6a^2+2a+6a+2=6a^2+8a+2$

79. Simplifying; $\left(-\frac{2}{5}\right)^2=\left(-\frac{2}{5}\right)\left(-\frac{2}{5}\right)=\frac{4}{25}$

81. Simplifying; $\left(3a^3\right)^2\left(2a^2\right)^3=\left(9a^6\right)\left(8a^6\right)=72a^{12}$

© 2012 Cengage Learning. All Rights Reserved. May not be scanned, copied or duplicated, or posted to a publicly accessible website, in whole or in part.

83. Simplifying; $\dfrac{(4x)^{-7}}{(4x)^{-5}}=(4x)^{-2}=\dfrac{1}{(4x)^2}=\dfrac{1}{16x^2}$

85. Simplifying; $\dfrac{12a^5b^3}{72a^2b^5}=\dfrac{1}{6}a^{5-2}b^{3-5}=\dfrac{1}{6}a^3b^{-2}=\dfrac{a^3}{6b^2}$

87. Simplifying; $\dfrac{15x^{-5}y^3}{45x^2y^5}=\dfrac{1}{3}x^{-5-2}y^{3-5}=\dfrac{1}{3}x^{-7}y^{-2}=\dfrac{1}{3x^7y^2}$

89. Simplifying; $(-7x^3y)(3xy^4)=-21x^{3+1}y^{1+4}=-21x^4y^5$

91. Simplifying; $(-5a^3b^{-1})(4a^{-2}b^4)=-20a^{3-2}b^{-1+4}=-20ab^3$

93. Simplifying; $(9a^2b^3)(-3a^3b^5)=-27a^{2+3}b^{3+5}=-27a^5b^8$

6.3 More Trinomials to Factor

1. Factoring the trinomial: $2x^2+7x+3=(2x+1)(x+3)$
3. Factoring the trinomial: $2a^2-a-3=(2a-3)(a+1)$
5. Factoring the trinomial: $3x^2+2x-5=(3x+5)(x-1)$
7. Factoring the trinomial: $3y^2-14y-5=(3y+1)(y-5)$
9. Factoring the trinomial: $6x^2+13x+6=(3x+2)(2x+3)$
11. Factoring the trinomial: $4x^2-12xy+9y^2=(2x-3y)(2x-3y)=(2x-3y)^2$
13. Factoring the trinomial: $4y^2-11y-3=(4y+1)(y-3)$
15. Factoring the trinomial: $20x^2-41x+20=(4x-5)(5x-4)$
17. Factoring the trinomial: $20a^2+48ab-5b^2=(10a-b)(2a+5b)$
19. Factoring the trinomial: $20x^2-21x-5=(4x-5)(5x+1)$
21. Factoring the trinomial: $12m^2+16m-3=(6m-1)(2m+3)$
23. Factoring the trinomial: $20x^2+37x+15=(4x+5)(5x+3)$
25. Factoring the trinomial: $12a^2-25ab+12b^2=(3a-4b)(4a-3b)$
27. Factoring the trinomial: $3x^2-xy-14y^2=(3x-7y)(x+2y)$
29. Factoring the trinomial: $14x^2+29x-15=(2x+5)(7x-3)$
31. Factoring the trinomial: $6x^2-43x+55=(3x-5)(2x-11)$
33. Factoring the trinomial: $15t^2-67t+38=(5t-19)(3t-2)$
35. Factoring the trinomial: $4x^2+2x-6=2(2x^2+x-3)=2(2x+3)(x-1)$
37. Factoring the trinomial: $24a^2-50a+24=2(12a^2-25a+12)=2(4a-3)(3a-4)$
39. Factoring the trinomial: $10x^3-23x^2+12x=x(10x^2-23x+12)=x(5x-4)(2x-3)$
41. Factoring the trinomial: $6x^4-11x^3-10x^2=x^2(6x^2-11x-10)=x^2(3x+2)(2x-5)$
43. Factoring the trinomial: $10a^3-6a^2-4a=2a(5a^2-3a-2)=2a(5a+2)(a-1)$
45. Factoring the trinomial: $15x^3-102x^2-21x=3x(5x^2-34x-7)=3x(5x+1)(x-7)$
47. Factoring the trinomial: $35y^3-60y^2-20y=5y(7y^2-12y-4)=5y(7y+2)(y-2)$

© 2012 Cengage Learning. All Rights Reserved. May not be scanned, copied or duplicated, or posted to a publicly accessible website, in whole or in part.

49. Factoring the trinomial: $15a^4 - 2a^3 - a^2 = a^2\left(15a^2 - 2a - 1\right) = a^2\left(5a+1\right)\left(3a-1\right)$

51. Factoring the trinomial: $12x^2y - 34xy^2 + 14y^3 = 2y\left(6x^2 - 17xy + 7y^2\right) = 2y(2x-y)(3x-7y)$

53. Evaluating each expression when $x = 2$:

$$2x^2 + 7x + 3 = 2(2)^2 + 7(2) + 3 = 8 + 14 + 3 = 25$$
$$(2x+1)(x+3) = (2 \bullet 2 + 1)(2+3) = (5)(5) = 25$$

Both expressions equal 25.

55. Multiplying using the difference of squares formula: $(2x+3)(2x-3) = (2x)^2 - (3)^2 = 4x^2 - 9$

57. Multiplying using the difference of squares formula:

$$(x+3)(x-3)\left(x^2+9\right) = \left(x^2-9\right)\left(x^2+9\right) = \left(x^2\right)^2 - (9)^2 = x^4 - 81$$

59. Factoring: $12x^2 - 71x + 105 = (x-3)(12x-35)$. The other factor is $12x - 35$.

61. Factoring: $54x^2 + 111x + 56 = (6x+7)(9x+8)$. The other factor is $9x + 8$.

63. Factoring: $16t^2 - 64t + 48 = 16\left(t^2 - 4t + 3\right) = 16(t-1)(t-3)$

65. Factoring: $h = 2{,}716 - 16t^2 = 4\left(679 - 4t^2\right)$

Substituting $t = 6$: $h = 4\left(679 - 4(6)^2\right) = 4(679 - 144) = 4(535) = 2{,}140$ feet

67. **a.** Factoring: $h = 8 + 62t - 16t^2 = 2\left(4 + 31t - 8t^2\right) = 2(4-t)(1+8t)$

b. Completing the table:

Time t (seconds)	Height h (feet)
0	8
1	54
2	68
3	50
4	0

69. Multiplying: $(x+3)(x-3) = x^2 - (3)^2 = x^2 - 9$

71. Multiplying: $(x+5)(x-5) = x^2 - (5)^2 = x^2 - 25$

73. Multiplying: $(x+7)(x-7) = x^2 - (7)^2 = x^2 - 49$

75. Multiplying: $(x+9)(x-9) = x^2 - (9)^2 = x^2 - 81$

77. Multiplying: $(2x-3y)(2x+3y) = (2x)^2 - (3y)^2 = 4x^2 - 9y^2$

79. Multiplying: $\left(x^2+4\right)(x+2)(x-2) = \left(x^2+4\right)\left(x^2-4\right) = \left(x^2\right)^2 - (4)^2 = x^4 - 16$

81. Multiplying: $(x+3)^2 = x^2 + 2(x)(3) + (3)^2 = x^2 + 6x + 9$

83. Multiplying: $(x+5)^2 = x^2 + 2(x)(5) + (5)^2 = x^2 + 10x + 25$

85. Multiplying: $(x+7)^2 = x^2 + 2(x)(7) + (7)^2 = x^2 + 14x + 49$

87. Multiplying: $(x+9)^2 = x^2 + 2(x)(9) + (9)^2 = x^2 + 18x + 81$

89. Multiplying: $(2x+3)^2 = (2x)^2 + 2(2x)(3) + (3)^2 = 4x^2 + 12x + 9$

91. Multiplying: $(4x-2y)^2 = (4x)^2 - 2(4x)(2y) + (2y)^2 = 16x^2 - 16xy + 4y^2$

93. Simplifying: $\left(6x^3 - 4x^2 + 2x\right) + \left(9x^2 - 6x + 3\right) = 6x^3 - 4x^2 + 2x + 9x^2 - 6x + 3 = 6x^3 + 5x^2 - 4x + 3$

© 2012 Cengage Learning. All Rights Reserved. May not be scanned, copied or duplicated, or posted to a publicly accessible website, in whole or in part.

95. Simplifying: $\left(-7x^4+4x^3-6x\right)+\left(8x^4+7x^3-9\right)=-7x^4+4x^3-6x+8x^4+7x^3-9=x^4+11x^3-6x-9$

97. Simplifying: $\left(2x^5+3x^3+4x\right)+\left(5x^3-6x-7\right)=2x^5+3x^3+4x+5x^3-6x-7=2x^5+8x^3-2x-7$

99. Simplifying: $\left(-8x^5-5x^4+7\right)+\left(7x^4+2x^2+5\right)=-8x^5-5x^4+7+7x^4+2x^2+5=-8x^5+2x^4+2x^2+12$

101. Simplifying: $\frac{24x^3y^7}{6x^{-2}y^4}+\frac{27x^{-2}y^{10}}{9x^{-7}y^7}=4x^5y^3+3x^5y^3=7x^5y^3$

103. Simplifying: $\frac{18a^5b^9}{3a^3b^6}-\frac{48a^{-3}b^{-1}}{16a^{-5}b^{-4}}=6a^2b^3-3a^2b^3=3a^2b^3$

6.4 The Difference of Two Squares

1. Factoring the binomial: $x^2-9=(x+3)(x-3)$

3. Factoring the binomial: $a^2-36=(a+6)(a-6)$

5. Factoring the binomial: $x^2-49=(x+7)(x-7)$

7. Factoring the binomial: $4a^2-16=4\left(a^2-4\right)=4(a+2)(a-2)$

9. The expression $9x^2+25$ cannot be factored.

11. Factoring the binomial: $25x^2-169=(5x+13)(5x-13)$

13. Factoring the binomial: $9a^2-16b^2=(3a+4b)(3a-4b)$

15. Factoring the binomial: $9-m^2=(3+m)(3-m)$

17. Factoring the binomial: $25-4x^2=(5+2x)(5-2x)$

19. Factoring the binomial: $2x^2-18=2\left(x^2-9\right)=2(x+3)(x-3)$

21. Factoring the binomial: $32a^2-128=32\left(a^2-4\right)=32(a+2)(a-2)$

23. Factoring the binomial: $8x^2y-18y=2y\left(4x^2-9\right)=2y(2x+3)(2x-3)$

25. Factoring the binomial: $a^4-b^4=\left(a^2+b^2\right)\left(a^2-b^2\right)=\left(a^2+b^2\right)(a+b)(a-b)$

27. Factoring the binomial: $16m^4-81=\left(4m^2+9\right)\left(4m^2-9\right)=\left(4m^2+9\right)(2m+3)(2m-3)$

29. Factoring the binomial: $3x^3y-75xy^3=3xy\left(x^2-25y^2\right)=3xy(x+5y)(x-5y)$

31. Factoring the trinomial: $x^2-2x+1=(x-1)(x-1)=(x-1)^2$

33. Factoring the trinomial: $x^2+2x+1=(x+1)(x+1)=(x+1)^2$

35. Factoring the trinomial: $a^2-10a+25=(a-5)(a-5)=(a-5)^2$

37. Factoring the trinomial: $y^2+4y+4=(y+2)(y+2)=(y+2)^2$

39. Factoring the trinomial: $x^2-4x+4=(x-2)(x-2)=(x-2)^2$

41. Factoring the trinomial: $m^2-12m+36=(m-6)(m-6)=(m-6)^2$

43. Factoring the trinomial: $4a^2+12a+9=(2a+3)(2a+3)=(2a+3)^2$

45. Factoring the trinomial: $49x^2-14x+1=(7x-1)(7x-1)=(7x-1)^2$

47. Factoring the trinomial: $9y^2-30y+25=(3y-5)(3y-5)=(3y-5)^2$

49. Factoring the trinomial: $x^2+10xy+25y^2=(x+5y)(x+5y)=(x+5y)^2$

51. Factoring the trinomial: $9a^2+6ab+b^2=(3a+b)(3a+b)=(3a+b)^2$

© 2012 Cengage Learning. All Rights Reserved. May not be scanned, copied or duplicated, or posted to a publicly accessible website, in whole or in part.

53. Factoring the trinomial: $y^2-3y+\frac{9}{4}=\left(y-\frac{3}{2}\right)\left(y-\frac{3}{2}\right)=\left(y-\frac{3}{2}\right)^2$

55. Factoring the trinomial: $a^2+a+\frac{1}{4}=\left(a+\frac{1}{2}\right)\left(a+\frac{1}{2}\right)=\left(a+\frac{1}{2}\right)^2$

57. Factoring the trinomial: $x^2-7x+\frac{49}{4}=\left(x-\frac{7}{2}\right)\left(x-\frac{7}{2}\right)=\left(x-\frac{7}{2}\right)^2$

59. Factoring the trinomial: $x^2-\frac{3}{4}x+\frac{9}{64}=\left(x-\frac{3}{8}\right)\left(x-\frac{3}{8}\right)=\left(x-\frac{3}{8}\right)^2$

61. Factoring the trinomial: $3a^2+18a+27=3\left(a^2+6a+9\right)=3(a+3)(a+3)=3(a+3)^2$

63. Factoring the trinomial: $2x^2+20xy+50y^2=2\left(x^2+10xy+25y^2\right)=2(x+5y)(x+5y)=2(x+5y)^2$

65. Factoring the trinomial: $x^3+4x^2+4x=x\left(x^2+4x+4\right)=x(x+2)^2$

67. Factoring the trinomial: $y^4-8y^3+16y^2=y^2\left(y^2-8y+16\right)=y^2(y-4)^2$

69. Factoring the trinomial: $5x^3+30x^2y+45xy^2=5x\left(x^2+6xy+9y^2\right)=5x(x+3y)(x+3y)=5x(x+3y)^2$

71. Factoring the trinomial: $12y^4-60y^3+75y^2=3y^2\left(4y^2-20y+25\right)=3y^2(2y-5)^2$

73. Factoring by grouping: $x^2+6x+9-y^2=(x+3)^2-y^2=(x+3+y)(x+3-y)$

75. Factoring by grouping: $x^2+2xy+y^2-9=(x+y)^2-9=(x+y+3)(x+y-3)$

77. Since $(x+7)^2=x^2+14x+49$, the value is $b=14$.

79. Since $(x+5)^2=x^2+10x+25$, the value is $c=25$.

81. a. Multiplying: $1^3=1$
b. Multiplying: $2^3=8$
c. Multiplying: $3^3=27$
d. Multiplying: $4^3=64$
e. Multiplying: $5^3=125$

83. a. Multiplying: $x\left(x^2-x+1\right)=x^3-x^2+x$
b. Multiplying: $1\left(x^2-x+1\right)=x^2-x+1$
c. Multiplying: $(x+1)\left(x^2-x+1\right)=x\left(x^2-x+1\right)+1\left(x^2-x+1\right)=x^3-x^2+x+x^2-x+1=x^3+1$

85. a. Multiplying: $x\left(x^2-2x+4\right)=x^3-2x^2+4x$
b. Multiplying: $2\left(x^2-2x+4\right)=2x^2-4x+8$
c. Multiplying: $(x+2)\left(x^2-2x+4\right)=x\left(x^2-2x+4\right)+2\left(x^2-2x+4\right)=x^3-2x^2+4x+2x^2-4x+8=x^3+8$

87. a. Multiplying: $x\left(x^2-3x+9\right)=x^3-3x^2+9x$
b. Multiplying: $3\left(x^2-3x+9\right)=3x^2-9x+27$
c. Multiplying:

$$(x+3)\left(x^2-3x+9\right)=x\left(x^2-3x+9\right)+3\left(x^2-3x+9\right)=x^3-3x^2+9x+3x^2-9x+27=x^3+27$$

89. a. Multiplying: $x\left(x^2-4x+16\right)=x^3-4x^2+16x$
b. Multiplying: $4\left(x^2-4x+16\right)=4x^2-16x+64$
c. Multiplying:

$$(x+4)\left(x^2-4x+16\right)=x\left(x^2-4x+16\right)+4\left(x^2-4x+16\right)=x^3-4x^2+16x+4x^2-16x+64=x^3+64$$

91. a. Multiplying: $x\left(x^2-5x+25\right)=x^3-5x^2+25x$
b. Multiplying: $5\left(x^2-5x+25\right)=5x^2-25x+125$
c. Multiplying:

$$(x+5)\left(x^2-5x+25\right)=x\left(x^2-5x+25\right)+5\left(x^2-5x+25\right)=x^3-5x^2+25x+5x^2-25x+125=x^3+125$$

© 2012 Cengage Learning. All Rights Reserved. May not be scanned, copied or duplicated, or posted to a publicly accessible website, in whole or in part.

93. Dividing: $\frac{\$6.05 \text{ million}}{32 \text{ million}} \approx \$0.19$ per message

95. Dividing: $\frac{24y^3-36y^2-18y}{6y}=\frac{24y^3}{6y}-\frac{36y^2}{6y}-\frac{18y}{6y}=4y^2-6y-3$

97. Dividing: $\frac{48x^7-36x^5+12x^2}{4x^2}=\frac{48x^7}{4x^2}-\frac{36x^5}{4x^2}+\frac{12x^2}{4x^2}=12x^5-9x^3+3$

99. Dividing: $\frac{18x^7+12x^6-6x^5}{-3x^4}=\frac{18x^7}{-3x^4}+\frac{12x^6}{-3x^4}-\frac{6x^5}{-3x^4}=-6x^3-4x^2+2x$

101. Dividing: $\frac{-42x^5+24x^4-66x^2}{6x^2}=\frac{-42x^5}{6x^2}+\frac{24x^4}{6x^2}-\frac{66x^2}{6x^2}=-7x^3+4x^2-11$

103. Using long division:

$$\begin{array}{r} x-2 \\ x-3\overline{)x^2-5x+8} \\ \underline{x^2-3x} \\ -2x+8 \\ \underline{-2x+6} \\ 2 \end{array}$$

The quotient is $x-2+\frac{2}{x-3}$.

105. Using long division:

$$\begin{array}{r} 3x-2 \\ 2x+3\overline{)6x^2+5x+3} \\ \underline{6x^2+9x} \\ -4x+3 \\ \underline{-4x-6} \\ 9 \end{array}$$

The quotient is $3x-2+\frac{9}{2x+3}$.

107. Factoring the trinomial: $t^2-\frac{2}{5}t+\frac{1}{25}=\left(t-\frac{1}{5}\right)\left(t-\frac{1}{5}\right)=\left(t-\frac{1}{5}\right)^2$

6.5 The Sum and Difference of Two Cubes

1. Factoring: $x^3-y^3=(x-y)(x^2+xy+y^2)$

3. Factoring: $a^3+8=(a+2)(a^2-2a+4)$

5. Factoring: $27+x^3=(3+x)(9-3x+x^2)$

7. Factoring: $y^3-1=(y-1)(y^2+y+1)$

9. Factoring: $y^3-64=(y-4)(y^2+4y+16)$

11. Factoring: $125h^3-t^3=(5h-t)(25h^2+5ht+t^2)$

13. Factoring: $x^3-216=(x-6)(x^2+6x+36)$

15. Factoring: $2y^3-54=2(y^3-27)=2(y-3)(y^2+3y+9)$

17. Factoring: $2a^3-128b^3=2(a^3-64b^3)=2(a-4b)(a^2+4ab+16b^2)$

19. Factoring: $2x^3+432y^3=2(x^3+216y^3)=2(x+6y)(x^2-6xy+36y^2)$

21. Factoring: $10a^3-640b^3=10(a^3-64b^3)=10(a-4b)(a^2+4ab+16b^2)$

23. Factoring: $10r^3-1250=10(r^3-125)=10(r-5)(r^2+5r+25)$

25. Factoring: $64+27a^3=(4+3a)(16-12a+9a^2)$

27. Factoring: $8x^3-27y^3=(2x-3y)(4x^2+6xy+9y^2)$

29. Factoring: $t^3+\frac{1}{27}=\left(t+\frac{1}{3}\right)\left(t^2-\frac{1}{3}t+\frac{1}{9}\right)$

31. Factoring: $27x^3-\frac{1}{27}=\left(3x-\frac{1}{3}\right)\left(9x^2+x+\frac{1}{9}\right)$

33. Factoring: $64a^3+125b^3=(4a+5b)(16a^2-20ab+25b^2)$

35. Factoring: $\frac{1}{8}x^3-\frac{1}{27}y^3=\left(\frac{1}{2}x-\frac{1}{3}y\right)\left(\frac{1}{4}x^2+\frac{1}{6}xy+\frac{1}{9}y^2\right)$

© 2012 Cengage Learning. All Rights Reserved. May not be scanned, copied or duplicated, or posted to a publicly accessible website, in whole or in part.

37. Factoring: $a^6 - b^6 = \left(a^3 + b^3\right)\left(a^3 - b^3\right) = (a+b)\left(a^2 - ab + b^2\right)(a-b)\left(a^2 + ab + b^2\right)$

39. Factoring: $64x^6 - y^6 = \left(8x^3 + y^3\right)\left(8x^3 - y^3\right) = (2x+y)\left(4x^2 - 2xy + y^2\right)(2x-y)\left(4x^2 + 2xy + y^2\right)$

41. Factoring: $x^6 - (5y)^6 = \left(x^3 + (5y)^3\right)\left(x^3 - (5y)^3\right) = (x+5y)\left(x^2 - 5xy + 25y^2\right)(x-5y)\left(x^2 + 5xy + 25y^2\right)$

43. Multiplying: $2x^3(x+2)(x-2) = 2x^3\left(x^2 - 4\right) = 2x^5 - 8x^3$

45. Multiplying: $3x^2(x-3)^2 = 3x^2\left(x^2 - 6x + 9\right) = 3x^4 - 18x^3 + 27x^2$

47. Multiplying: $y\left(y^2 + 25\right) = y^3 + 25y$

49. Multiplying: $(5a-2)(3a+1) = 15a^2 + 5a - 6a - 2 = 15a^2 - a - 2$

51. Multiplying: $4x^2(x-5)(x+2) = 4x^2\left(x^2 - 3x - 10\right) = 4x^4 - 12x^3 - 40x^2$

53. Multiplying: $2ab^3\left(b^2 - 4b + 1\right) = 2ab^5 - 8ab^4 + 2ab^3$

55. Finding the amount: $0.37(101.6) = 37.59$ quadrillion Btu

57. Solving for x:

$$\begin{aligned} 2x - 6y &= 8 \\ 2x &= 6y + 8 \\ x &= 3y + 4 \end{aligned}$$

59. Solving for x:

$$\begin{aligned} 4x - 6y &= 8 \\ 4x &= 6y + 8 \\ x &= \frac{3}{2}y + 2 \end{aligned}$$

61. Solving for y:

$$\begin{aligned} 3x - 6y &= -18 \\ -6y &= -3x - 18 \\ y &= \frac{1}{2}x + 3 \end{aligned}$$

63. Solving for y:

$$\begin{aligned} 4x - 6y &= 24 \\ -6y &= -4x + 24 \\ y &= \frac{2}{3}x - 4 \end{aligned}$$

6.6 Factoring: A General Review

1. Factoring the polynomial: $x^2 - 81 = (x+9)(x-9)$
3. Factoring the polynomial: $x^2 + 2x - 15 = (x+5)(x-3)$
5. Factoring the polynomial: $x^2 + 6x + 9 = (x+3)(x+3) = (x+3)^2$
7. Factoring the polynomial: $y^2 - 10y + 25 = (y-5)(y-5) = (y-5)^2$
9. Factoring the polynomial: $2a^3b + 6a^2b + 2ab = 2ab\left(a^2 + 3a + 1\right)$
11. The polynomial $x^2 + x + 1$ cannot be factored.
13. Factoring the polynomial: $12a^2 - 75 = 3\left(4a^2 - 25\right) = 3(2a+5)(2a-5)$
15. Factoring the polynomial: $9x^2 - 12xy + 4y^2 = (3x-2y)(3x-2y) = (3x-2y)^2$
17. Factoring the polynomial: $4x^3 + 16xy^2 = 4x\left(x^2 + 4y^2\right)$
19. Factoring the polynomial: $2y^3 + 20y^2 + 50y = 2y\left(y^2 + 10y + 25\right) = 2y(y+5)(y+5) = 2y(y+5)^2$
21. Factoring the polynomial: $a^6 + 4a^4b^2 = a^4\left(a^2 + 4b^2\right)$
23. Factoring the polynomial: $xy + 3x + 4y + 12 = x(y+3) + 4(y+3) = (y+3)(x+4)$
25. Factoring the polynomial: $x^3 - 27 = (x-3)\left(x^2 + 3x + 9\right)$
27. Factoring the polynomial: $xy - 5x + 2y - 10 = x(y-5) + 2(y-5) = (y-5)(x+2)$

© 2012 Cengage Learning. All Rights Reserved. May not be scanned, copied or duplicated, or posted to a publicly accessible website, in whole or in part.

29. Factoring the polynomial: $5a^2+10ab+5b^2=5(a^2+2ab+b^2)=5(a+b)(a+b)=5(a+b)^2$

31. The polynomial x^2+49 cannot be factored.

33. Factoring the polynomial: $3x^2+15xy+18y^2=3(x^2+5xy+6y^2)=3(x+2y)(x+3y)$

35. Factoring the polynomial: $2x^2+15x-38=(2x+19)(x-2)$

37. Factoring the polynomial: $100x^2-300x+200=100(x^2-3x+2)=100(x-2)(x-1)$

39. Factoring the polynomial: $x^2-64=(x+8)(x-8)$

41. Factoring the polynomial: $x^2+3x+ax+3a=x(x+3)+a(x+3)=(x+3)(x+a)$

43. Factoring the polynomial: $49a^7-9a^5=a^5(49a^2-9)=a^5(7a+3)(7a-3)$

45. The polynomial $49x^2+9y^2$ cannot be factored.

47. Factoring the polynomial: $25a^3+20a^2+3a=a(25a^2+20a+3)=a(5a+3)(5a+1)$

49. Factoring the polynomial: $xa-xb+ay-by=x(a-b)+y(a-b)=(a-b)(x+y)$

51. Factoring the polynomial: $48a^4b-3a^2b=3a^2b(16a^2-1)=3a^2b(4a+1)(4a-1)$

53. Factoring the polynomial: $20x^4-45x^2=5x^2(4x^2-9)=5x^2(2x+3)(2x-3)$

55. Factoring the polynomial: $3x^2+35xy-82y^2=(3x+41y)(x-2y)$

57. Factoring the polynomial: $16x^5-44x^4+30x^3=2x^3(8x^2-22x+15)=2x^3(2x-3)(4x-5)$

59. Factoring the polynomial: $2x^2+2ax+3x+3a=2x(x+a)+3(x+a)=(x+a)(2x+3)$

61. Factoring the polynomial: $y^4-1=(y^2+1)(y^2-1)=(y^2+1)(y+1)(y-1)$

63. Factoring the polynomial:

$$12x^4y^2+36x^3y^3+27x^2y^4=3x^2y^2(4x^2+12xy+9y^2)=3x^2y^2(2x+3y)(2x+3y)=3x^2y^2(2x+3y)^2$$

65. Factoring the polynomial: $16t^2-64t+48=16(t^2-4t+3)=16(t-1)(t-3)$

67. Factoring the polynomial: $54x^2+111x+56=(6x+7)(9x+8)$

69. Solving the equation:

$$\begin{aligned}3x-6&=9\\3x&=15\\x&=5\end{aligned}$$

71. Solving the equation:

$$\begin{aligned}2x+3&=0\\2x&=-3\\x&=-\frac{3}{2}\end{aligned}$$

73. Solving the equation:

$$\begin{aligned}4x+3&=0\\4x&=-3\\x&=-\frac{3}{4}\end{aligned}$$

75. Solving the equation:

$$\begin{aligned}-2(x+4)&=-10\\-2x-8&=-10\\-2x&=-2\\x&=1\end{aligned}$$

77. Solving the equation:

$$\begin{aligned}\frac{3}{5}x+4&=22\\\frac{3}{5}x&=18\\x&=\frac{5}{3}(18)=30\end{aligned}$$

79. Solving the equation:

$$\begin{aligned}6x-4(9-x)&=-96\\6x-36+4x&=-96\\10x-36&=-96\\10x&=-60\\x&=-6\end{aligned}$$

© 2012 Cengage Learning. All Rights Reserved. May not be scanned, copied or duplicated, or posted to a publicly accessible website, in whole or in part.

81. Solving the equation:

$$\begin{aligned} 2x-3(4x-7)&=-3x \\ 2x-12x+21&=-3x \\ -10x+21&=-3x \\ 21&=7x \\ x&=3 \end{aligned}$$

83. Solving the equation:

$$\begin{aligned} \frac{1}{2}x-\frac{5}{12}&=\frac{1}{12}x+\frac{5}{12} \\ 12\left(\frac{1}{2}x-\frac{5}{12}\right)&=12\left(\frac{1}{12}x+\frac{5}{12}\right) \\ 6x-5&=x+5 \\ 5x-5&=5 \\ 5x&=10 \\ x&=2 \end{aligned}$$

6.7 Solving Quadratic Equations by Factoring

1. Setting each factor equal to 0:

$x+2=0$, $x=-2$ $\qquad$ $x-1=0$, $x=1$

The solutions are –2 and 1.

3. Setting each factor equal to 0:

$a-4=0$, $a=4$ $\qquad$ $a-5=0$, $a=5$

The solutions are 4 and 5.

5. Setting each factor equal to 0:

$x=0$ $\qquad$ $x+1=0$, $x=-1$ $\qquad$ $x-3=0$, $x=3$

The solutions are 0, –1 and 3.

7. Setting each factor equal to 0:

$$3x+2=0 \quad 3x=-2 \quad x=-\frac{2}{3}$$

$$2x+3=0 \quad 2x=-3 \quad x=-\frac{3}{2}$$

The solutions are $-\frac{2}{3}$ and $-\frac{3}{2}$.

9. Setting each factor equal to 0:

$m=0$

$$3m+4=0 \quad 3m=-4 \quad m=-\frac{4}{3}$$

$$3m-4=0 \quad 3m=4 \quad m=\frac{4}{3}$$

The solutions are 0, $-\frac{4}{3}$ and $\frac{4}{3}$.

11. Setting each factor equal to 0:

$2y=0$, $y=0$

$$3y+1=0 \quad 3y=-1 \quad y=-\frac{1}{3}$$

$$5y+3=0 \quad 5y=-3 \quad y=-\frac{3}{5}$$

The solutions are 0, $-\frac{1}{3}$ and $-\frac{3}{5}$.

13. Solving by factoring:

$$\begin{aligned} x^2+3x+2&=0 \\ (x+2)(x+1)&=0 \\ x&=-2,-1 \end{aligned}$$

15. Solving by factoring:

$$\begin{aligned} x^2-9x+20&=0 \\ (x-4)(x-5)&=0 \\ x&=4,5 \end{aligned}$$

© 2012 Cengage Learning. All Rights Reserved. May not be scanned, copied or duplicated, or posted to a publicly accessible website, in whole or in part.

17. Solving by factoring:

$$a^2-2a-24=0$$
$$(a-6)(a+4)=0$$
$$a=6,-4$$

19. Solving by factoring:

$$100x^2-500x+600=0$$
$$100\left(x^2-5x+6\right)=0$$
$$100(x-2)(x-3)=0$$
$$x=2,3$$

21. Solving by factoring:

$$x^2=-6x-9$$
$$x^2+6x+9=0$$
$$(x+3)^2=0$$
$$x+3=0$$
$$x=-3$$

23. Solving by factoring:

$$a^2-16=0$$
$$(a+4)(a-4)=0$$
$$a=-4,4$$

25. Solving by factoring:

$$2x^2+5x-12=0$$
$$(2x-3)(x+4)=0$$
$$x=\frac{3}{2},-4$$

27. Solving by factoring:

$$9x^2+12x+4=0$$
$$(3x+2)^2=0$$
$$3x+2=0$$
$$x=-\frac{2}{3}$$

29. Solving by factoring:

$$a^2+25=10a$$
$$a^2-10a+25=0$$
$$(a-5)^2=0$$
$$a-5=0$$
$$a=5$$

31. Solving by factoring:

$$2x^2=3x+20$$
$$2x^2-3x-20=0$$
$$(2x+5)(x-4)=0$$
$$x=-\frac{5}{2},4$$

33. Solving by factoring:

$$3m^2=20-7m$$
$$3m^2+7m-20=0$$
$$(3m-5)(m+4)=0$$
$$m=\frac{5}{3},-4$$

35. Solving by factoring:

$$4x^2-49=0$$
$$(2x+7)(2x-7)=0$$
$$x=-\frac{7}{2},\frac{7}{2}$$

37. Solving by factoring:

$$x^2+6x=0$$
$$x(x+6)=0$$
$$x=0,-6$$

39. Solving by factoring:

$$x^2-3x=0$$
$$x(x-3)=0$$
$$x=0,3$$

41. Solving by factoring:

$$2x^2=8x$$
$$2x^2-8x=0$$
$$2x(x-4)=0$$
$$x=0,4$$

43. Solving by factoring:

$$3x^2=15x$$
$$3x^2-15x=0$$
$$3x(x-5)=0$$
$$x=0,5$$

45. Solving by factoring:

$$1400=400+700x-100x^2$$
$$100x^2-700x+1000=0$$
$$100\left(x^2-7x+10\right)=0$$
$$100(x-5)(x-2)=0$$
$$x=2,5$$

47. Solving by factoring:

$$6x^2=-5x+4$$
$$6x^2+5x-4=0$$
$$(3x+4)(2x-1)=0$$
$$x=-\frac{4}{3},\frac{1}{2}$$

© 2012 Cengage Learning. All Rights Reserved. May not be scanned, copied or duplicated, or posted to a publicly accessible website, in whole or in part.

49. Solving by factoring:

$$\begin{aligned} x(2x-3) &= 20 \\ 2x^2-3x &= 20 \\ 2x^2-3x-20 &= 0 \\ (2x+5)(x-4) &= 0 \\ x &= -\frac{5}{2},4 \end{aligned}$$

51. Solving by factoring:

$$\begin{aligned} t(t+2) &= 80 \\ t^2+2t &= 80 \\ t^2+2t-80 &= 0 \\ (t+10)(t-8) &= 0 \\ t &= -10,8 \end{aligned}$$

53. Solving by factoring:

$$\begin{aligned} 4000 &= (1300-100p)p \\ 4000 &= 1300p-100p^2 \\ 100p^2-1300p+4000 &= 0 \\ 100\left(p^2-13p+40\right) &= 0 \\ 100(p-8)(p-5) &= 0 \\ p &= 5,8 \end{aligned}$$

55. Solving by factoring:

$$\begin{aligned} x(14-x) &= 48 \\ 14x-x^2 &= 48 \\ -x^2+14x-48 &= 0 \\ x^2-14x+48 &= 0 \\ (x-6)(x-8) &= 0 \\ x &= 6,8 \end{aligned}$$

57. Solving by factoring:

$$\begin{aligned} (x+5)^2 &= 2x+9 \\ x^2+10x+25 &= 2x+9 \\ x^2+8x+16 &= 0 \\ (x+4)^2 &= 0 \\ x+4 &= 0 \\ x &= -4 \end{aligned}$$

59. Solving by factoring:

$$\begin{aligned} (y-6)^2 &= y-4 \\ y^2-12y+36 &= y-4 \\ y^2-13y+40 &= 0 \\ (y-5)(y-8) &= 0 \\ y &= 5,8 \end{aligned}$$

61. Solving by factoring:

$$\begin{aligned} 10^2 &= (x+2)^2+x^2 \\ 100 &= x^2+4x+4+x^2 \\ 100 &= 2x^2+4x+4 \\ 0 &= 2x^2+4x-96 \\ 0 &= 2\left(x^2+2x-48\right) \\ 0 &= 2(x+8)(x-6) \\ x &= -8,6 \end{aligned}$$

63. Solving by factoring:

$$\begin{aligned} 2x^3+11x^2+12x &= 0 \\ x\left(2x^2+11x+12\right) &= 0 \\ x(2x+3)(x+4) &= 0 \\ x &= 0,-\frac{3}{2},-4 \end{aligned}$$

65. Solving by factoring:

$$\begin{aligned} 4y^3-2y^2-30y &= 0 \\ 2y\left(2y^2-y-15\right) &= 0 \\ 2y(2y+5)(y-3) &= 0 \\ y &= 0,-\frac{5}{2},3 \end{aligned}$$

67. Solving by factoring:

$$\begin{aligned} 8x^3+16x^2 &= 10x \\ 8x^3+16x^2-10x &= 0 \\ 2x\left(4x^2+8x-5\right) &= 0 \\ 2x(2x-1)(2x+5) &= 0 \\ x &= 0,\frac{1}{2},-\frac{5}{2} \end{aligned}$$

69. Solving by factoring:

$$\begin{aligned} 20a^3 &= -18a^2+18a \\ 20a^3+18a^2-18a &= 0 \\ 2a\left(10a^2+9a-9\right) &= 0 \\ 2a(5a-3)(2a+3) &= 0 \\ a &= 0,\frac{3}{5},-\frac{3}{2} \end{aligned}$$

71. Solving by factoring:

$$\begin{aligned} 16t^2-32t+12 &= 0 \\ 4\left(4t^2-8t+3\right) &= 0 \\ 4(2t-1)(2t-3) &= 0 \\ t &= \frac{1}{2},\frac{3}{2} \end{aligned}$$

© 2012 Cengage Learning. All Rights Reserved. May not be scanned, copied or duplicated, or posted to a publicly accessible website, in whole or in part.

73. Solving the equation:

$$(a-5)(a+4)=-2a$$
$$a^2-a-20=-2a$$
$$a^2+a-20=0$$
$$(a+5)(a-4)=0$$
$$a=-5,4$$

75. Solving the equation:

$$3x(x+1)-2x(x-5)=-42$$
$$3x^2+3x-2x^2+10x=-42$$
$$x^2+13x+42=0$$
$$(x+7)(x+6)=0$$
$$x=-7,-6$$

77. Solving the equation:

$$2x(x+3)=x(x+2)-3$$
$$2x^2+6x=x^2+2x-3$$
$$x^2+4x+3=0$$
$$(x+3)(x+1)=0$$
$$x=-3,-1$$

79. Solving the equation:

$$a(a-3)+6=2a$$
$$a^2-3a+6=2a$$
$$a^2-5a+6=0$$
$$(a-2)(a-3)=0$$
$$a=2,3$$

81. Solving the equation:

$$15(x+20)+15x=2x(x+20)$$
$$15x+300+15x=2x^2+40x$$
$$30x+300=2x^2+40x$$
$$0=2x^2+10x-300$$
$$0=2(x^2+5x-150)$$
$$0=2(x+15)(x-10)$$
$$x=-15,10$$

83. Solving the equation:

$$15=a(a+2)$$
$$15=a^2+2a$$
$$0=a^2+2a-15$$
$$0=(a+5)(a-3)$$
$$a=-5,3$$

85. Solving by factoring:

$$x^3+3x^2-4x-12=0$$
$$x^2(x+3)-4(x+3)=0$$
$$(x+3)(x^2-4)=0$$
$$(x+3)(x+2)(x-2)=0$$
$$x=-3,-2,2$$

87. Solving by factoring:

$$x^3+x^2-16x-16=0$$
$$x^2(x+1)-16(x+1)=0$$
$$(x+1)(x^2-16)=0$$
$$(x+1)(x+4)(x-4)=0$$
$$x=-1,-4,4$$

89. Writing the equation:

$$(x-3)(x-5)=0$$
$$x^2-8x+15=0$$

91. **a.** Writing the equation:

$$(x-3)(x-2)=0$$
$$x^2-5x+6=0$$

b. Writing the equation:

$$(x-1)(x-6)=0$$
$$x^2-7x+6=0$$

c. Writing the equation:

$$(x-3)(x+2)=0$$
$$x^2-x-6=0$$

93. Let x and $x+1$ represent the two consecutive integers. The equation is: $x(x+1)=72$

95. Let x and $x+2$ represent the two consecutive odd integers. The equation is: $x(x+2)=99$

97. Let x and $x+2$ represent the two consecutive even integers. The equation is: $x(x+2)=5[x+(x+2)]-10$

99. Let x represent the cost of the suit and $5x$ represent the cost of the bicycle. The equation is:

$$x+5x=90$$
$$6x=90$$
$$x=15$$
$$5x=75$$

The suit costs \$15 and the bicycle costs \$75.

© 2012 Cengage Learning. All Rights Reserved. May not be scanned, copied or duplicated, or posted to a publicly accessible website, in whole or in part.

101. Let x represent the cost of the lot and $4x$ represent the cost of the house. The equation is:

$$\begin{aligned} x+4x&=3000\\ 5x&=3000\\ x&=600\\ 4x&=2400 \end{aligned}$$

The lot cost \$600 and the house cost \$2,400.

103. Simplifying using the properties of exponents: $2^{-3}=\frac{1}{2^3}=\frac{1}{8}$

105. Simplifying using the properties of exponents: $\frac{x^5}{x^{-3}}=x^{5-(-3)}=x^{5+3}=x^8$

107. Simplifying using the properties of exponents: $\frac{\left(x^2\right)^3}{\left(x^{-3}\right)^4}=\frac{x^6}{x^{-12}}=x^{6-(-12)}=x^{6+12}=x^{18}$

109. Writing in scientific notation: $0.0056=5.6\times10^{-3}$

111. Writing in scientific notation: $5,670,000,000=5.67\times10^9$

6.8 Applications of Quadratic Equations

1. Let x and $x + 2$ represent the two integers. The equation is:

$$\begin{aligned} x(x+2)&=80\\ x^2+2x&=80\\ x^2+2x-80&=0\\ (x+10)(x-8)&=0\\ x&=-10,8\\ x+2&=-8,10 \end{aligned}$$

The two numbers are either –10 and –8, or 8 and 10.

3. Let x and $x + 2$ represent the two integers. The equation is:

$$\begin{aligned} x(x+2)&=99\\ x^2+2x&=99\\ x^2+2x-99&=0\\ (x+11)(x-9)&=0\\ x&=-11,9\\ x+2&=-9,11 \end{aligned}$$

The two numbers are either –11 and –9, or 9 and 11.

5. Let x and $x + 2$ represent the two integers. The equation is:

$$\begin{aligned} x(x+2)&=5(x+x+2)-10\\ x^2+2x&=5(2x+2)-10\\ x^2+2x&=10x+10-10\\ x^2+2x&=10x\\ x^2-8x&=0\\ x(x-8)&=0\\ x&=0,8\\ x+2&=2,10 \end{aligned}$$

The two numbers are either 0 and 2, or 8 and 10.

© 2012 Cengage Learning. All Rights Reserved. May not be scanned, copied or duplicated, or posted to a publicly accessible website, in whole or in part.

7. Let x and $14 - x$ represent the two numbers. The equation is:

$$\begin{aligned} x(14-x) &= 48 \\ 14x - x^2 &= 48 \\ 0 &= x^2 - 14x + 48 \\ 0 &= (x-8)(x-6) \\ x &= 8, 6 \\ 14 - x &= 6, 8 \end{aligned}$$

The two numbers are 6 and 8.

9. Let x and $5x + 2$ represent the two numbers. The equation is:

$$\begin{aligned} x(5x+2) &= 24 \\ 5x^2 + 2x &= 24 \\ 5x^2 + 2x - 24 &= 0 \\ (5x+12)(x-2) &= 0 \\ x &= -\frac{12}{5}, 2 \\ 5x + 2 &= -10, 12 \end{aligned}$$

The two numbers are either $-\frac{12}{5}$ and –10, or 2 and 12.

11. Let x and $4x$ represent the two numbers. The equation is:

$$\begin{aligned} x(4x) &= 4(x+4x) \\ 4x^2 &= 4(5x) \\ 4x^2 &= 20x \\ 4x^2 - 20x &= 0 \\ 4x(x-5) &= 0 \\ x &= 0, 5 \\ 4x &= 0, 20 \end{aligned}$$

The two numbers are either 0 and 0, or 5 and 20.

13. Let w represent the width and $w + 1$ represent the length. The equation is:

$$\begin{aligned} w(w+1) &= 12 \\ w^2 + w &= 12 \\ w^2 + w - 12 &= 0 \\ (w+4)(w-3) &= 0 \\ w &= 3 \quad (w = -4 \text{ is impossible}) \\ w + 1 &= 4 \end{aligned}$$

The width is 3 inches and the length is 4 inches.

15. Let b represent the base and $2b$ represent the height. The equation is:

$$\begin{aligned} \frac{1}{2}b(2b) &= 9 \\ b^2 &= 9 \\ b^2 - 9 &= 0 \\ (b+3)(b-3) &= 0 \\ b &= 3 \quad (b = -3 \text{ is impossible}) \end{aligned}$$

The base is 3 inches.

© 2012 Cengage Learning. All Rights Reserved. May not be scanned, copied or duplicated, or posted to a publicly accessible website, in whole or in part.

17. Let x and $x + 2$ represent the two legs. The equation is:

$$\begin{aligned} x^2+(x+2)^2 &= 10^2 \\ x^2+x^2+4x+4 &= 100 \\ 2x^2+4x+4 &= 100 \\ 2x^2+4x-96 &= 0 \\ 2(x^2+2x-48) &= 0 \\ 2(x+8)(x-6) &= 0 \\ x &= 6 \quad (x=-8 \text{ is impossible}) \\ x+2 &= 8 \end{aligned}$$

The legs are 6 inches and 8 inches.

19. Let x represent the longer leg and $x + 1$ represent the hypotenuse. The equation is:

$$\begin{aligned} 5^2+x^2 &= (x+1)^2 \\ 25+x^2 &= x^2+2x+1 \\ 25 &= 2x+1 \\ 24 &= 2x \\ x &= 12 \end{aligned}$$

The longer leg is 12 meters.

21. Setting $C = \$1,400$:

$$\begin{aligned} 1400 &= 400+700x-100x^2 \\ 100x^2-700x+1000 &= 0 \\ 100(x^2-7x+10) &= 0 \\ 100(x-5)(x-2) &= 0 \\ x &= 2,5 \end{aligned}$$

The company can manufacture either 200 items or 500 items.

23. Setting $C = \$2,200$:

$$\begin{aligned} 2200 &= 600+1000x-100x^2 \\ 100x^2-1000x+1600 &= 0 \\ 100(x^2-10x+16) &= 0 \\ 100(x-2)(x-8) &= 0 \\ x &= 2,8 \end{aligned}$$

The company can manufacture either 200 DVD's or 800 DVD's.

25. The revenue is given by: $R = xp = (1200-100p)p$. Setting $R = \$3,200$:

$$\begin{aligned} 3200 &= (1200-100p)p \\ 3200 &= 1200p-100p^2 \\ 100p^2-1200p+3200 &= 0 \\ 100(p^2-12p+32) &= 0 \\ 100(p-4)(p-8) &= 0 \\ p &= 4,8 \end{aligned}$$

The company should sell the ribbons for either \$4 or \$8.

© 2012 Cengage Learning. All Rights Reserved. May not be scanned, copied or duplicated, or posted to a publicly accessible website, in whole or in part.

27. The revenue is given by: $R = xp = (1700 - 100p)p$. Setting $R = \$7,000$:

$$\begin{aligned} 7000 &= (1700 - 100p)p \\ 7000 &= 1700p - 100p^2 \\ 100p^2 - 1700p + 7000 &= 0 \\ 100(p^2 - 17p + 70) &= 0 \\ 100(p-7)(p-10) &= 0 \\ p &= 7, 10 \end{aligned}$$

The calculators should be sold for either \$7 or \$10.

29. **a.** Let x represent the distance from the base to the wall, and $2x + 2$ represent the height on the wall. Using the Pythagorean theorem:

$$\begin{aligned} x^2 + (2x+2)^2 &= 13^2 \\ x^2 + 4x^2 + 8x + 4 &= 169 \\ 5x^2 + 8x - 165 &= 0 \\ (5x+33)(x-5) &= 0 \\ x &= 5 \quad \left(x = -\frac{33}{5} \text{ is impossible}\right) \end{aligned}$$

The base of the ladder is 5 feet from the wall.

b. Since $2x + 2 = 2 \bullet 5 + 2 = 12$, the ladder reaches a height of 12 feet.

31. **a.** Finding when $h = 0$:

$$\begin{aligned} 0 &= -16t^2 + 396t + 100 \\ 16t^2 - 396t - 100 &= 0 \\ 4t^2 - 99t - 25 &= 0 \\ (4t+1)(t-25) &= 0 \\ t &= 25 \quad \left(t = -\frac{1}{4} \text{ is impossible}\right) \end{aligned}$$

The bullet will land on the ground after 25 seconds.

b. Completing the table:

t (seconds)	h (feet)
0	100
5	1,680
10	2,460
15	2,440
20	1,620
25	0

33. **a.** Setting $h = 1{,}550$:

$$\begin{aligned} 1{,}550 &= 1{,}614 - 16t^2 \\ 16t^2 &= 64 \\ t^2 &= 4 \\ t^2 - 4 &= 0 \\ (t+2)(t-2) &= 0 \\ t &= 2 \quad (t = -2 \text{ is impossible}) \end{aligned}$$

The object is 1,550 feet above the ground after 2 seconds

b. Substituting $t = 4$: $h = 1{,}614 - 16(4)^2 = 1{,}614 - 256 = 1{,}358$. Since the object has fallen to a height of 1,358 feet, it has fallen 256 feet after 4 seconds.

35. Simplifying the expression: $(5x^3)^2(2x^6)^3 = 25x^6 \bullet 8x^{18} = 200x^{24}$

© 2012 Cengage Learning. All Rights Reserved. May not be scanned, copied or duplicated, or posted to a publicly accessible website, in whole or in part.

37. Simplifying the expression: $\frac{x^4}{x^{-3}} = x^{4-(-3)} = x^{4+3} = x^7$

39. Simplifying the expression: $(2\times10^{-4})(4\times10^5) = 8\times10^1 = 80$

41. Simplifying the expression: $20ab^2 - 16ab^2 + 6ab^2 = 10ab^2$

43. Multiplying using the distributive property:

$$2x^2(3x^2+3x-1) = 2x^2(3x^2) + 2x^2(3x) - 2x^2(1) = 6x^4 + 6x^3 - 2x^2$$

45. Multiplying using the square of binomial formula: $(3y-5)^2 = (3y)^2 - 2(3y)(5) + (5)^2 = 9y^2 - 30y + 25$

47. Multiplying using the difference of squares formula: $(2a^2+7)(2a^2-7) = (2a^2)^2 - (7)^2 = 4a^4 - 49$

Chapter 6 Review

1. Factoring the polynomial: $10x - 20 = 10(x-2)$

3. Factoring the polynomial: $5x - 5y = 5(x-y)$

5. Factoring the polynomial: $49a^3 - 14b^3 = 7(7a^3 - 2b^3)$

7. Factoring the polynomial: $xy + bx + ay + ab = x(y+b) + a(y+b) = (y+b)(x+a)$

9. Factoring the polynomial: $2xy + 10x - 3y - 15 = 2x(y+5) - 3(y+5) = (y+5)(2x-3)$

11. Factoring the polynomial: $y^2 + 9y + 14 = (y+7)(y+2)$

13. Factoring the polynomial: $y^2 + 20y + 99 = (y+9)(y+11)$

15. Factoring the polynomial: $2x^2 + 13x + 15 = (2x+3)(x+5)$

17. Factoring the polynomial: $6r^2 + 5rt - 6t^2 = (3r-2t)(2r+3t)$

19. Factoring the polynomial: $n^2 - 81 = (n+9)(n-9)$

21. The expression $x^2 + 49$ cannot be factored.

23. Factoring the polynomial: $64t^2 + 16t + 1 = (8t+1)(8t+1) = (8t+1)^2$

25. Factoring the polynomial: $4r^2 - 12rt + 9t^2 = (2r-3t)(2r-3t) = (2r-3t)^2$

27. Factoring the polynomial: $2x^2 + 20x + 48 = 2(x^2 + 10x + 24) = 2(x+4)(x+6)$

29. Factoring the polynomial: $3m^3 - 18m^2 - 21m = 3m(m^2 - 6m - 7) = 3m(m-7)(m+1)$

31. Factoring the polynomial: $8x^2 + 16x + 6 = 2(4x^2 + 8x + 3) = 2(2x+1)(2x+3)$

33. Factoring the polynomial: $20m^3 - 34m^2 + 6m = 2m(10m^2 - 17m + 3) = 2m(5m-1)(2m-3)$

35. Factoring the polynomial: $4x^2 + 40x + 100 = 4(x^2 + 10x + 25) = 4(x+5)(x+5) = 4(x+5)^2$

37. Factoring the polynomial: $5x^2 - 45 = 5(x^2 - 9) = 5(x+3)(x-3)$

39. Factoring the polynomial: $27x^3 + 8y^3 = (3x+2y)(9x^2 - 6xy + 4y^2)$

41. Factoring the polynomial: $6a^3b + 33a^2b^2 + 15ab^3 = 3ab(2a^2 + 11ab + 5b^2) = 3ab(2a+b)(a+5b)$

43. Factoring the polynomial: $4y^6 + 9y^4 = y^4(4y^2 + 9)$

© 2012 Cengage Learning. All Rights Reserved. May not be scanned, copied or duplicated, or posted to a publicly accessible website, in whole or in part.

45. Setting each factor equal to 0:

$$x-5=0 \qquad\qquad x+2=0$$
$$x=5 \qquad\qquad x=-2$$

The solutions are –2, 5.

47. Solving the equation by factoring:

$$\begin{aligned}6y^2&=-13y-6\\6y^2+13y+6&=0\\(3y+2)(2y+3)&=0\\y&=-\frac{2}{3},-\frac{3}{2}\end{aligned}$$

49. Let x and $x+2$ represent the two integers. The equation is:

$$\begin{aligned}x(x+2)&=120\\x^2+2x&=120\\x^2+2x-120&=0\\(x+12)(x-10)&=0\\x&=-12,10\\x+2&=-10,12\end{aligned}$$

The two integers are either –12 and –10, or 10 and 12.

51. Let x and $20-x$ represent the two numbers. The equation is:

$$\begin{aligned}x(20-x)&=75\\20x-x^2&=75\\0&=x^2-20x+75\\0&=(x-15)(x-5)\\x&=15,5\\20-x&=5,15\end{aligned}$$

The two numbers are 5 and 15.

Chapter 6 Cumulative Review

1. Simplifying the expression: $9+(-7)+(-8)=9+(-15)=-6$

3. Simplifying the expression: $\frac{-63}{-7}=9$

5. Simplifying the expression: $(-4)^3=(-4)(-4)(-4)=-64$

7. Simplifying the expression: $\frac{-3(4-7)-5(7-2)}{-5-2-1}=\frac{-3(-3)-5(5)}{-8}=\frac{9-25}{-8}=\frac{-16}{-8}=2$

9. Simplifying the expression: $-a+3+6a-8=(-a+6a)+(3-8)=5a-5$

11. Simplifying using the rules of exponents: $\left(x^4\right)^{10}=x^{4\bullet 10}=x^{40}$

13. Multiplying using the FOIL method: $(5x-2)(3x+4)=15x^2-6x+20x-8=15x^2+14x-8$

15. Solving the equation:

$$\begin{aligned}3x&=-18\\\frac{1}{3}(3x)&=\frac{1}{3}(-18)\\x&=-6\end{aligned}$$

17. Solving the equation:

$$\begin{aligned}-\frac{x}{3}&=7\\-3\left(-\frac{x}{3}\right)&=-3(7)\\x&=-21\end{aligned}$$

© 2012 Cengage Learning. All Rights Reserved. May not be scanned, copied or duplicated, or posted to a publicly accessible website, in whole or in part.

19. Setting each factor equal to 0:

$$4m = 0 \qquad m - 7 = 0 \qquad 2m - 7 = 0$$
$$m = 0 \qquad m = 7 \qquad 2m = 7$$
$$m = \frac{7}{2}$$

The solutions are $0, 7,$ and $\frac{7}{2}$.

21. Solving the inequality:

$$-2x > -8$$
$$-\frac{1}{2}(-2x) < -\frac{1}{2}(-8)$$
$$x < 4$$

23. Graphing the line:

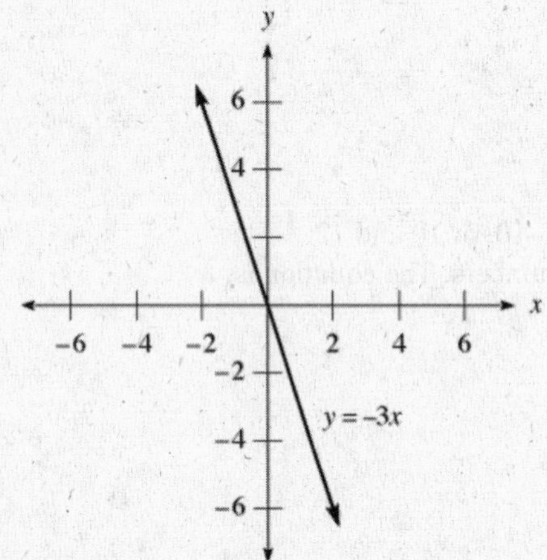

25. Checking the point $(0,0)$: $2(0) + 3(0) = 0 \geq 6$ (false)
Graphing the linear inequality:

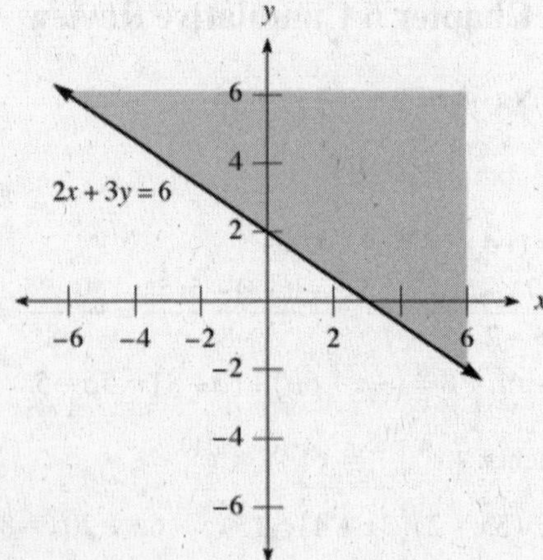

27. Substituting the points:

$(0,3)$: $3(0) - 4(3) = 0 - 12 = -12 \neq 12$

$(4,0)$: $3(4) - 4(0) = 12 - 0 = 12$

$\left(\frac{16}{3}, 1\right)$: $3\left(\frac{16}{3}\right) - 4(1) = 16 - 4 = 12$

The ordered pairs $(4,0)$ and $\left(\frac{16}{3}, 1\right)$ are solutions to the equation.

29. Finding the slope: $m = \frac{-4-3}{-2-7} = \frac{-7}{-9} = \frac{7}{9}$

© 2012 Cengage Learning. All Rights Reserved. May not be scanned, copied or duplicated, or posted to a publicly accessible website, in whole or in part.

31. The slope-intercept form is $y=-\frac{2}{5}x-\frac{2}{3}$.

33. Graphing both lines:

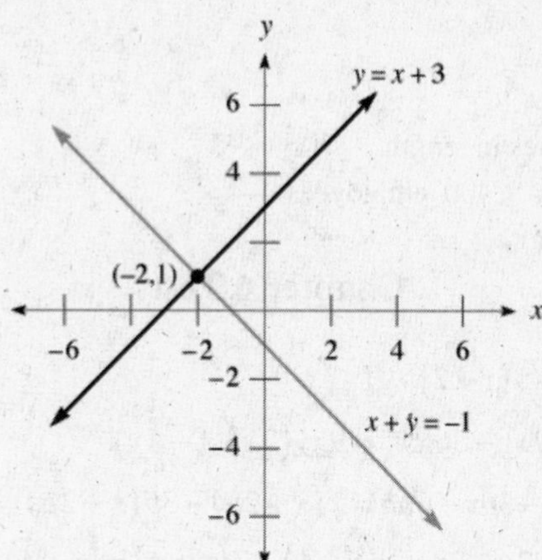

The intersection point is $(-2,1)$.

35. Adding the two equations:

$$2y=10$$
$$y=5$$

Substituting into the second equation:

$$x+5=7$$
$$x=2$$

The solution is $(2,5)$.

37. Substituting into the first equation:

$$2(y-1)+y=4$$
$$2y-2+y=4$$
$$3y-2=4$$
$$3y=6$$
$$y=2$$
$$x=2-1=1$$

The solution is $(1,2)$.

39. Factoring the polynomial: $n^2-5n-36=(n-9)(n+4)$

41. Factoring the polynomial: $16-a^2=(4+a)(4-a)$

43. Factoring the polynomial: $45x^2y-30xy^2+5y^3=5y(9x^2-6xy+y^2)=5y(3x-y)(3x-y)=5y(3x-y)^2$

45. Factoring the polynomial: $3xy+15x-2y-10=3x(y+5)-2(y+5)=(y+5)(3x-2)$

47. The opposite of –2 is 2, the reciprocal is $-\frac{1}{2}$, and the absolute value is 2.

49. Using long division:

$$\begin{array}{r} 4x+5 \\ x-3\overline{)4x^2-7x-13} \\ \underline{4x^2-12x} \\ 5x-13 \\ \underline{5x-15} \\ 2 \end{array}$$

The quotient is $4x+5+\frac{2}{x-3}$.

© 2012 Cengage Learning. All Rights Reserved. May not be scanned, copied or duplicated, or posted to a publicly accessible website, in whole or in part.

51. Let x and $x+4$ represent the length of each piece. The equation is:
$$\begin{aligned} x+x+4&=72\\ 2x+4&=72\\ 2x&=68\\ x&=34\\ x+4&=38 \end{aligned}$$
The pieces are 34 inches and 38 inches in length.

53. Finding the amount: $0.36(15{,}000)=5{,}400$ employees

Chapter 6 Test

1. Factoring the polynomial: $5x-10=5(x-2)$
2. Factoring the polynomial: $18x^2y-9xy-36xy^2=9xy(2x-1-4y)$
3. Factoring the polynomial: $x^2+2ax-3bx-6ab=x(x+2a)-3b(x+2a)=(x+2a)(x-3b)$
4. Factoring the polynomial: $xy+4x-7y-28=x(y+4)-7(y+4)=(y+4)(x-7)$
5. Factoring the polynomial: $x^2-5x+6=(x-2)(x-3)$
6. Factoring the polynomial: $x^2-x-6=(x-3)(x+2)$
7. Factoring the polynomial: $a^2-16=(a+4)(a-4)$
8. The polynomial x^2+25 cannot be factored.
9. Factoring the polynomial: $x^4-81=(x^2+9)(x^2-9)=(x^2+9)(x+3)(x-3)$
10. Factoring the polynomial: $27x^2-75y^2=3(9x^2-25y^2)=3(3x+5y)(3x-5y)$
11. Factoring the polynomial: $x^3+5x^2-9x-45=x^2(x+5)-9(x+5)=(x+5)(x^2-9)=(x+5)(x+3)(x-3)$
12. Factoring the polynomial: $x^2-bx+5x-5b=x(x-b)+5(x-b)=(x-b)(x+5)$
13. Factoring the polynomial: $4a^2+22a+10=2(2a^2+11a+5)=2(2a+1)(a+5)$
14. Factoring the polynomial: $3m^2-3m-18=3(m^2-m-6)=3(m-3)(m+2)$
15. Factoring the polynomial: $6y^2+7y-5=(3y+5)(2y-1)$
16. Factoring the polynomial: $12x^3-14x^2-10x=2x(6x^2-7x-5)=2x(3x-5)(2x+1)$
17. Factoring the polynomial: $a^3+64b^3=(a+4b)(a^2-4ab+16b^2)$
18. Factoring the polynomial: $54x^3-16=2(27x^3-8)=2(3x-2)(9x^2+6x+4)$
19. Solving the equation by factoring:
$$\begin{aligned} x^2+7x+12&=0\\ (x+4)(x+3)&=0\\ x&=-4,-3 \end{aligned}$$
20. Solving the equation by factoring:
$$\begin{aligned} x^2-4x+4&=0\\ (x-2)^2&=0\\ x-2&=0\\ x&=2 \end{aligned}$$
21. Solving the equation by factoring:
$$\begin{aligned} x^2-36&=0\\ (x+6)(x-6)&=0\\ x&=-6,6 \end{aligned}$$
22. Solving the equation by factoring:
$$\begin{aligned} x^2&=x+20\\ x^2-x-20&=0\\ (x+4)(x-5)&=0\\ x&=-4,5 \end{aligned}$$

© 2012 Cengage Learning. All Rights Reserved. May not be scanned, copied or duplicated, or posted to a publicly accessible website, in whole or in part.

23. Solving the equation by factoring:

$$\begin{aligned} x^2-11x&=-30\\ x^2-11x+30&=0\\ (x-6)(x-5)&=0\\ x&=5,6 \end{aligned}$$

24. Solving the equation by factoring:

$$\begin{aligned} y^3&=16y\\ y^3-16y&=0\\ y\left(y^2-16\right)&=0\\ y(y+4)(y-4)&=0\\ y&=0,-4,4 \end{aligned}$$

25. Solving the equation by factoring:

$$\begin{aligned} 2a^2&=a+15\\ 2a^2-a-15&=0\\ (2a+5)(a-3)&=0\\ a&=-\frac{5}{2},3 \end{aligned}$$

26. Solving the equation by factoring:

$$\begin{aligned} 30x^3-20x^2&=10x\\ 30x^3-20x^2-10x&=0\\ 10x\left(3x^2-2x-1\right)&=0\\ 10x(3x+1)(x-1)&=0\\ x&=0,-\frac{1}{3},1 \end{aligned}$$

27. Let x and $20-x$ represent the numbers. The equation is:

$$\begin{aligned} x(20-x)&=64\\ 20x-x^2&=64\\ 0&=x^2-20x+64\\ 0&=(x-16)(x-4)\\ x&=4,16\\ 20-x&=16,4 \end{aligned}$$

The numbers are 4 and 16.

28. Let x and $x+2$ represent the two integers. The equation is:

$$\begin{aligned} x(x+2)&=x+x+2+7\\ x^2+2x&=2x+9\\ x^2-9&=0\\ (x+3)(x-3)&=0\\ x&=-3,3\\ x+2&=-1,5 \end{aligned}$$

The integers are either –3 and –1, or 3 and 5.

29. Let w represent the width and $3w+5$ represent the length. The equation is:

$$\begin{aligned} w(3w+5)&=42\\ 3w^2+5w&=42\\ 3w^2+5w-42&=0\\ (3w+14)(w-3)&=0\\ w&=3 \qquad \left(w=-\frac{14}{3}\text{ is impossible}\right)\\ 3w+5&=14 \end{aligned}$$

The width is 3 feet and the length is 14 feet.

30. Let x and $2x+2$ represent the two legs. The equation is:

$$\begin{aligned} x^2+(2x+2)^2&=13^2\\ x^2+4x^2+8x+4&=169\\ 5x^2+8x-165&=0\\ (5x+33)(x-5)&=0\\ x&=5 \qquad \left(x=-\frac{33}{5}\text{ is impossible}\right)\\ 2x+2&=12 \end{aligned}$$

The two legs are 5 meters and 12 meters in length.

© 2012 Cengage Learning. All Rights Reserved. May not be scanned, copied or duplicated, or posted to a publicly accessible website, in whole or in part.

31. Setting $C = \$800$:

$$800 = 200 + 500x - 100x^2$$
$$100x^2 - 500x + 600 = 0$$
$$100\left(x^2 - 5x + 6\right) = 0$$
$$100(x-2)(x-3) = 0$$
$$x = 2, 3$$

The company can manufacture either 200 items or 300 items.

32. The revenue is given by: $R = xp = (900 - 100p)p$. Setting $R = \$1{,}800$:

$$1800 = (900 - 100p)p$$
$$1800 = 900p - 100p^2$$
$$100p^2 - 900p + 1800 = 0$$
$$100\left(p^2 - 9p + 18\right) = 0$$
$$100(p-6)(p-3) = 0$$
$$p = 3, 6$$

The manufacturer should sell the items at either \$3 or \$6.

33. Let x represent the distance from the ladder to the wall, and $2x - 1$ represent the height on the wall the ladder reaches. Using the Pythagorean theorem:

$$(x)^2 + (2x-1)^2 = 17^2$$
$$x^2 + 4x^2 - 4x + 1 = 289$$
$$5x^2 - 4x - 288 = 0$$
$$(5x+36)(x-8) = 0$$
$$x = 8 \quad \left(x = -\frac{36}{5} \text{ is impossible}\right)$$

a. The ladder is 8 feet from the wall.

b. The ladder reaches a height of 15 feet up the wall.

34. **a.** Setting $h = 40$:

$$40 = 24 + 40t - 16t^2$$
$$16t^2 - 40t + 16 = 0$$
$$2t^2 - 5t + 2 = 0$$
$$(2t-1)(t-2) = 0$$
$$t = \frac{1}{2}, 2$$

The ball will be 40 feet above the ground after $\frac{1}{2}$ or 2 seconds.

b. Setting $h = 0$:

$$0 = 24 + 40t - 16t^2$$
$$16t^2 - 40t - 24 = 0$$
$$2t^2 - 5t - 3 = 0$$
$$(2t+1)(t-3) = 0$$
$$t = 3 \quad \left(t = -\frac{1}{2} \text{ is impossible}\right)$$

The ball will hit the ground after 3 seconds.

© 2012 Cengage Learning. All Rights Reserved. May not be scanned, copied or duplicated, or posted to a publicly accessible website, in whole or in part.

Chapter 7
Rational Expressions

Getting Ready for Chapter 7

1. Simplifying: $\frac{2}{9} \bullet \frac{15}{22} = \frac{1}{3} \bullet \frac{5}{11} = \frac{5}{33}$

2. Simplifying: $\frac{3}{5} \div \frac{15}{7} = \frac{3}{5} \bullet \frac{7}{15} = \frac{1}{5} \bullet \frac{7}{5} = \frac{7}{25}$

3. Simplifying: $1 - \frac{5}{3} = 1 \bullet \frac{3}{3} - \frac{5}{3} = \frac{3}{3} - \frac{5}{3} = -\frac{2}{3}$

4. Simplifying: $1 + \frac{3}{5} = 1 \bullet \frac{5}{5} + \frac{3}{5} = \frac{5}{5} + \frac{3}{5} = \frac{8}{5}$

5. The expression $\frac{5}{0}$ is undefined.

6. Simplifying: $\frac{1}{10} + \frac{3}{14} = \frac{1}{10} \bullet \frac{7}{7} + \frac{3}{14} \bullet \frac{5}{5} = \frac{7}{70} + \frac{15}{70} = \frac{22}{70} = \frac{11}{35}$

7. Simplifying: $\frac{1}{21} + \frac{4}{15} = \frac{1}{21} \bullet \frac{5}{5} + \frac{4}{15} \bullet \frac{7}{7} = \frac{5}{105} + \frac{28}{105} = \frac{33}{105} = \frac{11}{35}$

8. Multiplying: $x(x+2) = x^2 + 2x$

9. Multiplying: $(x+3)(x-4) = x^2 + 3x - 4x - 12 = x^2 - x - 12$

10. Factoring: $x^2 - 25 = (x+5)(x-5)$

11. Factoring: $2x - 4 = 2(x-2)$

12. Factoring: $x^2 + 7x + 12 = (x+4)(x+3)$

13. Factoring: $a^2 - 4a = a(a-4)$

14. Factoring: $xy^2 + y = y(xy+1)$

15. Solving the equation:
$$\begin{aligned} 72 &= 2x \\ x &= 36 \end{aligned}$$

16. Solving the equation:
$$\begin{aligned} 15 &= 3x - 3 \\ 18 &= 3x \\ x &= 6 \end{aligned}$$

17. Solving the equation:
$$\begin{aligned} a^2 - a - 20 &= -2a \\ a^2 + a - 20 &= 0 \\ (a+5)(a-4) &= 0 \\ a &= -5, 4 \end{aligned}$$

18. Solving the equation:
$$\begin{aligned} x^2 + 2x &= 8 \\ x^2 + 2x - 8 &= 0 \\ (x+4)(x-2) &= 0 \\ x &= -4, 2 \end{aligned}$$

19. Substituting $x = 10$: $y = \frac{20}{10} = 2$

20. Substituting $y = 72$:
$$\begin{aligned} 72 &= 2x^2 \\ x^2 &= 36 \\ x^2 - 36 &= 0 \\ (x+6)(x-6) &= 0 \\ x &= -6, 6 \end{aligned}$$

© 2012 Cengage Learning. All Rights Reserved. May not be scanned, copied or duplicated, or posted to a publicly accessible website, in whole or in part.

7.1 Reducing Rational Expressions to Lowest Terms

1. a. Simplifying: $\frac{5+1}{25-1}=\frac{6}{24}=\frac{1}{4}$

b. Simplifying: $\frac{x+1}{x^2-1}=\frac{1(x+1)}{(x+1)(x-1)}=\frac{1}{x-1}$. The variable restriction is $x \neq -1, 1$.

c. Simplifying: $\frac{x^2-x}{x^2-1}=\frac{x(x-1)}{(x+1)(x-1)}=\frac{x}{x+1}$. The variable restriction is $x \neq -1, 1$.

d. Simplifying: $\frac{x^3-1}{x^2-1}=\frac{(x-1)(x^2+x+1)}{(x+1)(x-1)}=\frac{x^2+x+1}{x+1}$. The variable restriction is $x \neq -1, 1$.

e. Simplifying: $\frac{x^3-1}{x^3-x^2}=\frac{(x-1)(x^2+x+1)}{x^2(x-1)}=\frac{x^2+x+1}{x^2}$. The variable restriction is $x \neq 0, 1$.

3. Reducing the rational expression: $\frac{5}{5x-10}=\frac{5}{5(x-2)}=\frac{1}{x-2}$. The variable restriction is $x \neq 2$.

5. Reducing the rational expression: $\frac{a-3}{a^2-9}=\frac{1(a-3)}{(a+3)(a-3)}=\frac{1}{a+3}$. The variable restriction is $a \neq -3, 3$.

7. Reducing the rational expression: $\frac{x+5}{x^2-25}=\frac{1(x+5)}{(x+5)(x-5)}=\frac{1}{x-5}$. The variable restriction is $x \neq -5, 5$.

9. Reducing the rational expression: $\frac{2x^2-8}{4}=\frac{2(x^2-4)}{4}=\frac{2(x+2)(x-2)}{4}=\frac{(x+2)(x-2)}{2}$

There are no variable restrictions.

11. Reducing the rational expression: $\frac{2x-10}{3x-6}=\frac{2(x-5)}{3(x-2)}$. The variable restriction is $x \neq 2$.

13. Reducing the rational expression: $\frac{10a+20}{5a+10}=\frac{10(a+2)}{5(a+2)}=\frac{2}{1}=2$

15. Reducing the rational expression: $\frac{5x^2-5}{4x+4}=\frac{5(x^2-1)}{4(x+1)}=\frac{5(x+1)(x-1)}{4(x+1)}=\frac{5(x-1)}{4}$

17. Reducing the rational expression: $\frac{x-3}{x^2-6x+9}=\frac{1(x-3)}{(x-3)^2}=\frac{1}{x-3}$

19. Reducing the rational expression: $\frac{3x+15}{3x^2+24x+45}=\frac{3(x+5)}{3(x^2+8x+15)}=\frac{3(x+5)}{3(x+5)(x+3)}=\frac{1}{x+3}$

21. Reducing the rational expression: $\frac{a^2-3a}{a^3-8a^2+15a}=\frac{a(a-3)}{a(a^2-8a+15)}=\frac{a(a-3)}{a(a-3)(a-5)}=\frac{1}{a-5}$

23. Reducing the rational expression: $\frac{3x-2}{9x^2-4}=\frac{1(3x-2)}{(3x+2)(3x-2)}=\frac{1}{3x+2}$

25. Reducing the rational expression: $\frac{x^2+8x+15}{x^2+5x+6}=\frac{(x+5)(x+3)}{(x+2)(x+3)}=\frac{x+5}{x+2}$

27. Reducing the rational expression: $\frac{2m^3-2m^2-12m}{m^2-5m+6}=\frac{2m(m^2-m-6)}{(m-2)(m-3)}=\frac{2m(m-3)(m+2)}{(m-2)(m-3)}=\frac{2m(m+2)}{m-2}$

© 2012 Cengage Learning. All Rights Reserved. May not be scanned, copied or duplicated, or posted to a publicly accessible website, in whole or in part.

29. Reducing the rational expression: $\frac{x^3+3x^2-4x}{x^3-16x}=\frac{x(x^2+3x-4)}{x(x^2-16)}=\frac{x(x+4)(x-1)}{x(x+4)(x-4)}=\frac{x-1}{x-4}$

31. Reducing the rational expression: $\frac{4x^3-10x^2+6x}{2x^3+x^2-3x}=\frac{2x(2x^2-5x+3)}{x(2x^2+x-3)}=\frac{2x(2x-3)(x-1)}{x(2x+3)(x-1)}=\frac{2(2x-3)}{2x+3}$

33. Reducing the rational expression: $\frac{4x^2-12x+9}{4x^2-9}=\frac{(2x-3)^2}{(2x+3)(2x-3)}=\frac{2x-3}{2x+3}$

35. Reducing the rational expression: $\frac{x+3}{x^4-81}=\frac{x+3}{(x^2+9)(x^2-9)}=\frac{x+3}{(x^2+9)(x+3)(x-3)}=\frac{1}{(x^2+9)(x-3)}$

37. Reducing the rational expression: $\frac{3x^2+x-10}{x^4-16}=\frac{(3x-5)(x+2)}{(x^2+4)(x^2-4)}=\frac{(3x-5)(x+2)}{(x^2+4)(x+2)(x-2)}=\frac{3x-5}{(x^2+4)(x-2)}$

39. Reducing the rational expression: $\frac{42x^3-20x^2-48x}{6x^2-5x-4}=\frac{2x(21x^2-10x-24)}{(3x-4)(2x+1)}=\frac{2x(7x+6)(3x-4)}{(3x-4)(2x+1)}=\frac{2x(7x+6)}{2x+1}$

41. Reducing the rational expression: $\frac{x^3-y^3}{x^2-y^2}=\frac{(x-y)(x^2+xy+y^2)}{(x+y)(x-y)}=\frac{x^2+xy+y^2}{x+y}$

43. Reducing the rational expression: $\frac{x^3+8}{x^2-4}=\frac{(x+2)(x^2-2x+4)}{(x+2)(x-2)}=\frac{x^2-2x+4}{x-2}$

45. Reducing the rational expression: $\frac{x^3+8}{x^2+x-2}=\frac{(x+2)(x^2-2x+4)}{(x+2)(x-1)}=\frac{x^2-2x+4}{x-1}$

47. Reducing the rational expression: $\frac{xy+3x+2y+6}{xy+3x+5y+15}=\frac{x(y+3)+2(y+3)}{x(y+3)+5(y+3)}=\frac{(y+3)(x+2)}{(y+3)(x+5)}=\frac{x+2}{x+5}$

49. Reducing the rational expression: $\frac{x^2-3x+ax-3a}{x^2-3x+bx-3b}=\frac{x(x-3)+a(x-3)}{x(x-3)+b(x-3)}=\frac{(x-3)(x+a)}{(x-3)(x+b)}=\frac{x+a}{x+b}$

51. Reducing the rational expression: $\frac{xy+bx+ay+ab}{xy+bx+3y+3b}=\frac{x(y+b)+a(y+b)}{x(y+b)+3(y+b)}=\frac{(y+b)(x+a)}{(y+b)(x+3)}=\frac{x+a}{x+3}$

53. **a.** Adding: $(x^2-4x)+(4x-16)=x^2-4x+4x-16=x^2-16$

b. Subtracting: $(x^2-4x)-(4x-16)=x^2-4x-4x+16=x^2-8x+16$

c. Multiplying: $(x^2-4x)(4x-16)=4x^3-16x^2-16x^2+64x=4x^3-32x^2+64x$

d. Reducing: $\frac{x^2-4x}{4x-16}=\frac{x(x-4)}{4(x-4)}=\frac{x}{4}$

55. Writing as a ratio: $\frac{8}{6}=\frac{4}{3}$

57. Writing as a ratio: $\frac{200}{250}=\frac{4}{5}$

59. Writing as a ratio: $\frac{32}{4}=\frac{8}{1}$

© 2012 Cengage Learning. All Rights Reserved. May not be scanned, copied or duplicated, or posted to a publicly accessible website, in whole or in part.

61. Finding the distance for each car:

Koenigsegg CCX: $245(1.5) = 367.5 < 375$

Saleen S7 Twin Turbo: $248(1.5) = 372 < 375$

Bugatti Veyron: $253(1.5) = 379.5 > 375$

SSC Ultimate Aero: $257(1.5) = 385.5 > 375$

The Bugatti Veyron and the SSC Ultimate Aero could travel 375 miles in 1.5 hours.

63. The average speed is: $\frac{122 \text{ miles}}{3 \text{ hours}} \approx 40.7$ miles/hour

65. The average speed is: $\frac{785 \text{ feet}}{20 \text{ minutes}} = 39.25$ feet/minute

67. The average speed is: $\frac{188 \text{ feet}}{30 \text{ seconds}} \approx 6.3$ feet/second

69. The average fuel consumption is: $\frac{168 \text{ miles}}{3.5 \text{ gallons}} = 48$ miles/gallon

71. Substituting $x = 5$ and $y = 4$: $\frac{x^2 - y^2}{x - y} = \frac{5^2 - 4^2}{5 - 4} = \frac{25 - 16}{5 - 4} = \frac{9}{1} = 9$. The result is equal to $5 + 4 = 9$.

73. Completing the table:

x	$\frac{x-3}{3-x}$
−2	−1
−1	−1
0	−1
1	−1
2	−1

Simplifying: $\frac{x-3}{3-x} = \frac{-(3-x)}{3-x} = -1$

75. Completing the table:

x	$\frac{x-5}{x^2-25}$	$\frac{1}{x+5}$
0	$1/5$	$1/5$
2	$1/7$	$1/7$
−2	$1/3$	$1/3$
5	undefined	$1/10$
−5	undefined	undefined

77. Simplifying: $\frac{3}{4} \cdot \frac{10}{21} = \frac{3 \cdot 2 \cdot 5}{2 \cdot 2 \cdot 3 \cdot 7} = \frac{5}{14}$

79. Simplifying: $\frac{4}{5} \div \frac{8}{9} = \frac{4}{5} \cdot \frac{9}{8} = \frac{2 \cdot 2 \cdot 9}{5 \cdot 2 \cdot 2 \cdot 2} = \frac{9}{10}$

81. Factoring: $x^2 - 9 = (x+3)(x-3)$

83. Factoring: $3x - 9 = 3(x-3)$

85. Factoring: $x^2 - x - 20 = (x-5)(x+4)$

87. Factoring: $a^2 + 5a = a(a+5)$

89. Simplifying: $\frac{a(a+5)(a-5)(a+4)}{a^2 + 5a} = \frac{a(a+5)(a-5)(a+4)}{a(a+5)} = (a-5)(a+4)$

91. Multiplying: $\frac{5,603}{11} \cdot \frac{1}{5,280} \cdot \frac{60}{1} \approx 5.8$

93. Simplifying the expression: $\frac{27x^5}{9x^2} - \frac{45x^8}{15x^5} = 3x^3 - 3x^3 = 0$

95. Simplifying the expression: $\frac{72a^3b^7}{9ab^5} + \frac{64a^5b^3}{8a^3b} = 8a^2b^2 + 8a^2b^2 = 16a^2b^2$

97. Dividing by the monomial: $\frac{38x^7 + 42x^5 - 84x^3}{2x^3} = \frac{38x^7}{2x^3} + \frac{42x^5}{2x^3} - \frac{84x^3}{2x^3} = 19x^4 + 21x^2 - 42$

© 2012 Cengage Learning. All Rights Reserved. May not be scanned, copied or duplicated, or posted to a publicly accessible website, in whole or in part.

99. Dividing by the monomial: $\frac{28a^5b^5+36ab^4-44a^4b}{4ab}=\frac{28a^5b^5}{4ab}+\frac{36ab^4}{4ab}-\frac{44a^4b}{4ab}=7a^4b^4+9b^3-11a^3$

101. Completing the table:

Stock	Price	Earnings per Share	P/E
Yahoo	37.80	1.07	35.33
Google	381.24	4.51	84.53
Disney	24.96	1.34	18.63
Nike	85.46	4.88	17.51
Ebay	40.96	0.73	56.11

The stocks Disney and Nike appear to be undervalued.

7.2 Multiplication and Division of Rational Expressions

1. Simplifying the expression: $\frac{x+y}{3}\bullet\frac{6}{x+y}=\frac{6(x+y)}{3(x+y)}=2$

3. Simplifying the expression: $\frac{2x+10}{x^2}\bullet\frac{x^3}{4x+20}=\frac{2(x+5)}{x^2}\bullet\frac{x^3}{4(x+5)}=\frac{2x^3(x+5)}{4x^2(x+5)}=\frac{x}{2}$

5. Simplifying the expression: $\frac{9}{2a-8}\div\frac{3}{a-4}=\frac{9}{2a-8}\bullet\frac{a-4}{3}=\frac{9}{2(a-4)}\bullet\frac{a-4}{3}=\frac{9(a-4)}{6(a-4)}=\frac{3}{2}$

7. Simplifying the expression:

$$\frac{x+1}{x^2-9}\div\frac{2x+2}{x+3}=\frac{x+1}{x^2-9}\bullet\frac{x+3}{2x+2}=\frac{x+1}{(x+3)(x-3)}\bullet\frac{x+3}{2(x+1)}=\frac{(x+1)(x+3)}{2(x+3)(x-3)(x+1)}=\frac{1}{2(x-3)}$$

9. Simplifying the expression: $\frac{a^2+5a}{7a}\bullet\frac{4a^2}{a^2+4a}=\frac{a(a+5)}{7a}\bullet\frac{4a^2}{a(a+4)}=\frac{4a^3(a+5)}{7a^2(a+4)}=\frac{4a(a+5)}{7(a+4)}$

11. Simplifying the expression:

$$\frac{y^2-5y+6}{2y+4}\div\frac{2y-6}{y+2}=\frac{y^2-5y+6}{2y+4}\bullet\frac{y+2}{2y-6}=\frac{(y-2)(y-3)}{2(y+2)}\bullet\frac{y+2}{2(y-3)}=\frac{(y-2)(y-3)(y+2)}{4(y+2)(y-3)}=\frac{y-2}{4}$$

13. Simplifying the expression:

$$\frac{2x-8}{x^2-4}\bullet\frac{x^2+6x+8}{x-4}=\frac{2(x-4)}{(x+2)(x-2)}\bullet\frac{(x+4)(x+2)}{x-4}=\frac{2(x-4)(x+4)(x+2)}{(x+2)(x-2)(x-4)}=\frac{2(x+4)}{x-2}$$

15. Simplifying the expression:

$$\frac{x-1}{x^2-x-6}\bullet\frac{x^2+5x+6}{x^2-1}=\frac{x-1}{(x-3)(x+2)}\bullet\frac{(x+2)(x+3)}{(x+1)(x-1)}=\frac{(x-1)(x+2)(x+3)}{(x-3)(x+2)(x+1)(x-1)}=\frac{x+3}{(x-3)(x+1)}$$

17. Simplifying the expression:

$$\frac{a^2+10a+25}{a+5}\div\frac{a^2-25}{a-5}=\frac{a^2+10a+25}{a+5}\bullet\frac{a-5}{a^2-25}=\frac{(a+5)^2}{a+5}\bullet\frac{a-5}{(a+5)(a-5)}=\frac{(a+5)^2(a-5)}{(a+5)^2(a-5)}=1$$

© 2012 Cengage Learning. All Rights Reserved. May not be scanned, copied or duplicated, or posted to a publicly accessible website, in whole or in part.

19. Simplifying the expression:

$$\begin{aligned}\frac{y^3-5y^2}{y^4+3y^3+2y^2}\div\frac{y^2-5y+6}{y^2-2y-3}&=\frac{y^3-5y^2}{y^4+3y^3+2y^2}\bullet\frac{y^2-2y-3}{y^2-5y+6}\\&=\frac{y^2(y-5)}{y^2(y+2)(y+1)}\bullet\frac{(y-3)(y+1)}{(y-2)(y-3)}\\&=\frac{y^2(y-5)(y-3)(y+1)}{y^2(y+2)(y+1)(y-2)(y-3)}\\&=\frac{y-5}{(y+2)(y-2)}\end{aligned}$$

21. Simplifying the expression:

$$\begin{aligned}\frac{2x^2+17x+21}{x^2+2x-35}\bullet\frac{x^3-125}{2x^2-7x-15}&=\frac{(2x+3)(x+7)}{(x+7)(x-5)}\bullet\frac{(x-5)\left(x^2+5x+25\right)}{(2x+3)(x-5)}\\&=\frac{(2x+3)(x+7)(x-5)\left(x^2+5x+25\right)}{(x+7)(x-5)^2(2x+3)}\\&=\frac{x^2+5x+25}{x-5}\end{aligned}$$

23. Simplifying the expression:

$$\begin{aligned}\frac{2x^2+10x+12}{4x^2+24x+32}\bullet\frac{2x^2+18x+40}{x^2+8x+15}&=\frac{2\left(x^2+5x+6\right)}{4\left(x^2+6x+8\right)}\bullet\frac{2\left(x^2+9x+20\right)}{x^2+8x+15}\\&=\frac{2(x+2)(x+3)}{4(x+4)(x+2)}\bullet\frac{2(x+5)(x+4)}{(x+5)(x+3)}\\&=\frac{4(x+2)(x+3)(x+4)(x+5)}{4(x+2)(x+3)(x+4)(x+5)}\\&=1\end{aligned}$$

25. Simplifying the expression:

$$\begin{aligned}\frac{2a^2+7a+3}{a^2-16}\div\frac{4a^2+8a+3}{2a^2-5a-12}&=\frac{2a^2+7a+3}{a^2-16}\bullet\frac{2a^2-5a-12}{4a^2+8a+3}\\&=\frac{(2a+1)(a+3)}{(a+4)(a-4)}\bullet\frac{(2a+3)(a-4)}{(2a+1)(2a+3)}\\&=\frac{(2a+1)(a+3)(2a+3)(a-4)}{(a+4)(a-4)(2a+1)(2a+3)}\\&=\frac{a+3}{a+4}\end{aligned}$$

27. Simplifying the expression:

$$\begin{aligned}\frac{4y^2-12y+9}{y^2-36}\div\frac{2y^2-5y+3}{y^2+5y-6}&=\frac{4y^2-12y+9}{y^2-36}\bullet\frac{y^2+5y-6}{2y^2-5y+3}\\&=\frac{(2y-3)^2}{(y+6)(y-6)}\bullet\frac{(y+6)(y-1)}{(2y-3)(y-1)}\\&=\frac{(2y-3)^2(y+6)(y-1)}{(y+6)(y-6)(2y-3)(y-1)}\\&=\frac{2y-3}{y-6}\end{aligned}$$

© 2012 Cengage Learning. All Rights Reserved. May not be scanned, copied or duplicated, or posted to a publicly accessible website, in whole or in part.

29. Simplifying the expression:

$$\frac{x^2-1}{6x^2+42x+60}\bullet\frac{7x^2+17x+6}{x^3+1}\bullet\frac{6x+30}{7x^2-11x-6}=\frac{(x+1)(x-1)}{6(x+5)(x+2)}\bullet\frac{(7x+3)(x+2)}{(x+1)(x^2-x+1)}\bullet\frac{6(x+5)}{(7x+3)(x-2)}$$

$$=\frac{6(x+1)(x-1)(7x+3)(x+2)(x+5)}{6(x+5)(x+2)(x+1)(7x+3)(x-2)(x^2-x+1)}$$

$$=\frac{x-1}{(x-2)(x^2-x+1)}$$

31. Simplifying the expression:

$$\frac{18x^3+21x^2-60x}{21x^2-25x-4}\bullet\frac{28x^2-17x-3}{16x^3+28x^2-30x}=\frac{3x(6x^2+7x-20)}{21x^2-25x-4}\bullet\frac{28x^2-17x-3}{2x(8x^2+14x-15)}$$

$$=\frac{3x(3x-4)(2x+5)}{(7x+1)(3x-4)}\bullet\frac{(7x+1)(4x-3)}{2x(4x-3)(2x+5)}$$

$$=\frac{3x(3x-4)(2x+5)(7x+1)(4x-3)}{2x(7x+1)(3x-4)(4x-3)(2x+5)}$$

$$=\frac{3}{2}$$

33. **a.** Simplifying: $\frac{9-1}{27-1}=\frac{8}{26}=\frac{4}{13}$

b. Reducing: $\frac{x^2-1}{x^3-1}=\frac{(x+1)(x-1)}{(x-1)(x^2+x+1)}=\frac{x+1}{x^2+x+1}$

c. Multiplying: $\frac{x^2-1}{x^3-1}\bullet\frac{x-1}{x+1}=\frac{(x+1)(x-1)}{(x-1)(x^2+x+1)}\bullet\frac{x-1}{x+1}=\frac{x-1}{x^2+x+1}$

d. Dividing: $\frac{x^2-1}{x^3-1}\div\frac{x-1}{x^2+x+1}=\frac{(x+1)(x-1)}{(x-1)(x^2+x+1)}\bullet\frac{x^2+x+1}{x-1}=\frac{x+1}{x-1}$

35. Simplifying the expression: $(x^2-9)\left(\frac{2}{x+3}\right)=\frac{(x+3)(x-3)}{1}\bullet\frac{2}{x+3}=\frac{2(x+3)(x-3)}{x+3}=2(x-3)$

37. Simplifying the expression: $a(a+5)(a-5)\left(\frac{2}{a^2-25}\right)=\frac{a(a+5)(a-5)}{1}\bullet\frac{2}{(a+5)(a-5)}=\frac{2a(a+5)(a-5)}{(a+5)(a-5)}=2a$

39. Simplifying the expression: $(x^2-x-6)\left(\frac{x+1}{x-3}\right)=\frac{(x-3)(x+2)}{1}\bullet\frac{x+1}{x-3}=\frac{(x-3)(x+2)(x+1)}{x-3}=(x+2)(x+1)$

41. Simplifying the expression: $(x^2-4x-5)\left(\frac{-2x}{x+1}\right)=\frac{(x-5)(x+1)}{1}\bullet\frac{-2x}{x+1}=\frac{-2x(x-5)(x+1)}{x+1}=-2x(x-5)$

43. Simplifying the expression:

$$\frac{x^2-9}{x^2-3x}\bullet\frac{2x+10}{xy+5x+3y+15}=\frac{(x+3)(x-3)}{x(x-3)}\bullet\frac{2(x+5)}{x(y+5)+3(y+5)}=\frac{2(x+3)(x-3)(x+5)}{x(x-3)(y+5)(x+3)}=\frac{2(x+5)}{x(y+5)}$$

45. Simplifying the expression:

$$\frac{2x^2+4x}{x^2-y^2}\bullet\frac{x^2+3x+xy+3y}{x^2+5x+6}=\frac{2x(x+2)}{(x+y)(x-y)}\bullet\frac{x(x+3)+y(x+3)}{(x+2)(x+3)}=\frac{2x(x+2)(x+3)(x+y)}{(x+y)(x-y)(x+2)(x+3)}=\frac{2x}{x-y}$$

© 2012 Cengage Learning. All Rights Reserved. May not be scanned, copied or duplicated, or posted to a publicly accessible website, in whole or in part.

47. Simplifying the expression:

$$\frac{x^3-3x^2+4x-12}{x^4-16}\bullet\frac{3x^2+5x-2}{3x^2-10x+3}=\frac{x^2(x-3)+4(x-3)}{(x^2+4)(x^2-4)}\bullet\frac{(3x-1)(x+2)}{(3x-1)(x-3)}$$

$$=\frac{(x-3)(x^2+4)}{(x^2+4)(x+2)(x-2)}\bullet\frac{(3x-1)(x+2)}{(3x-1)(x-3)}$$

$$=\frac{(x-3)(x^2+4)(3x-1)(x+2)}{(x^2+4)(x+2)(x-2)(3x-1)(x-3)}$$

$$=\frac{1}{x-2}$$

49. Simplifying the expression: $\left(1-\frac{1}{2}\right)\left(1-\frac{1}{3}\right)\left(1-\frac{1}{4}\right)\left(1-\frac{1}{5}\right)=\left(\frac{2}{2}-\frac{1}{2}\right)\left(\frac{3}{3}-\frac{1}{3}\right)\left(\frac{4}{4}-\frac{1}{4}\right)\left(\frac{5}{5}-\frac{1}{5}\right)=\frac{1}{2}\bullet\frac{2}{3}\bullet\frac{3}{4}\bullet\frac{4}{5}=\frac{1}{5}$

51. Simplifying the expression: $\left(1-\frac{1}{2}\right)\left(1-\frac{1}{3}\right)\left(1-\frac{1}{4}\right)\cdots\left(1-\frac{1}{99}\right)\left(1-\frac{1}{100}\right)=\frac{1}{2}\bullet\frac{2}{3}\bullet\frac{3}{4}\cdots\frac{98}{99}\bullet\frac{99}{100}=\frac{1}{100}$

53. Converting the speed: $\frac{245 \text{ miles}}{1 \text{ hour}}\bullet\frac{5{,}280 \text{ feet}}{1 \text{ mile}}\bullet\frac{1 \text{ hour}}{3{,}600 \text{ seconds}}\approx 359.3$ feet/second

55. Since 5,280 feet = 1 mile, the height is: $\frac{14{,}494 \text{ feet}}{5{,}280 \text{ feet/mile}}\approx 2.7$ miles

57. Converting to miles per hour: $\frac{1088 \text{ feet}}{1 \text{ second}}\bullet\frac{1 \text{ mile}}{5280 \text{ feet}}\bullet\frac{60 \text{ seconds}}{1 \text{ minute}}\bullet\frac{60 \text{ minutes}}{1 \text{ hour}}\approx 742$ miles/hour

59. Converting to miles per hour: $\frac{785 \text{ feet}}{20 \text{ minutes}}\bullet\frac{60 \text{ minutes}}{1 \text{ hour}}\bullet\frac{1 \text{ mile}}{5280 \text{ feet}}\approx 0.45$ miles/hour

61. Converting to miles per hour: $\frac{518 \text{ feet}}{40 \text{ seconds}}\bullet\frac{60 \text{ seconds}}{1 \text{ minute}}\bullet\frac{60 \text{ minutes}}{1 \text{ hour}}\bullet\frac{1 \text{ mile}}{5280 \text{ feet}}\approx 8.8$ miles/hour

63. Adding the fractions: $\frac{1}{5}+\frac{3}{5}=\frac{4}{5}$

65. Adding the fractions: $\frac{1}{10}+\frac{3}{14}=\frac{1}{10}\bullet\frac{7}{7}+\frac{3}{14}\bullet\frac{5}{5}=\frac{7}{70}+\frac{15}{70}=\frac{22}{70}=\frac{11}{35}$

67. Subtracting the fractions: $\frac{1}{10}-\frac{3}{14}=\frac{1}{10}\bullet\frac{7}{7}-\frac{3}{14}\bullet\frac{5}{5}=\frac{7}{70}-\frac{15}{70}=-\frac{8}{70}=-\frac{4}{35}$

69. Multiplying: $2(x-3)=2x-6$

71. Multiplying: $(x+4)(x-5)=x^2-5x+4x-20=x^2-x-20$

73. Reducing the fraction: $\frac{x+3}{x^2-9}=\frac{x+3}{(x+3)(x-3)}=\frac{1}{x-3}$

75. Reducing the fraction: $\frac{x^2-x-30}{2(x+5)(x-5)}=\frac{(x+5)(x-6)}{2(x+5)(x-5)}=\frac{x-6}{2(x-5)}$

77. Simplifying: $(x+4)(x-5)-10=x^2-5x+4x-20-10=x^2-x-30$

79. Adding the fractions: $\frac{1}{2}+\frac{5}{2}=\frac{6}{2}=3$

81. Adding the fractions: $2+\frac{3}{4}=\frac{2\bullet 4}{1\bullet 4}+\frac{3}{4}=\frac{8}{4}+\frac{3}{4}=\frac{11}{4}$

83. Simplifying the expression: $\frac{10x^4}{2x^2}+\frac{12x^6}{3x^4}=5x^2+4x^2=9x^2$

85. Simplifying the expression: $\frac{12a^2b^5}{3ab^3}+\frac{14a^4b^7}{7a^3b^5}=4ab^2+2ab^2=6ab^2$

© 2012 Cengage Learning. All Rights Reserved. May not be scanned, copied or duplicated, or posted to a publicly accessible website, in whole or in part.

7.3 Addition and Subtraction of Rational Expressions

1. Combining the fractions: $\frac{3}{x}+\frac{4}{x}=\frac{7}{x}$

3. Combining the fractions: $\frac{9}{a}-\frac{5}{a}=\frac{4}{a}$

5. Combining the fractions: $\frac{1}{x+1}+\frac{x}{x+1}=\frac{1+x}{x+1}=1$

7. Combining the fractions: $\frac{y^2}{y-1}-\frac{1}{y-1}=\frac{y^2-1}{y-1}=\frac{(y+1)(y-1)}{y-1}=y+1$

9. Combining the fractions: $\frac{x^2}{x+2}+\frac{4x+4}{x+2}=\frac{x^2+4x+4}{x+2}=\frac{(x+2)^2}{x+2}=x+2$

11. Combining the fractions: $\frac{x^2}{x-2}-\frac{4x-4}{x-2}=\frac{x^2-4x+4}{x-2}=\frac{(x-2)^2}{x-2}=x-2$

13. Combining the fractions: $\frac{x+2}{x+6}-\frac{x-4}{x+6}=\frac{x+2-x+4}{x+6}=\frac{6}{x+6}$

15. Combining the fractions: $\frac{y}{2}-\frac{2}{y}=\frac{y\bullet y}{2\bullet y}-\frac{2\bullet 2}{y\bullet 2}=\frac{y^2}{2y}-\frac{4}{2y}=\frac{y^2-4}{2y}=\frac{(y+2)(y-2)}{2y}$

17. Combining the fractions: $\frac{1}{2}+\frac{a}{3}=\frac{1\bullet 3}{2\bullet 3}+\frac{a\bullet 2}{3\bullet 2}=\frac{3}{6}+\frac{2a}{6}=\frac{2a+3}{6}$

19. Combining the fractions: $\frac{x}{x+1}+\frac{3}{4}=\frac{x\bullet 4}{(x+1)\bullet 4}+\frac{3\bullet(x+1)}{4\bullet(x+1)}=\frac{4x}{4(x+1)}+\frac{3x+3}{4(x+1)}=\frac{4x+3x+3}{4(x+1)}=\frac{7x+3}{4(x+1)}$

21. Combining the fractions:

$$\frac{x+1}{x-2}-\frac{4x+7}{5x-10}=\frac{(x+1)\bullet 5}{(x-2)\bullet 5}-\frac{4x+7}{5(x-2)}=\frac{5x+5}{5(x-2)}-\frac{4x+7}{5(x-2)}=\frac{5x+5-4x-7}{5(x-2)}=\frac{x-2}{5(x-2)}=\frac{1}{5}$$

23. Combining the fractions:

$$\frac{4x-2}{3x+12}-\frac{x-2}{x+4}=\frac{4x-2}{3(x+4)}-\frac{(x-2)\bullet 3}{(x+4)\bullet 3}=\frac{4x-2}{3(x+4)}-\frac{3x-6}{3(x+4)}=\frac{4x-2-3x+6}{3(x+4)}=\frac{x+4}{3(x+4)}=\frac{1}{3}$$

25. Combining the fractions:

$$\frac{6}{x(x-2)}+\frac{3}{x}=\frac{6}{x(x-2)}+\frac{3\bullet(x-2)}{x\bullet(x-2)}=\frac{6}{x(x-2)}+\frac{3x-6}{x(x-2)}=\frac{6+3x-6}{x(x-2)}=\frac{3x}{x(x-2)}=\frac{3}{x-2}$$

27. Combining the fractions:

$$\frac{4}{a}-\frac{12}{a^2+3a}=\frac{4\bullet(a+3)}{a\bullet(a+3)}-\frac{12}{a(a+3)}=\frac{4a+12}{a(a+3)}-\frac{12}{a(a+3)}=\frac{4a+12-12}{a(a+3)}=\frac{4a}{a(a+3)}=\frac{4}{a+3}$$

29. Combining the fractions:

$$\begin{aligned}\frac{2}{x+5}-\frac{10}{x^2-25}&=\frac{2\bullet(x-5)}{(x+5)\bullet(x-5)}-\frac{10}{(x+5)(x-5)}\\&=\frac{2x-10}{(x+5)(x-5)}-\frac{10}{(x+5)(x-5)}\\&=\frac{2x-10-10}{(x+5)(x-5)}\\&=\frac{2x-20}{(x+5)(x-5)}\\&=\frac{2(x-10)}{(x+5)(x-5)}\end{aligned}$$

31. Combining the fractions:

$$\begin{aligned}\frac{x-4}{x-3}+\frac{6}{x^2-9}&=\frac{(x-4)\bullet(x+3)}{(x-3)\bullet(x+3)}+\frac{6}{(x+3)(x-3)}\\&=\frac{x^2-x-12}{(x+3)(x-3)}+\frac{6}{(x+3)(x-3)}\\&=\frac{x^2-x-12+6}{(x+3)(x-3)}\\&=\frac{x^2-x-6}{(x+3)(x-3)}\\&=\frac{(x-3)(x+2)}{(x+3)(x-3)}\\&=\frac{x+2}{x+3}\end{aligned}$$

© 2012 Cengage Learning. All Rights Reserved. May not be scanned, copied or duplicated, or posted to a publicly accessible website, in whole or in part.

33. Combining the fractions:

$$\begin{aligned}\frac{a-4}{a-3}+\frac{5}{a^2-a-6}&=\frac{(a-4)\bullet(a+2)}{(a-3)\bullet(a+2)}+\frac{5}{(a-3)(a+2)}\\&=\frac{a^2-2a-8}{(a-3)(a+2)}+\frac{5}{(a-3)(a+2)}\\&=\frac{a^2-2a-8+5}{(a-3)(a+2)}\\&=\frac{a^2-2a-3}{(a-3)(a+2)}\\&=\frac{(a-3)(a+1)}{(a-3)(a+2)}\\&=\frac{a+1}{a+2}\end{aligned}$$

35. Combining the fractions:

$$\begin{aligned}\frac{8}{x^2-16}-\frac{7}{x^2-x-12}&=\frac{8}{(x+4)(x-4)}-\frac{7}{(x-4)(x+3)}\\&=\frac{8(x+3)}{(x+4)(x-4)(x+3)}-\frac{7(x+4)}{(x+4)(x-4)(x+3)}\\&=\frac{8x+24}{(x+4)(x-4)(x+3)}-\frac{7x+28}{(x+4)(x-4)(x+3)}\\&=\frac{8x+24-7x-28}{(x+4)(x-4)(x+3)}\\&=\frac{x-4}{(x+4)(x-4)(x+3)}\\&=\frac{1}{(x+4)(x+3)}\end{aligned}$$

37. Combining the fractions:

$$\begin{aligned}\frac{4y}{y^2+6y+5}-\frac{3y}{y^2+5y+4}&=\frac{4y}{(y+5)(y+1)}-\frac{3y}{(y+4)(y+1)}\\&=\frac{4y(y+4)}{(y+5)(y+1)(y+4)}-\frac{3y(y+5)}{(y+5)(y+1)(y+4)}\\&=\frac{4y^2+16y}{(y+5)(y+1)(y+4)}-\frac{3y^2+15y}{(y+5)(y+1)(y+4)}\\&=\frac{4y^2+16y-3y^2-15y}{(y+5)(y+1)(y+4)}\\&=\frac{y^2+y}{(y+5)(y+1)(y+4)}\\&=\frac{y(y+1)}{(y+5)(y+1)(y+4)}\\&=\frac{y}{(y+5)(y+4)}\end{aligned}$$

© 2012 Cengage Learning. All Rights Reserved. May not be scanned, copied or duplicated, or posted to a publicly accessible website, in whole or in part.

39. Combining the fractions:

$$\frac{4x+1}{x^2+5x+4}-\frac{x+3}{x^2+4x+3}=\frac{4x+1}{(x+4)(x+1)}-\frac{x+3}{(x+3)(x+1)}$$
$$=\frac{(4x+1)(x+3)}{(x+4)(x+1)(x+3)}-\frac{(x+3)(x+4)}{(x+4)(x+1)(x+3)}$$
$$=\frac{4x^2+13x+3}{(x+4)(x+1)(x+3)}-\frac{x^2+7x+12}{(x+4)(x+1)(x+3)}$$
$$=\frac{4x^2+13x+3-x^2-7x-12}{(x+4)(x+1)(x+3)}$$
$$=\frac{3x^2+6x-9}{(x+4)(x+1)(x+3)}$$
$$=\frac{3(x+3)(x-1)}{(x+4)(x+1)(x+3)}$$
$$=\frac{3(x-1)}{(x+4)(x+1)}$$

41. Combining the fractions:

$$\frac{1}{x}+\frac{x}{3x+9}-\frac{3}{x^2+3x}=\frac{1}{x}+\frac{x}{3(x+3)}-\frac{3}{x(x+3)}$$
$$=\frac{1\bullet 3(x+3)}{x\bullet 3(x+3)}+\frac{x\bullet x}{3(x+3)\bullet x}-\frac{3\bullet 3}{x(x+3)\bullet 3}$$
$$=\frac{3x+9}{3x(x+3)}+\frac{x^2}{3x(x+3)}-\frac{9}{3x(x+3)}$$
$$=\frac{3x+9+x^2-9}{3x(x+3)}$$
$$=\frac{x^2+3x}{3x(x+3)}$$
$$=\frac{x(x+3)}{3x(x+3)}$$
$$=\frac{1}{3}$$

43. **a.** Multiplying: $\frac{4}{9}\bullet\frac{1}{6}=\frac{2\bullet 2}{3\bullet 3}\bullet\frac{1}{2\bullet 3}=\frac{2}{3\bullet 3\bullet 3}=\frac{2}{27}$

b. Dividing: $\frac{4}{9}\div\frac{1}{6}=\frac{4}{9}\bullet\frac{6}{1}=\frac{2\bullet 2}{3\bullet 3}\bullet\frac{2\bullet 3}{1}=\frac{2\bullet 2\bullet 2}{3}=\frac{8}{3}$

c. Adding: $\frac{4}{9}+\frac{1}{6}=\frac{4}{9}\bullet\frac{2}{2}+\frac{1}{6}\bullet\frac{3}{3}=\frac{8}{18}+\frac{3}{18}=\frac{11}{18}$

d. Multiplying: $\frac{x+2}{x-2}\bullet\frac{3x+10}{x^2-4}=\frac{x+2}{x-2}\bullet\frac{3x+10}{(x+2)(x-2)}=\frac{3x+10}{(x-2)^2}$

e. Dividing: $\frac{x+2}{x-2}\div\frac{3x+10}{x^2-4}=\frac{x+2}{x-2}\bullet\frac{(x+2)(x-2)}{3x+10}=\frac{(x+2)^2}{3x+10}$

© 2012 Cengage Learning. All Rights Reserved. May not be scanned, copied or duplicated, or posted to a publicly accessible website, in whole or in part.

f. Subtracting:

$$\frac{x+2}{x-2}-\frac{3x+10}{x^2-4}=\frac{x+2}{x-2}\bullet\frac{x+2}{x+2}-\frac{3x+10}{(x+2)(x-2)}$$
$$=\frac{x^2+4x+4}{(x+2)(x-2)}-\frac{3x+10}{(x+2)(x-2)}$$
$$=\frac{x^2+x-6}{(x+2)(x-2)}$$
$$=\frac{(x+3)(x-2)}{(x+2)(x-2)}$$
$$=\frac{x+3}{x+2}$$

45. Completing the table:

Number x	Reciprocal $\frac{1}{x}$	Sum $1+\frac{1}{x}$	Sum $\frac{x+1}{x}$
1	1	2	2
2	$\frac{1}{2}$	$\frac{3}{2}$	$\frac{3}{2}$
3	$\frac{1}{3}$	$\frac{4}{3}$	$\frac{4}{3}$
4	$\frac{1}{4}$	$\frac{5}{4}$	$\frac{5}{4}$

47. Completing the table:

x	$x+\frac{4}{x}$	$\frac{x^2+4}{x}$	$x+4$
1	5	5	5
2	4	4	6
3	$\frac{13}{3}$	$\frac{13}{3}$	7
4	5	5	8

49. Combining the fractions: $1+\frac{1}{x+2}=\frac{1\bullet(x+2)}{1\bullet(x+2)}+\frac{1}{x+2}=\frac{x+2}{x+2}+\frac{1}{x+2}=\frac{x+2+1}{x+2}=\frac{x+3}{x+2}$

51. Combining the fractions: $1-\frac{1}{x+3}=\frac{1\bullet(x+3)}{1\bullet(x+3)}-\frac{1}{x+3}=\frac{x+3}{x+3}-\frac{1}{x+3}=\frac{x+3-1}{x+3}=\frac{x+2}{x+3}$

53. Simplifying: $6\left(\frac{1}{2}\right)=3$

55. Simplifying: $\frac{0}{5}=0$

57. Simplifying: $\frac{5}{0}$ is undefined

59. Simplifying: $1-\frac{5}{2}=\frac{2}{2}-\frac{5}{2}=-\frac{3}{2}$

61. Simplifying: $6\left(\frac{x}{3}+\frac{5}{2}\right)=6\bullet\frac{x}{3}+6\bullet\frac{5}{2}=2x+15$

63. Simplifying: $x^2\left(1-\frac{5}{x}\right)=x^2\bullet 1-x^2\bullet\frac{5}{x}=x^2-5x$

65. Solving the equation:

$$\begin{aligned}2x+15&=3\\2x&=-12\\x&=-6\end{aligned}$$

67. Solving the equation:

$$\begin{aligned}-2x-9&=x-3\\-3x-9&=-3\\-3x&=6\\x&=-2\end{aligned}$$

69. Solving the equation:

$$\begin{aligned}2x+3(x-3)&=6\\2x+3x-9&=6\\5x-9&=6\\5x&=15\\x&=3\end{aligned}$$

71. Solving the equation:

$$\begin{aligned}x-3(x+3)&=x-3\\x-3x-9&=x-3\\-2x-9&=x-3\\-3x-9&=-3\\-3x&=6\\x&=-2\end{aligned}$$

© 2012 Cengage Learning. All Rights Reserved. May not be scanned, copied or duplicated, or posted to a publicly accessible website, in whole or in part.

73. Solving the equation:

$$\begin{aligned} 7-2(3x+1)&=4x+3 \\ 7-6x-2&=4x+3 \\ -6x+5&=4x+3 \\ -10x+5&=3 \\ -10x&=-2 \\ x&=\frac{1}{5} \end{aligned}$$

75. Solving the equation:

$$\begin{aligned} x^2+5x+6&=0 \\ (x+2)(x+3)&=0 \\ x&=-2,-3 \end{aligned}$$

77. Solving the quadratic equation:

$$\begin{aligned} x^2-x&=6 \\ x^2-x-6&=0 \\ (x-3)(x+2)&=0 \\ x&=-2,3 \end{aligned}$$

79. Solving the quadratic equation:

$$\begin{aligned} x^2-5x&=0 \\ x(x-5)&=0 \\ x&=0,5 \end{aligned}$$

81. Converting the temperature: $C=\frac{5}{9}(136.4-32)=\frac{5}{9}(104.4)=58°\text{C}$

7.4 Equations Involving Rational Expressions

1. Multiplying both sides of the equation by 6:

$$\begin{aligned} 6\left(\frac{x}{3}+\frac{1}{2}\right)&=6\left(-\frac{1}{2}\right) \\ 2x+3&=-3 \\ 2x&=-6 \\ x&=-3 \end{aligned}$$

Since $x=-3$ checks in the original equation, the solution is $x=-3$.

3. Multiplying both sides of the equation by $5a$:

$$\begin{aligned} 5a\left(\frac{4}{a}\right)&=5a\left(\frac{1}{5}\right) \\ 20&=a \end{aligned}$$

Since $a=20$ checks in the original equation, the solution is $a=20$.

5. Multiplying both sides of the equation by x:

$$\begin{aligned} x\left(\frac{3}{x}+1\right)&=x\left(\frac{2}{x}\right) \\ 3+x&=2 \\ x&=-1 \end{aligned}$$

Since $x=-1$ checks in the original equation, the solution is $x=-1$.

7. Multiplying both sides of the equation by $5a$:

$$\begin{aligned} 5a\left(\frac{3}{a}-\frac{2}{a}\right)&=5a\left(\tfrac{1}{5}\right) \\ 15-10&=a \\ a&=5 \end{aligned}$$

Since $a=5$ checks in the original equation, the solution is $a=5$.

9. Multiplying both sides of the equation by $2x$:

$$\begin{aligned} 2x\left(\frac{3}{x}+2\right)&=2x\left(\frac{1}{2}\right) \\ 6+4x&=x \\ 6&=-3x \\ x&=-2 \end{aligned}$$

Since $x=-2$ checks in the original equation, the solution is $x=-2$.

© 2012 Cengage Learning. All Rights Reserved. May not be scanned, copied or duplicated, or posted to a publicly accessible website, in whole or in part.

11. Multiplying both sides of the equation by $4y$:

$$\begin{aligned} 4y\left(\frac{1}{y}-\frac{1}{2}\right)&=4y\left(-\frac{1}{4}\right)\\ 4-2y&=-y\\ 4&=y \end{aligned}$$

Since $y=4$ checks in the original equation, the solution is $y=4$.

13. Multiplying both sides of the equation by x^2:

$$\begin{aligned} x^2\left(1-\frac{8}{x}\right)&=x^2\left(-\frac{15}{x^2}\right)\\ x^2-8x&=-15\\ x^2-8x+15&=0\\ (x-3)(x-5)&=0\\ x&=3,5 \end{aligned}$$

Both $x=3$ and $x=5$ check in the original equation.

15. Multiplying both sides of the equation by $2x$:

$$\begin{aligned} 2x\left(\frac{x}{2}-\frac{4}{x}\right)&=2x\left(-\frac{7}{2}\right)\\ x^2-8&=-7x\\ x^2+7x-8&=0\\ (x+8)(x-1)&=0\\ x&=-8,1 \end{aligned}$$

Both $x=-8$ and $x=1$ check in the original equation.

17. Multiplying both sides of the equation by 6:

$$\begin{aligned} 6\left(\frac{x-3}{2}+\frac{2x}{3}\right)&=6\left(\frac{5}{6}\right)\\ 3(x-3)+2(2x)&=5\\ 3x-9+4x&=5\\ 7x-9&=5\\ 7x&=14\\ x&=2 \end{aligned}$$

Since $x=2$ checks in the original equation, the solution is $x=2$.

19. Multiplying both sides of the equation by 12:

$$\begin{aligned} 12\left(\frac{x+1}{3}+\frac{x-3}{4}\right)&=12\left(\frac{1}{6}\right)\\ 4(x+1)+3(x-3)&=2\\ 4x+4+3x-9&=2\\ 7x-5&=2\\ 7x&=7\\ x&=1 \end{aligned}$$

Since $x=1$ checks in the original equation, the solution is $x=1$.

21. Multiplying both sides of the equation by $5(x+2)$:

$$\begin{aligned} 5(x+2)\bullet\frac{6}{x+2}&=5(x+2)\bullet\frac{3}{5}\\ 30&=3x+6\\ 24&=3x\\ x&=8 \end{aligned}$$

Since $x=8$ checks in the original equation, the solution is $x=8$.

23. Multiplying both sides of the equation by $(y-2)(y-3)$:

$$\begin{aligned} (y-2)(y-3)\bullet\frac{3}{y-2}&=(y-2)(y-3)\bullet\frac{2}{y-3}\\ 3(y-3)&=2(y-2)\\ 3y-9&=2y-4\\ y&=5 \end{aligned}$$

Since $y=5$ checks in the original equation, the solution is $y=5$.

© 2012 Cengage Learning. All Rights Reserved. May not be scanned, copied or duplicated, or posted to a publicly accessible website, in whole or in part.

25. Multiplying both sides of the equation by $3(x-2)$:

$$\begin{aligned} 3(x-2)\left(\frac{x}{x-2}+\frac{2}{3}\right)&=3(x-2)\left(\frac{2}{x-2}\right)\\ 3x+2(x-2)&=6\\ 3x+2x-4&=6\\ 5x-4&=6\\ 5x&=10\\ x&=2 \end{aligned}$$

Since $x=2$ does not check in the original equation, there is no solution.

27. Multiplying both sides of the equation by $2(x-2)$:

$$\begin{aligned} 2(x-2)\left(\frac{x}{x-2}+\frac{3}{2}\right)&=2(x-2)\bullet\frac{9}{2(x-2)}\\ 2x+3(x-2)&=9\\ 2x+3x-6&=9\\ 5x-6&=9\\ 5x&=15\\ x&=3 \end{aligned}$$

Since $x=3$ checks in the original equation, the solution is $x=3$.

29. Multiplying both sides of the equation by $x^2+5x+6=(x+2)(x+3)$:

$$\begin{aligned} (x+2)(x+3)\left(\frac{5}{x+2}+\frac{1}{x+3}\right)&=(x+2)(x+3)\bullet\frac{-1}{(x+2)(x+3)}\\ 5(x+3)+1(x+2)&=-1\\ 5x+15+x+2&=-1\\ 6x+17&=-1\\ 6x&=-18\\ x&=-3 \end{aligned}$$

Since $x=-3$ does not check in the original equation, there is no solution.

31. Multiplying both sides of the equation by $x^2-4=(x+2)(x-2)$:

$$\begin{aligned} (x+2)(x-2)\left(\frac{8}{x^2-4}+\frac{3}{x+2}\right)&=(x+2)(x-2)\bullet\frac{1}{x-2}\\ 8+3(x-2)&=1(x+2)\\ 8+3x-6&=x+2\\ 3x+2&=x+2\\ 2x&=0\\ x&=0 \end{aligned}$$

Since $x=0$ checks in the original equation, the solution is $x=0$.

33. Multiplying both sides of the equation by $2(a-3)$:

$$\begin{aligned} 2(a-3)\left(\frac{a}{2}+\frac{3}{a-3}\right)&=2(a-3)\bullet\frac{a}{a-3}\\ a(a-3)+6&=2a\\ a^2-3a+6&=2a\\ a^2-5a+6&=0\\ (a-2)(a-3)&=0\\ a&=2,3 \end{aligned}$$

Since $a=3$ does not check in the original equation, the solution is $a=2$.

© 2012 Cengage Learning. All Rights Reserved. May not be scanned, copied or duplicated, or posted to a publicly accessible website, in whole or in part.

35. Since $y^2-4=(y+2)(y-2)$ and $y^2+2y=y(y+2)$, the LCD is $y(y+2)(y-2)$. Multiplying by the LCD:

$$\begin{aligned}y(y+2)(y-2)\bullet\frac{6}{(y+2)(y-2)}&=y(y+2)(y-2)\bullet\frac{4}{y(y+2)}\\6y&=4(y-2)\\6y&=4y-8\\2y&=-8\\y&=-4\end{aligned}$$

Since $y=-4$ checks in the original equation, the solution is $y=-4$.

37. Since $a^2-9=(a+3)(a-3)$ and $a^2+a-12=(a+4)(a-3)$, the LCD is $(a+3)(a-3)(a+4)$. Multiplying by the LCD:

$$\begin{aligned}(a+3)(a-3)(a+4)\bullet\frac{2}{(a+3)(a-3)}&=(a+3)(a-3)(a+4)\bullet\frac{3}{(a+4)(a-3)}\\2(a+4)&=3(a+3)\\2a+8&=3a+9\\-a+8&=9\\-a&=1\\a&=-1\end{aligned}$$

Since $a=-1$ checks in the original equation, the solution is $a=-1$.

39. Multiplying both sides of the equation by $x^2-4x-5=(x-5)(x+1)$:

$$\begin{aligned}(x-5)(x+1)\left(\frac{3x}{x-5}-\frac{2x}{x+1}\right)&=(x-5)(x+1)\bullet\frac{-42}{(x-5)(x+1)}\\3x(x+1)-2x(x-5)&=-42\\3x^2+3x-2x^2+10x&=-42\\x^2+13x+42&=0\\(x+7)(x+6)&=0\\x&=-7,-6\end{aligned}$$

Both $x=-7$ and $x=-6$ check in the original equation.

41. Multiplying both sides of the equation by $x^2+5x+6=(x+2)(x+3)$:

$$\begin{aligned}(x+2)(x+3)\left(\frac{2x}{x+2}\right)&=(x+2)(x+3)\left(\frac{x}{x+3}-\frac{3}{x^2+5x+6}\right)\\2x(x+3)&=x(x+2)-3\\2x^2+6x&=x^2+2x-3\\x^2+4x+3&=0\\(x+3)(x+1)&=0\\x&=-3,-1\end{aligned}$$

Since $x=-3$ does not check in the original equation, the solution is $x=-1$.

© 2012 Cengage Learning. All Rights Reserved. May not be scanned, copied or duplicated, or posted to a publicly accessible website, in whole or in part.

43. **a.** Solving the equation:

$$5x-1=0$$
$$5x=1$$
$$x=\frac{1}{5}$$

b. Solving the equation:

$$\frac{5}{x}-1=0$$
$$x\left(\frac{5}{x}-1\right)=x(0)$$
$$5-x=0$$
$$x=5$$

c. Solving the equation:

$$\frac{x}{5}-1=\frac{2}{3}$$
$$15\left(\frac{x}{5}-1\right)=15\left(\frac{2}{3}\right)$$
$$3x-15=10$$
$$3x=25$$
$$x=\frac{25}{3}$$

d. Solving the equation:

$$\frac{5}{x}-1=\frac{2}{3}$$
$$3x\left(\frac{5}{x}-1\right)=3x\left(\frac{2}{3}\right)$$
$$15-3x=2x$$
$$15=5x$$
$$x=3$$

e. Solving the equation:

$$\frac{5}{x^2}+5=\frac{26}{x}$$
$$x^2\left(\frac{5}{x^2}+5\right)=x^2\left(\frac{26}{x}\right)$$
$$5+5x^2=26x$$
$$5x^2-26x+5=0$$
$$(5x-1)(x-5)=0$$
$$x=\frac{1}{5},5$$

45. **a.** Dividing: $\frac{7}{a^2-5a-6}\div\frac{a+2}{a+1}=\frac{7}{(a-6)(a+1)}\bullet\frac{a+1}{a+2}=\frac{7}{(a-6)(a+2)}$

b. Adding:

$$\frac{7}{a^2-5a-6}+\frac{a+2}{a+1}=\frac{7}{(a-6)(a+1)}+\frac{a+2}{a+1}\bullet\frac{a-6}{a-6}$$
$$=\frac{7}{(a-6)(a+1)}+\frac{a^2-4a-12}{(a-6)(a+1)}$$
$$=\frac{a^2-4a-5}{(a-6)(a+1)}$$
$$=\frac{(a-5)(a+1)}{(a-6)(a+1)}$$
$$=\frac{a-5}{a-6}$$

© 2012 Cengage Learning. All Rights Reserved. May not be scanned, copied or duplicated, or posted to a publicly accessible website, in whole or in part.

c. Solving the equation:

$$\frac{7}{a^2-5a-6}+\frac{a+2}{a+1}=2$$
$$\frac{7}{(a-6)(a+1)}+\frac{a+2}{a+1}=2$$
$$(a-6)(a+1)\left(\frac{7}{(a-6)(a+1)}+\frac{a+2}{a+1}\right)=2(a-6)(a+1)$$
$$7+(a-6)(a+2)=2(a-6)(a+1)$$
$$7+a^2-4a-12=2a^2-10a-12$$
$$0=a^2-6a-7$$
$$0=(a-7)(a+1)$$
$$a=-1,7$$

Upon checking, only $a = 7$ checks in the original equation.

47. Multiplying both sides of the equation by $2x$:

$$2x\left(\frac{1}{x}+\frac{1}{2x}\right)=2x\left(\frac{9}{2}\right)$$
$$2+1=9x$$
$$9x=3$$
$$x=\frac{1}{3}$$

49. Multiplying both sides of the equation by $30x$:

$$30x\left(\frac{1}{10}-\frac{1}{15}\right)=30x\left(\frac{1}{x}\right)$$
$$3x-2x=30$$
$$x=30$$

51. Evaluating when $x = -6$: $y=-\frac{6}{-6}=1$

53. Evaluating when $x = 2$: $y=-\frac{6}{2}=-3$

55. Finding the speed: $v=\frac{420\text{ meters}}{27\text{ minutes}}\bullet\frac{100\text{ cm}}{1\text{ meter}}\bullet\frac{1\text{ minute}}{60\text{ seconds}}\approx 26\text{ cm/second}$

57. Let x represent the number. The equation is:

$$2(x-3)-5=3$$
$$2x-6-5=3$$
$$2x-11=3$$
$$2x=14$$
$$x=7$$

The number is 7.

59. Let w represent the width, and $2w + 5$ represent the length. Using the perimeter formula:

$$2w+2(2w+5)=34$$
$$2w+4w+10=34$$
$$6w+10=34$$
$$6w=24$$
$$w=4$$
$$2w+5=13$$

The length is 13 inches and the width is 4 inches.

© 2012 Cengage Learning. All Rights Reserved. May not be scanned, copied or duplicated, or posted to a publicly accessible website, in whole or in part.

61. Let x and $x + 2$ represent the two integers. The equation is:

$$\begin{aligned} x(x+2) &= 48 \\ x^2+2x &= 48 \\ x^2+2x-48 &= 0 \\ (x+8)(x-6) &= 0 \\ x &= -8, 6 \\ x+2 &= -6, 8 \end{aligned}$$

The two integers are either –8 and –6, or 6 and 8.

63. Let x and $x + 2$ represent the two legs. The equation is:

$$\begin{aligned} x^2+(x+2)^2 &= 10^2 \\ x^2+x^2+4x+4 &= 100 \\ 2x^2+4x-96 &= 0 \\ x^2+2x-48 &= 0 \\ (x+8)(x-6) &= 0 \\ x &= 6 \quad (x=-8 \text{ is impossible}) \\ x+2 &= 8 \end{aligned}$$

The legs are 6 inches and 8 inches.

7.5 Applications of Rational Expressions

1. Let x and $3x$ represent the two numbers. The equation is:

$$\begin{aligned} \frac{1}{x}+\frac{1}{3x} &= \frac{16}{3} \\ 3x\left(\frac{1}{x}+\frac{1}{3x}\right) &= 3x\left(\frac{16}{3}\right) \\ 3+1 &= 16x \\ 16x &= 4 \\ x &= \frac{1}{4} \\ 3x &= \frac{3}{4} \end{aligned}$$

The numbers are $\frac{1}{4}$ and $\frac{3}{4}$.

3. Let x represent the number. The equation is:

$$\begin{aligned} x+\frac{1}{x} &= \frac{13}{6} \\ 6x\left(x+\frac{1}{x}\right) &= 6x\left(\frac{13}{6}\right) \\ 6x^2+6 &= 13x \\ 6x^2-13x+6 &= 0 \\ (3x-2)(2x-3) &= 0 \\ x &= \frac{2}{3}, \frac{3}{2} \end{aligned}$$

The number is either $\frac{2}{3}$ or $\frac{3}{2}$.

5. Let x represent the number. The equation is:

$$\begin{aligned} \frac{7+x}{9+x} &= \frac{5}{7} \\ 7(9+x)\bullet\frac{7+x}{9+x} &= 7(9+x)\bullet\frac{5}{7} \\ 7(7+x) &= 5(9+x) \\ 49+7x &= 45+5x \\ 49+2x &= 45 \\ 2x &= -4 \\ x &= -2 \end{aligned}$$

The number is –2.

© 2012 Cengage Learning. All Rights Reserved. May not be scanned, copied or duplicated, or posted to a publicly accessible website, in whole or in part.

7. Let x and $x + 2$ represent the two integers. The equation is:

$$\frac{1}{x}+\frac{1}{x+2}=\frac{5}{12}$$
$$12x(x+2)\left(\frac{1}{x}+\frac{1}{x+2}\right)=12x(x+2)\left(\frac{5}{12}\right)$$
$$12(x+2)+12x=5x(x+2)$$
$$12x+24+12x=5x^2+10x$$
$$0=5x^2-14x-24$$
$$(5x+6)(x-4)=0$$
$$x=4 \quad \left(x=-\frac{6}{5}\text{ is impossible}\right)$$
$$x+2=6$$

The integers are 4 and 6.

9. Let x represent the rate of the boat in still water:

	d	r	t
Upstream	26	$x-3$	$\frac{26}{x-3}$
Downstream	38	$x+3$	$\frac{38}{x+3}$

The equation is:

$$\frac{26}{x-3}=\frac{38}{x+3}$$
$$(x+3)(x-3)\bullet\frac{26}{x-3}=(x+3)(x-3)\bullet\frac{38}{x+3}$$
$$26(x+3)=38(x-3)$$
$$26x+78=38x-114$$
$$-12x+78=-114$$
$$-12x=-192$$
$$x=16$$

The speed of the boat in still water is 16 mph.

11. Let x represent the plane speed in still air:

	d	r	t
Against Wind	140	$x-20$	$\frac{140}{x-20}$
With Wind	160	$x+20$	$\frac{160}{x+20}$

The equation is:

$$\frac{140}{x-20}=\frac{160}{x+20}$$
$$(x+20)(x-20)\bullet\frac{140}{x-20}=(x+20)(x-20)\bullet\frac{160}{x+20}$$
$$140(x+20)=160(x-20)$$
$$140x+2800=160x-3200$$
$$-20x+2800=-3200$$
$$-20x=-6000$$
$$x=300$$

The plane speed in still air is 300 mph.

13. Let x and $x + 20$ represent the rates of each plane:

	d	r	t
Plane 1	285	$x+20$	$\frac{285}{x+20}$
Plane 2	255	x	$\frac{255}{x}$

The equation is:

$$\frac{285}{x+20}=\frac{255}{x}$$
$$x(x+20)\bullet\frac{285}{x+20}=x(x+20)\bullet\frac{255}{x}$$
$$285x=255(x+20)$$
$$285x=255x+5100$$
$$30x=5100$$
$$x=170$$
$$x+20=190$$

The plane speeds are 170 mph and 190 mph.

15. Let x represent her rate downhill:

	d	r	t
Level Ground	2	$x-3$	$\frac{2}{x-3}$
Downhill	6	x	$\frac{6}{x}$

The equation is:

$$\frac{2}{x-3}+\frac{6}{x}=1$$
$$x(x-3)\left(\frac{2}{x-3}+\frac{6}{x}\right)=x(x-3)\bullet 1$$
$$2x+6(x-3)=x(x-3)$$
$$2x+6x-18=x^2-3x$$
$$8x-18=x^2-3x$$
$$0=x^2-11x+18$$
$$0=(x-2)(x-9)$$
$$x=9 \quad (x=2\text{ is impossible})$$

Tina runs 9 mph on the downhill part of the course.

© 2012 Cengage Learning. All Rights Reserved. May not be scanned, copied or duplicated, or posted to a publicly accessible website, in whole or in part.

17. Let x represent her rate on level ground:

	d	r	t
Level Ground	4	x	$\frac{4}{x}$
Downhill	5	$x+2$	$\frac{5}{x+2}$

The equation is:

$$\begin{aligned}\frac{4}{x}+\frac{5}{x+2}&=1\\ x(x+2)\left(\frac{4}{x}+\frac{5}{x+2}\right)&=x(x+2)\bullet 1\\ 4(x+2)+5x&=x(x+2)\\ 4x+8+5x&=x^2+2x\\ 9x+8&=x^2+2x\\ 0&=x^2-7x-8\\ 0&=(x-8)(x+1)\\ x&=8 \quad (x=-1 \text{ is impossible})\end{aligned}$$

Jerri jogs 8 mph on level ground.

19. Let t represent the time to fill the pool with both pipes left open. The equation is:

$$\begin{aligned}\frac{1}{12}-\frac{1}{15}&=\frac{1}{t}\\ 60t\left(\frac{1}{12}-\frac{1}{15}\right)&=60t\bullet\frac{1}{t}\\ 5t-4t&=60\\ t&=60\end{aligned}$$

It will take 60 hours to fill the pool with both pipes left open.

21. Let t represent the time to fill the bathtub with both faucets open. The equation is:

$$\begin{aligned}\frac{1}{10}+\frac{1}{12}&=\frac{1}{t}\\ 60t\left(\frac{1}{10}+\frac{1}{12}\right)&=60t\bullet\frac{1}{t}\\ 6t+5t&=60\\ 11t&=60\\ t&=\frac{60}{11}=5\frac{5}{11}\end{aligned}$$

It will take $5\frac{5}{11}$ minutes to fill the tub with both faucets open.

23. Let t represent the time to fill the sink with both the faucet and the drain left open. The equation is:

$$\begin{aligned}\frac{1}{3}-\frac{1}{4}&=\frac{1}{t}\\ 12t\left(\frac{1}{3}-\frac{1}{4}\right)&=12t\bullet\frac{1}{t}\\ 4t-3t&=12\\ t&=12\end{aligned}$$

It will take 12 minutes for the sink to overflow with both the faucet and drain left open.

© 2012 Cengage Learning. All Rights Reserved. May not be scanned, copied or duplicated, or posted to a publicly accessible website, in whole or in part.

25. Sketching the graph:

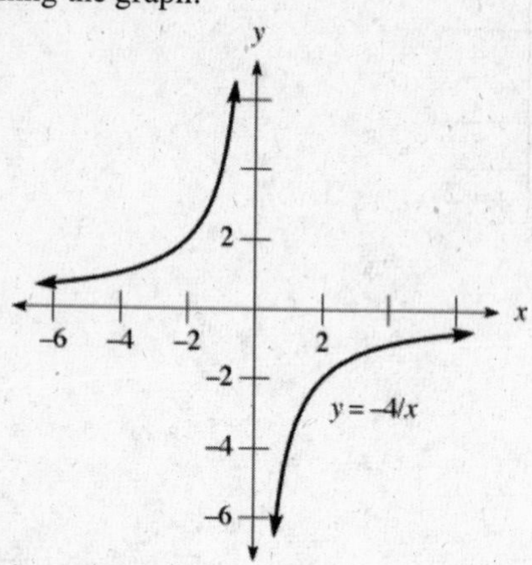

27. Sketching the graph:

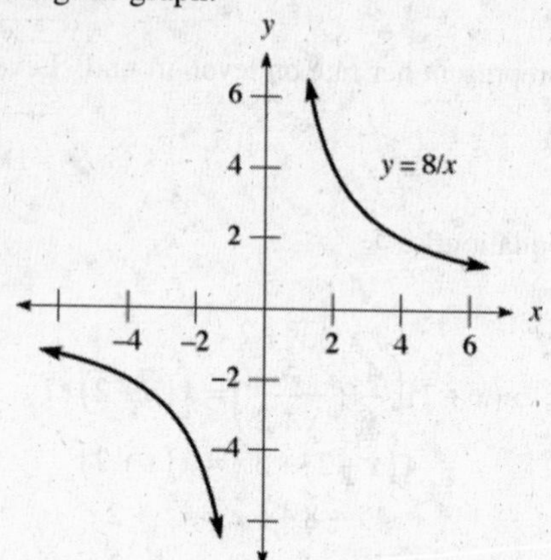

29. Sketching the graph:

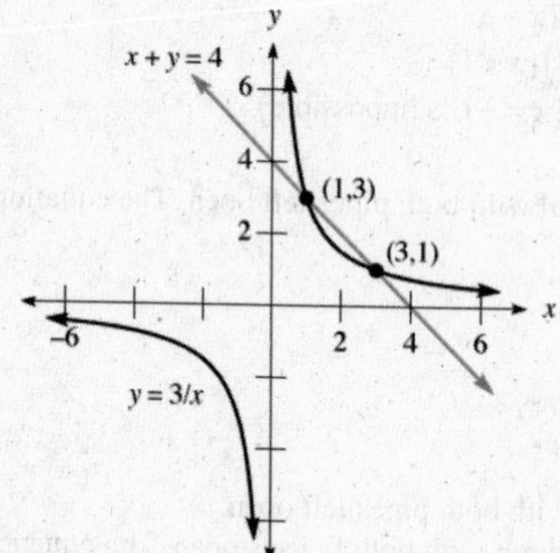

The intersection points are (1,3) and (3,1).

31. Simplifying: $\frac{1}{2} \div \frac{2}{3} = \frac{1}{2} \bullet \frac{3}{2} = \frac{3}{4}$

33. Simplifying: $1 + \frac{1}{2} = \frac{2}{2} + \frac{1}{2} = \frac{3}{2}$

35. Simplifying: $y^5 \bullet \frac{2x^3}{y^2} = \frac{2x^3y^5}{y^2} = 2x^3y^3$

37. Simplifying: $\frac{2x^3}{y^2} \bullet \frac{y^5}{4x} = \frac{2x^3y^5}{4xy^2} = \frac{x^2y^3}{2}$

39. Factoring: $x^2y + x = x(xy+1)$

41. Reducing the fraction: $\frac{2x^3y^2}{4x} = \frac{x^2y^2}{2}$

43. Reducing the fraction: $\frac{x^2-4}{x^2-x-6} = \frac{(x+2)(x-2)}{(x+2)(x-3)} = \frac{x-2}{x-3}$

45. Finding the amount:

$$0.42(x) = 8.36$$

$$x = \frac{8.36}{0.42} \approx 19.9 \text{ million}$$

47. Factoring the polynomial: $15a^3b^3 - 20a^2b - 35ab^2 = 5ab\left(3a^2b^2 - 4a - 7b\right)$

49. Factoring the polynomial: $x^2 - 4x - 12 = (x-6)(x+2)$

51. Factoring the polynomial: $x^4 - 16 = \left(x^2+4\right)\left(x^2-4\right) = \left(x^2+4\right)(x+2)(x-2)$

53. Factoring the polynomial: $5x^3 - 25x^2 - 30x = 5x\left(x^2-5x-6\right) = 5x(x-6)(x+1)$

© 2012 Cengage Learning. All Rights Reserved. May not be scanned, copied or duplicated, or posted to a publicly accessible website, in whole or in part.

7.6 Complex Fractions

1. Simplifying the complex fraction: $\dfrac{\frac{3}{4}}{\frac{1}{8}}=\dfrac{\frac{3}{4}\bullet 8}{\frac{1}{8}\bullet 8}=\dfrac{6}{1}=6$

3. Simplifying the complex fraction: $\dfrac{\frac{2}{3}}{4}=\dfrac{\frac{2}{3}\bullet 3}{4\bullet 3}=\dfrac{2}{12}=\dfrac{1}{6}$

5. Simplifying the complex fraction: $\dfrac{\frac{x^2}{y}}{\frac{x}{y^3}}=\dfrac{\frac{x^2}{y}\bullet y^3}{\frac{x}{y^3}\bullet y^3}=\dfrac{x^2y^2}{x}=xy^2$

7. Simplifying the complex fraction: $\dfrac{\frac{4x^3}{y^6}}{\frac{8x^2}{y^7}}=\dfrac{\frac{4x^3}{y^6}\bullet y^7}{\frac{8x^2}{y^7}\bullet y^7}=\dfrac{4x^3y}{8x^2}=\dfrac{xy}{2}$

9. Simplifying the complex fraction: $\dfrac{y+\frac{1}{x}}{x+\frac{1}{y}}=\dfrac{\left(y+\frac{1}{x}\right)\bullet xy}{\left(x+\frac{1}{y}\right)\bullet xy}=\dfrac{xy^2+y}{x^2y+x}=\dfrac{y(xy+1)}{x(xy+1)}=\dfrac{y}{x}$

11. Simplifying the complex fraction: $\dfrac{1+\frac{1}{a}}{1-\frac{1}{a}}=\dfrac{\left(1+\frac{1}{a}\right)\bullet a}{\left(1-\frac{1}{a}\right)\bullet a}=\dfrac{a+1}{a-1}$

13. Simplifying the complex fraction: $\dfrac{\frac{x+1}{x^2-9}}{\frac{2}{x+3}}=\dfrac{\frac{x+1}{(x+3)(x-3)}\bullet(x+3)(x-3)}{\frac{2}{x+3}\bullet(x+3)(x-3)}=\dfrac{x+1}{2(x-3)}$

15. Simplifying the complex fraction: $\dfrac{\frac{1}{a+2}}{\frac{1}{a^2-a-6}}=\dfrac{\frac{1}{a+2}\bullet(a-3)(a+2)}{\frac{1}{(a-3)(a+2)}\bullet(a-3)(a+2)}=\dfrac{a-3}{1}=a-3$

17. Simplifying the complex fraction: $\dfrac{1-\frac{9}{y^2}}{1-\frac{1}{y}-\frac{6}{y^2}}=\dfrac{\left(1-\frac{9}{y^2}\right)\bullet y^2}{\left(1-\frac{1}{y}-\frac{6}{y^2}\right)\bullet y^2}=\dfrac{y^2-9}{y^2-y-6}=\dfrac{(y+3)(y-3)}{(y+2)(y-3)}=\dfrac{y+3}{y+2}$

19. Simplifying the complex fraction: $\dfrac{\frac{1}{y}+\frac{1}{x}}{\frac{1}{xy}}=\dfrac{\left(\frac{1}{y}+\frac{1}{x}\right)\bullet xy}{\left(\frac{1}{xy}\right)\bullet xy}=\dfrac{x+y}{1}=x+y$

21. Simplifying the complex fraction: $\dfrac{1-\frac{1}{a^2}}{1-\frac{1}{a}}=\dfrac{\left(1-\frac{1}{a^2}\right)\bullet a^2}{\left(1-\frac{1}{a}\right)\bullet a^2}=\dfrac{a^2-1}{a^2-a}=\dfrac{(a+1)(a-1)}{a(a-1)}=\dfrac{a+1}{a}$

© 2012 Cengage Learning. All Rights Reserved. May not be scanned, copied or duplicated, or posted to a publicly accessible website, in whole or in part.

23. Simplifying the complex fraction: $\dfrac{\dfrac{1}{10x}-\dfrac{y}{10x^2}}{\dfrac{1}{10}-\dfrac{y}{10x}}=\dfrac{\left(\dfrac{1}{10x}-\dfrac{y}{10x^2}\right)\bullet 10x^2}{\left(\dfrac{1}{10}-\dfrac{y}{10x}\right)\bullet 10x^2}=\dfrac{x-y}{x^2-xy}=\dfrac{1(x-y)}{x(x-y)}=\dfrac{1}{x}$

25. Simplifying the complex fraction: $\dfrac{\dfrac{1}{a+1}+2}{\dfrac{1}{a+1}+3}=\dfrac{\left(\dfrac{1}{a+1}+2\right)\bullet(a+1)}{\left(\dfrac{1}{a+1}+3\right)\bullet(a+1)}=\dfrac{1+2(a+1)}{1+3(a+1)}=\dfrac{1+2a+2}{1+3a+3}=\dfrac{2a+3}{3a+4}$

27. Simplifying each parenthesis first:

$$1-\frac{1}{x}=\frac{x}{x}-\frac{1}{x}=\frac{x-1}{x}$$

$$1-\frac{1}{x+1}=\frac{x+1}{x+1}-\frac{1}{x+1}=\frac{x}{x+1}$$

$$1-\frac{1}{x+2}=\frac{x+2}{x+2}-\frac{1}{x+2}=\frac{x+1}{x+2}$$

Now performing the multiplication: $\left(1-\dfrac{1}{x}\right)\left(1-\dfrac{1}{x+1}\right)\left(1-\dfrac{1}{x+2}\right)=\dfrac{x-1}{x}\bullet\dfrac{x}{x+1}\bullet\dfrac{x+1}{x+2}=\dfrac{x-1}{x+2}$

29. Simplifying each parenthesis first:

$$1+\frac{1}{x+3}=\frac{x+3}{x+3}+\frac{1}{x+3}=\frac{x+4}{x+3}$$

$$1+\frac{1}{x+2}=\frac{x+2}{x+2}+\frac{1}{x+2}=\frac{x+3}{x+2}$$

$$1+\frac{1}{x+1}=\frac{x+1}{x+1}+\frac{1}{x+1}=\frac{x+2}{x+1}$$

Now performing the multiplication: $\left(1+\dfrac{1}{x+3}\right)\left(1+\dfrac{1}{x+2}\right)\left(1+\dfrac{1}{x+1}\right)=\dfrac{x+4}{x+3}\bullet\dfrac{x+3}{x+2}\bullet\dfrac{x+2}{x+1}=\dfrac{x+4}{x+1}$

31. Simplifying each term in the sequence:

$$2+\frac{1}{2+1}=2+\frac{1}{3}=\frac{6}{3}+\frac{1}{3}=\frac{7}{3}$$

$$2+\frac{1}{2+\dfrac{1}{2+1}}=2+\frac{1}{\dfrac{7}{3}}=2+\frac{3}{7}=\frac{14}{7}+\frac{3}{7}=\frac{17}{7}$$

$$2+\frac{2}{2+\dfrac{1}{2+\dfrac{1}{2+1}}}=2+\frac{1}{\dfrac{17}{7}}=2+\frac{7}{17}=\frac{34}{17}+\frac{7}{17}=\frac{41}{17}$$

33. Completing the table:

Number x	Reciprocal $\frac{1}{x}$	Quotient $\frac{x}{1/x}$	Square x^2
1	1	1	1
2	$\frac{1}{2}$	4	4
3	$\frac{1}{3}$	9	9
4	$\frac{1}{4}$	16	16

35. Completing the table:

Number x	Reciprocal $\frac{1}{x}$	Sum $1+\frac{1}{x}$	Quotient $\dfrac{1+\frac{1}{x}}{\frac{1}{x}}$
1	1	2	2
2	$\frac{1}{2}$	$\frac{3}{2}$	3
3	$\frac{1}{3}$	$\frac{4}{3}$	4
4	$\frac{1}{4}$	$\frac{5}{4}$	5

© 2012 Cengage Learning. All Rights Reserved. May not be scanned, copied or duplicated, or posted to a publicly accessible website, in whole or in part.

37. Converting the speed: $v = \frac{253}{\frac{100}{161}} = 253 \bullet \frac{161}{100} \approx 407$ kilometers per hour

39. Solving the equation:
$$21 = 6x$$
$$x = \frac{21}{6} = \frac{7}{2}$$

41. Solving the equation:
$$x^2 + x = 6$$
$$x^2 + x - 6 = 0$$
$$(x+3)(x-2) = 0$$
$$x = -3, 2$$

43. Solving the inequality:
$$2x + 3 < 5$$
$$2x + 3 - 3 < 5 - 3$$
$$2x < 2$$
$$\frac{1}{2}(2x) < \frac{1}{2}(2)$$
$$x < 1$$

45. Solving the inequality:
$$-3x \le 21$$
$$-\frac{1}{3}(-3x) \ge -\frac{1}{3}(21)$$
$$x \ge -7$$

47. Solving the inequality:
$$-2x + 8 > -4$$
$$-2x + 8 - 8 > -4 - 8$$
$$-2x > -12$$
$$-\frac{1}{2}(-2x) < -\frac{1}{2}(-12)$$
$$x < 6$$

49. Solving the inequality:
$$4 - 2(x+1) \ge -2$$
$$4 - 2x - 2 \ge -2$$
$$-2x + 2 \ge -2$$
$$-2x + 2 - 2 \ge -2 - 2$$
$$-2x \ge -4$$
$$-\frac{1}{2}(-2x) \le -\frac{1}{2}(-4)$$
$$x \le 2$$

7.7 Proportions

1. Solving the proportion:
$$\frac{x}{2} = \frac{6}{12}$$
$$12x = 12$$
$$x = 1$$

3. Solving the proportion:
$$\frac{2}{5} = \frac{4}{x}$$
$$2x = 20$$
$$x = 10$$

5. Solving the proportion:
$$\frac{10}{20} = \frac{20}{x}$$
$$10x = 400$$
$$x = 40$$

7. Solving the proportion:
$$\frac{a}{3} = \frac{5}{12}$$
$$12a = 15$$
$$a = \frac{15}{12} = \frac{5}{4}$$

9. Solving the proportion:
$$\frac{2}{x} = \frac{6}{7}$$
$$6x = 14$$
$$x = \frac{14}{6} = \frac{7}{3}$$

11. Solving the proportion:
$$\frac{x+1}{3} = \frac{4}{x}$$
$$x^2 + x = 12$$
$$x^2 + x - 12 = 0$$
$$(x+4)(x-3) = 0$$
$$x = -4, 3$$

13. Solving the proportion:
$$\frac{x}{2} = \frac{8}{x}$$
$$x^2 = 16$$
$$x^2 - 16 = 0$$
$$(x+4)(x-4) = 0$$
$$x = -4, 4$$

15. Solving the proportion:
$$\frac{4}{a+2} = \frac{a}{2}$$
$$a^2 + 2a = 8$$
$$a^2 + 2a - 8 = 0$$
$$(a+4)(a-2) = 0$$
$$a = -4, 2$$

© 2012 Cengage Learning. All Rights Reserved. May not be scanned, copied or duplicated, or posted to a publicly accessible website, in whole or in part.

17. Solving the proportion:

$$\frac{1}{x}=\frac{x-5}{6}$$
$$x^2-5x=6$$
$$x^2-5x-6=0$$
$$(x-6)(x+1)=0$$
$$x=-1,6$$

19. Solving the proportion:

$$\frac{9}{7}=\frac{x}{9.8}$$
$$7x=88.2$$
$$x=\frac{88.2}{7}=12.6$$

The consumption was 12.6 million metric tons.

21. Comparing hits to games, the proportion is:

$$\frac{6}{18}=\frac{x}{45}$$
$$18x=270$$
$$x=15$$

He will get 15 hits in 45 games.

23. Comparing ml alcohol to ml water, the proportion is:

$$\frac{12}{16}=\frac{x}{28}$$
$$16x=336$$
$$x=21$$

The solution will have 21 ml of alcohol.

25. Comparing grams of fat to total grams, the proportion is:

$$\frac{13}{100}=\frac{x}{350}$$
$$100x=4550$$
$$x=45.5$$

There are 45.5 grams of fat in 350 grams of ice cream.

27. Comparing inches on the map to actual miles, the proportion is:

$$\frac{3.5}{100}=\frac{x}{420}$$
$$100x=1470$$
$$x=14.7$$

They are 14.7 inches apart on the map.

29. Comparing inches on the map to actual miles, the proportion is:

$$\frac{0.5}{5}=\frac{1.25}{x}$$
$$0.5x=6.25$$
$$x=12.5$$

The actual distance is 12.5 miles.

31. Comparing driving time to distance:

$$\frac{46}{34}=\frac{x}{12.5}$$
$$34x=575$$
$$x\approx 17$$

The driving time is approximately 17 minutes.

33. Comparing miles to hours, the proportion is:

$$\frac{245}{5}=\frac{x}{7}$$
$$5x=1715$$
$$x=343$$

He will travel 343 miles.

35. Solving the proportion:

$$\frac{h}{6}=\frac{6}{4}$$
$$4h=36$$
$$h=9$$

37. Solving the proportion:

$$\frac{y}{21}=\frac{8}{12}$$
$$12y=168$$
$$y=14$$

39. Substituting $y = 50$:

$$50=2x^2$$
$$2x^2-50=0$$
$$2(x^2-25)=0$$
$$2(x+5)(x-5)=0$$
$$x=-5,5$$

41. Substituting $y = 15$ and $x = 3$:

$$15=K\bullet 3$$
$$K=5$$

© 2012 Cengage Learning. All Rights Reserved. May not be scanned, copied or duplicated, or posted to a publicly accessible website, in whole or in part.

43. Substituting $y = 32$ and $x = 4$:

$$32 = K(4)^2$$
$$32 = 16K$$
$$K = 2$$

45. Reducing the fraction: $\frac{x^2 - x - 6}{x^2 - 9} = \frac{(x-3)(x+2)}{(x+3)(x-3)} = \frac{x+2}{x+3}$

47. Multiplying the fractions:

$$\frac{x^2 - 25}{x+4} \bullet \frac{2x+8}{x^2 - 9x + 20} = \frac{(x+5)(x-5)}{x+4} \bullet \frac{2(x+4)}{(x-5)(x-4)} = \frac{2(x+5)(x-5)(x+4)}{(x+4)(x-5)(x-4)} = \frac{2(x+5)}{x-4}$$

49. Adding the fractions: $\frac{x}{x^2 - 16} + \frac{4}{x^2 - 16} = \frac{x+4}{x^2 - 16} = \frac{1(x+4)}{(x+4)(x-4)} = \frac{1}{x-4}$

51. Answers will vary.

7.8 Direct and Inverse Variation

1. The variation equation is $y = Kx$. Finding K:

$$10 = K \bullet 5$$
$$K = 2$$

So $y = 2x$. Substituting $x = 4$: $y = 2 \bullet 4 = 8$

3. The variation equation is $y = Kx$. Finding K:

$$39 = K \bullet 3$$
$$K = 13$$

So $y = 13x$. Substituting $x = 10$: $y = 13 \bullet 10 = 130$

5. The variation equation is $y = Kx$. Finding K:

$$-24 = K \bullet 4$$
$$K = -6$$

So $y = -6x$. Substituting $y = -30$:

$$-6x = -30$$
$$x = 5$$

7. The variation equation is $y = Kx$. Finding K:

$$-7 = K \bullet (-1)$$
$$K = 7$$

So $y = 7x$. Substituting $y = -21$:

$$7x = -21$$
$$x = -3$$

9. The variation equation is $y = Kx^2$. Finding K:

$$75 = K \bullet 5^2$$
$$75 = 25K$$
$$K = 3$$

So $y = 3x^2$. Substituting $x = 1$: $y = 3 \bullet 1^2 = 3 \bullet 1 = 3$

11. The variation equation is $y = Kx^2$. Finding K:

$$48 = K \bullet 4^2$$
$$48 = 16K$$
$$K = 3$$

So $y = 3x^2$. Substituting $x = 9$: $y = 3 \bullet 9^2 = 3 \bullet 81 = 243$

13. The variation equation is $y = \frac{K}{x}$. Finding K:

$$5 = \frac{K}{2}$$
$$K = 10$$

So $y = \frac{10}{x}$. Substituting $x = 5$: $y = \frac{10}{5} = 2$

15. The variation equation is $y = \frac{K}{x}$. Finding K:

$$2 = \frac{K}{1}$$
$$K = 2$$

So $y = \frac{2}{x}$. Substituting $x = 4$: $y = \frac{2}{4} = \frac{1}{2}$

© 2012 Cengage Learning. All Rights Reserved. May not be scanned, copied or duplicated, or posted to a publicly accessible website, in whole or in part.

17. The variation equation is $y=\frac{K}{x}$. Finding K:

$$5=\frac{K}{3}$$
$$K=15$$

So $y=\frac{15}{x}$. Substituting $y=15$:

$$\frac{15}{x}=15$$
$$15=15x$$
$$x=1$$

19. The variation equation is $y=\frac{K}{x}$. Finding K:

$$10=\frac{K}{10}$$
$$K=100$$

So $y=\frac{100}{x}$. Substituting $y=100$:

$$\frac{100}{x}=20$$
$$100=20x$$
$$x=5$$

21. The variation equation is $y=\frac{K}{x^2}$. Finding K:

$$4=\frac{K}{5^2}$$
$$4=\frac{K}{25}$$
$$K=100$$

So $y=\frac{100}{x^2}$. Substituting $x=2$: $y=\frac{100}{2^2}=\frac{100}{4}=25$

23. The variation equation is $y=\frac{K}{x^2}$. Finding K:

$$4=\frac{K}{3^2}$$
$$4=\frac{K}{9}$$
$$K=36$$

So $y=\frac{36}{x^2}$. Substituting $x=2$: $y=\frac{36}{2^2}=\frac{36}{4}=9$

25. The variation equation is $t=\frac{K}{s}$. Finding K:

$$2=\frac{K}{257}$$
$$K=514$$

So $t=\frac{514}{s}$. Substituting $s=248$: $t=\frac{514}{248}=\frac{257}{124}\approx 2.1$ hours.

27. The variation equation is $t=Kd$. Finding K:

$$42=K\bullet 2$$
$$K=21$$

So $t=21d$. Substituting $d=4$: $t=21\bullet 4=84$ pounds

29. The variation equation is $P=KI^2$. Finding K:

$$30=K\bullet 2^2$$
$$30=4K$$
$$K=\frac{15}{2}$$

So $P=\frac{15}{2}I^2$. Substituting $I=7$: $P=\frac{15}{2}\bullet 7^2=\frac{15}{2}\bullet 49=\frac{735}{2}=367.5$

31. The variation equation is $M=Kh$. Finding K:

$$157=K\bullet 20$$
$$K=7.85$$

So $M=7.85h$. Substituting $h=30$: $M=7.85\bullet 30=\$235.50$

© 2012 Cengage Learning. All Rights Reserved. May not be scanned, copied or duplicated, or posted to a publicly accessible website, in whole or in part.

33. The variation equation is $F=\frac{K}{d^2}$. Finding *K*:

$$150=\frac{K}{4000^2}$$
$$150=\frac{K}{1.6\times10^7}$$
$$K=2.4\times10^9$$

So $F=\frac{2.4\times10^9}{d^2}$. Substituting $d=5000$: $F=\frac{2.4\times10^9}{(5000)^2}=\frac{2.4\times10^9}{2.5\times10^7}=96$ pounds

35. The variation equation is $I=\frac{K}{R}$. Finding *K*:

$$30=\frac{K}{2}$$
$$K=60$$

So $I=\frac{60}{R}$. Substituting $R=5$: $I=\frac{60}{5}=12$ amps

37. Adding the two equations:

$$5x=10$$
$$x=2$$

Substituting into the first equation:

$$2(2)+y=3$$
$$4+y=3$$
$$y=-1$$

The solution is $(2,-1)$.

39. Multiplying the second equation by –4:

$$4x-5y=1$$
$$-4x+8y=8$$

Adding the two equations:

$$3y=9$$
$$y=3$$

Substituting into the second equation:

$$x-2(3)=-2$$
$$x-6=-2$$
$$x=4$$

The solution is $(4,3)$.

41. Substituting into the first equation:

$$5x+2(3x-2)=7$$
$$5x+6x-4=7$$
$$11x-4=7$$
$$11x=11$$
$$x=1$$

Substituting into the second equation: $y=3(1)-2=3-2=1$. The solution is $(1,1)$.

43. Substituting into the first equation:

$$2(2y+1)-3y=4$$
$$4y+2-3y=4$$
$$y+2=4$$
$$y=2$$

Substituting into the second equation: $x=2(2)+1=4+1=5$. The solution is $(5,2)$.

© 2012 Cengage Learning. All Rights Reserved. May not be scanned, copied or duplicated, or posted to a publicly accessible website, in whole or in part.

Chapter 7 Review

1. Reducing the rational expression: $\frac{7}{14x-28}=\frac{7}{14(x-2)}=\frac{1}{2(x-2)}$. The variable restriction is $x \neq 2$.

3. Reducing the rational expression: $\frac{8x-4}{4x+12}=\frac{4(2x-1)}{4(x+3)}=\frac{2x-1}{x+3}$. The variable restriction is $x \neq -3$.

5. Reducing the rational expression: $\frac{3x^3+16x^2-12x}{2x^3+9x^2-18x}=\frac{x\left(3x^2+16x-12\right)}{x\left(2x^2+9x-18\right)}=\frac{x(3x-2)(x+6)}{x(2x-3)(x+6)}=\frac{3x-2}{2x-3}$

The variable restriction is $x \neq -6, 0, \frac{3}{2}$.

7. Reducing the rational expression: $\frac{x^2+5x-14}{x+7}=\frac{(x+7)(x-2)}{x+7}=x-2$. The variable restriction is $x \neq -7$.

9. Reducing the rational expression: $\frac{xy+bx+ay+ab}{xy+5x+ay+5a}=\frac{x(y+b)+a(y+b)}{x(y+5)+a(y+5)}=\frac{(y+b)(x+a)}{(y+5)(x+a)}=\frac{y+b}{y+5}$

The variable restriction is $y \neq -5, x \neq -a$.

11. Performing the operations:

$$\begin{aligned}\frac{x^2+8x+16}{x^2+x-12} \div \frac{x^2-16}{x^2-x-6} &= \frac{x^2+8x+16}{x^2+x-12} \bullet \frac{x^2-x-6}{x^2-16} \\ &= \frac{(x+4)^2}{(x+4)(x-3)} \bullet \frac{(x+2)(x-3)}{(x+4)(x-4)} \\ &= \frac{(x+4)^2(x+2)(x-3)}{(x+4)^2(x-3)(x-4)} \\ &= \frac{x+2}{x-4}\end{aligned}$$

13. Performing the operations:

$$\begin{aligned}\frac{3x^2-2x-1}{x^2+6x+8} \div \frac{3x^2+13x+4}{x^2+8x+16} &= \frac{3x^2-2x-1}{x^2+6x+8} \bullet \frac{x^2+8x+16}{3x^2+13x+4} \\ &= \frac{(3x+1)(x-1)}{(x+4)(x+2)} \bullet \frac{(x+4)^2}{(3x+1)(x+4)} \\ &= \frac{(x+4)^2(3x+1)(x-1)}{(x+4)^2(x+2)(3x+1)} \\ &= \frac{x-1}{x+2}\end{aligned}$$

15. Performing the operations: $\frac{x^2}{x-9}-\frac{18x-81}{x-9}=\frac{x^2-18x+81}{x-9}=\frac{(x-9)^2}{x-9}=x-9$

17. Performing the operations: $\frac{x}{x+9}+\frac{5}{x}=\frac{x \bullet x}{(x+9)\bullet x}+\frac{5 \bullet (x+9)}{x \bullet (x+9)}=\frac{x^2}{x(x+9)}+\frac{5x+45}{x(x+9)}=\frac{x^2+5x+45}{x(x+9)}$

© 2012 Cengage Learning. All Rights Reserved. May not be scanned, copied or duplicated, or posted to a publicly accessible website, in whole or in part.

19. Performing the operations:

$$\begin{aligned}\frac{3}{x^2-36}-\frac{2}{x^2-4x-12}&=\frac{3}{(x+6)(x-6)}-\frac{2}{(x-6)(x+2)}\\&=\frac{3(x+2)}{(x+6)(x-6)(x+2)}-\frac{2(x+6)}{(x+6)(x-6)(x+2)}\\&=\frac{3x+6}{(x+6)(x-6)(x+2)}-\frac{2x+12}{(x+6)(x-6)(x+2)}\\&=\frac{3x+6-2x-12}{(x+6)(x-6)(x+2)}\\&=\frac{x-6}{(x+6)(x-6)(x+2)}\\&=\frac{1}{(x+6)(x+2)}\end{aligned}$$

21. Multiplying both sides of the equation by $2x$:

$$\begin{aligned}2x\left(\frac{3}{x}+\frac{1}{2}\right)&=2x\left(\frac{5}{x}\right)\\6+x&=10\\x&=4\end{aligned}$$

Since $x=4$ checks in the original equation, the solution is $x=4$.

23. Multiplying both sides of the equation by x^2:

$$\begin{aligned}x^2\left(1-\frac{7}{x}\right)&=x^2\left(\frac{-6}{x^2}\right)\\x^2-7x&=-6\\x^2-7x+6&=0\\(x-6)(x-1)&=0\\x&=1,6\end{aligned}$$

Both $x=1$ and $x=6$ check in the original equation.

25. Since $y^2-16=(y+4)(y-4)$ and $y^2+4y=y(y+4)$, multiply each side of the equation by $y(y+4)(y-4)$:

$$\begin{aligned}y(y+4)(y-4)\bullet\frac{2}{(y+4)(y-4)}&=y(y+4)(y-4)\bullet\frac{10}{y(y+4)}\\2y&=10(y-4)\\2y&=10y-40\\-8y&=-40\\y&=5\end{aligned}$$

Since $y=5$ checks in the original equation, the solution is $y=5$.

© 2012 Cengage Learning. All Rights Reserved. May not be scanned, copied or duplicated, or posted to a publicly accessible website, in whole or in part.

27. Let x represent the speed of the boat in still water. Completing the table:

	d	r	t
Upstream	48	$x-3$	$\frac{48}{x-3}$
Downstream	72	$x+3$	$\frac{72}{x+3}$

The equation is:

$$\begin{aligned} \frac{48}{x-3} &= \frac{72}{x+3} \\ (x+3)(x-3) \bullet \frac{48}{x-3} &= (x+3)(x-3) \bullet \frac{72}{x+3} \\ 48(x+3) &= 72(x-3) \\ 48x+144 &= 72x-216 \\ -24x+144 &= -216 \\ -24x &= -360 \\ x &= 15 \end{aligned}$$

The speed of the boat in still water is 15 mph.

29. Simplifying the complex fraction: $\dfrac{\frac{x+4}{x^2-16}}{\frac{2}{x-4}} = \dfrac{\frac{x+4}{(x+4)(x-4)}}{\frac{2}{x-4}} = \dfrac{\frac{1}{x-4} \bullet (x-4)}{\frac{2}{x-4} \bullet (x-4)} = \dfrac{1}{2}$

31. Simplifying the complex fraction: $\dfrac{\frac{1}{a-2}+4}{\frac{1}{a-2}+1} = \dfrac{\left(\frac{1}{a-2}+4\right)(a-2)}{\left(\frac{1}{a-2}+1\right)(a-2)} = \dfrac{1+4(a-2)}{1+1(a-2)} = \dfrac{1+4a-8}{1+a-2} = \dfrac{4a-7}{a-1}$

33. Writing as a fraction: $\dfrac{40 \text{ seconds}}{3 \text{ minutes}} = \dfrac{40 \text{ seconds}}{180 \text{ seconds}} = \dfrac{2}{9}$

35. Solving the proportion:

$$\begin{aligned} \frac{a}{3} &= \frac{12}{a} \\ a^2 &= 36 \\ a^2 - 36 &= 0 \\ (a+6)(a-6) &= 0 \\ a &= -6, 6 \end{aligned}$$

37. The variation equation is $y = Kx$. Finding K:

$$\begin{aligned} -20 &= K \bullet 4 \\ K &= -5 \end{aligned}$$

So $y = -5x$. Substituting $x = 7$: $y = -5 \bullet 7 = -35$

© 2012 Cengage Learning. All Rights Reserved. May not be scanned, copied or duplicated, or posted to a publicly accessible website, in whole or in part.

Chapter 7 Cumulative Review

1. Simplifying the expression: $8-11=8+(-11)=-3$
3. Simplifying the expression: $\frac{-48}{12}=-4$
5. Simplifying the expression: $5x-4-9x=5x-9x-4=-4x-4$
7. Simplifying the expression: $\frac{x^{-9}}{x^{-13}}=x^{-9+13}=x^4$
9. Simplifying the expression: $4^1+9^0+(-7)^0=4+1+1=6$
11. Simplifying the expression: $4x-7x=-3x$
13. Simplifying the expression: $\frac{x^2}{x-7}-\frac{14x-49}{x-7}=\frac{x^2-14x+49}{x-7}=\frac{(x-7)^2}{x-7}=x-7$
15. Solving the equation:
$$\begin{aligned}4x-3&=8x+5\\-4x-3&=5\\-4x&=8\\x&=-2\end{aligned}$$
17. Solving the equation:
$$\begin{aligned}98r^2-18&=0\\2\left(49r^2-9\right)&=0\\2(7r+3)(7r-3)&=0\\r&=-\frac{3}{7},\frac{3}{7}\end{aligned}$$
19. Multiplying each side of the equation by $3x$:
$$\begin{aligned}3x\left(\frac{5}{x}-\frac{1}{3}\right)&=3x\left(\frac{3}{x}\right)\\15-x&=9\\-x&=-6\\x&=6\end{aligned}$$
Since $x=6$ checks in the original equation, the solution is $x=6$.
21. Multiplying the second equation by 2:
$$\begin{aligned}x+2y&=1\\2x-2y&=8\end{aligned}$$
Adding the two equations:
$$\begin{aligned}3x&=9\\x&=3\end{aligned}$$
Substituting into the first equation:
$$\begin{aligned}3+2y&=1\\2y&=-2\\y&=-1\end{aligned}$$
The solution is $(3,-1)$.
23. Substituting into the second equation:
$$\begin{aligned}2(4y-3)-5y&=9\\8y-6-5y&=9\\3y-6&=9\\3y&=15\\y&=5\end{aligned}$$
Substituting into the first equation: $x=4(5)-3=20-3=17$. The solution is $(17,5)$.

© 2012 Cengage Learning. All Rights Reserved. May not be scanned, copied or duplicated, or posted to a publicly accessible website, in whole or in part.

25. Graphing the equation:

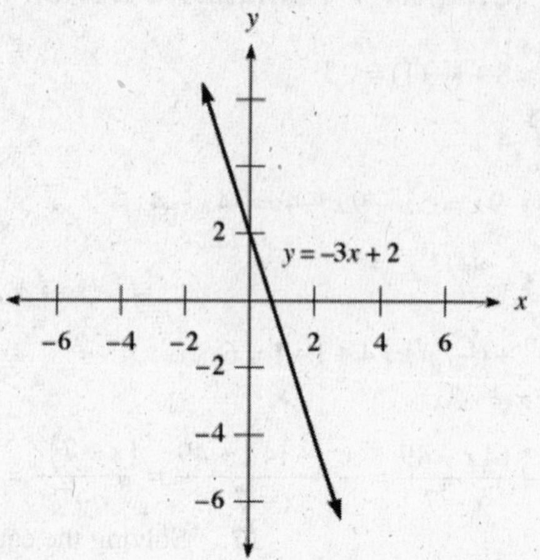

27. Solving the inequality:

$$-\frac{a}{3} \le -2$$
$$-3\left(-\frac{a}{3}\right) \ge -3(-2)$$
$$a \ge 6$$

29. Factoring the polynomial: $xy + 5x + ay + 5a = x(y+5) + a(y+5) = (y+5)(x+a)$

31. Factoring the polynomial: $20y^2 - 27y + 9 = (5y-3)(4y-3)$

33. Factoring the polynomial: $3x^2 + 12y^2 = 3(x^2 + 4y^2)$

35. Graphing each equation:

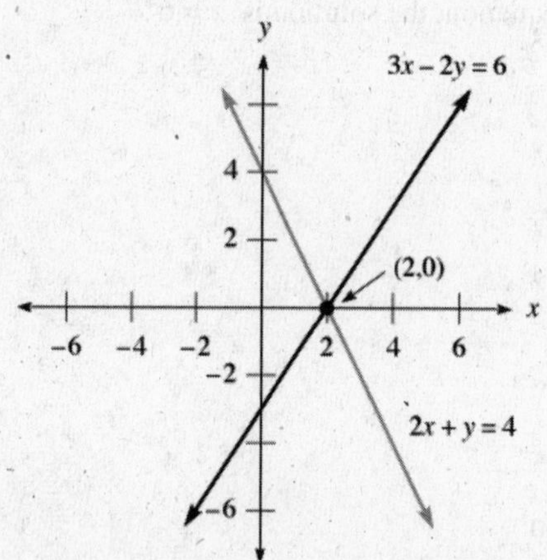

The intersection point is $(2,0)$.

37. To find the x-intercept, let $y = 0$:

$$2x + 5(0) = 10$$
$$2x = 10$$
$$x = 5$$

To find the y-intercept, let $x = 0$:

$$2(0) + 5y = 10$$
$$5y = 10$$
$$y = 2$$

39. The slope-intercept form is $y = -5x - 1$.

41. Subtracting: $6 - (-2) = 6 + 2 = 8$

43. Associative property of addition

© 2012 Cengage Learning. All Rights Reserved. May not be scanned, copied or duplicated, or posted to a publicly accessible website, in whole or in part.

45. Using long division:

$$\begin{array}{r} x-7 \\ x+4\overline{)x^2-3x-28} \\ \underline{x^2+4x} \\ -7x-28 \\ \underline{-7x-28} \\ 0 \end{array}$$

The quotient is $x-7$.

47. Simplifying the expression: $\frac{5a+10}{10a+20}=\frac{5(a+2)}{10(a+2)}=\frac{1}{2}$

49. The variation equation is $y=Kx^3$. Finding K:

$$\begin{aligned} -32 &= K\bullet 2^3 \\ -32 &= 8K \\ K &= -4 \end{aligned}$$

So $y=-4x^3$. Substituting $x=3$: $y=-4\bullet 3^3=-108$

51. Converting the units: $\frac{2{,}600{,}000 \text{ sq. km}}{2.59 \text{ sq. km/sq. mi}}\approx 1{,}003{,}861 \text{ sq. mi}$

Chapter 7 Test

1. Reducing the rational expression: $\frac{x^2-16}{x^2-8x+16}=\frac{(x+4)(x-4)}{(x-4)^2}=\frac{x+4}{x-4}$

2. Reducing the rational expression: $\frac{10a+20}{5a^2+20a+20}=\frac{10(a+2)}{5(a^2+4a+4)}=\frac{10(a+2)}{5(a+2)^2}=\frac{2}{a+2}$

3. Reducing the rational expression: $\frac{xy+7x+5y+35}{x^2+ax+5x+5a}=\frac{x(y+7)+5(y+7)}{x(x+a)+5(x+a)}=\frac{(y+7)(x+5)}{(x+a)(x+5)}=\frac{y+7}{x+a}$

4. Performing the operations: $\frac{3x-12}{4}\bullet\frac{8}{2x-8}=\frac{3(x-4)}{4}\bullet\frac{8}{2(x-4)}=\frac{24(x-4)}{8(x-4)}=3$

5. Performing the operations:

$$\begin{aligned} \frac{x^2-49}{x+1}\div\frac{x+7}{x^2-1} &= \frac{x^2-49}{x+1}\bullet\frac{x^2-1}{x+7} \\ &= \frac{(x+7)(x-7)}{x+1}\bullet\frac{(x+1)(x-1)}{x+7} \\ &= \frac{(x+7)(x-7)(x+1)(x-1)}{(x+1)(x+7)} \\ &= (x-7)(x-1) \end{aligned}$$

6. Performing the operations:

$$\begin{aligned} \frac{x^2-3x-10}{x^2-8x+15}\div\frac{3x^2+2x-8}{x^2+x-12} &= \frac{x^2-3x-10}{x^2-8x+15}\bullet\frac{x^2+x-12}{3x^2+2x-8} \\ &= \frac{(x-5)(x+2)}{(x-5)(x-3)}\bullet\frac{(x+4)(x-3)}{(3x-4)(x+2)} \\ &= \frac{(x-5)(x+2)(x+4)(x-3)}{(x-5)(x-3)(3x-4)(x+2)} \\ &= \frac{x+4}{3x-4} \end{aligned}$$

© 2012 Cengage Learning. All Rights Reserved. May not be scanned, copied or duplicated, or posted to a publicly accessible website, in whole or in part.

7. Performing the operations: $\left(x^2-9\right)\left(\frac{x+2}{x+3}\right)=\frac{(x+3)(x-3)}{1}\cdot\frac{x+2}{x+3}=\frac{(x+3)(x-3)(x+2)}{x+3}=(x-3)(x+2)$

8. Performing the operations: $\frac{3}{x-2}-\frac{6}{x-2}=\frac{3-6}{x-2}=-\frac{3}{x-2}$

9. Performing the operations:

$$\begin{aligned}\frac{x}{x^2-9}+\frac{4}{4x-12}&=\frac{x}{(x+3)(x-3)}+\frac{4}{4(x-3)}\\&=\frac{x}{(x+3)(x-3)}+\frac{1\cdot(x+3)}{(x+3)(x-3)}\\&=\frac{x}{(x+3)(x-3)}+\frac{x+3}{(x+3)(x-3)}\\&=\frac{2x+3}{(x+3)(x-3)}\end{aligned}$$

10. Performing the operations:

$$\begin{aligned}\frac{2x}{x^2-1}+\frac{x}{x^2-3x+2}&=\frac{2x}{(x+1)(x-1)}+\frac{x}{(x-1)(x-2)}\\&=\frac{2x\cdot(x-2)}{(x+1)(x-1)(x-2)}+\frac{x\cdot(x+1)}{(x+1)(x-1)(x-2)}\\&=\frac{2x^2-4x}{(x+1)(x-1)(x-2)}+\frac{x^2+x}{(x+1)(x-1)(x-2)}\\&=\frac{3x^2-3x}{(x+1)(x-1)(x-2)}\\&=\frac{3x(x-1)}{(x+1)(x-1)(x-2)}\\&=\frac{3x}{(x+1)(x-2)}\end{aligned}$$

11. Multiplying both sides of the equation by 15:

$$\begin{aligned}15\cdot\frac{7}{5}&=15\cdot\frac{x+2}{3}\\21&=5(x+2)\\21&=5x+10\\11&=5x\\x&=\frac{11}{5}\end{aligned}$$

Since $x=\frac{11}{5}$ checks in the original equation, the solution is $x=\frac{11}{5}$.

12. Multiplying both sides of the equation by $x(x+4)$:

$$\begin{aligned}x(x+4)\cdot\frac{10}{x+4}&=x(x+4)\cdot\left(\frac{6}{x}-\frac{4}{x}\right)\\10x&=6(x+4)-4(x+4)\\10x&=6x+24-4x-16\\10x&=2x+8\\8x&=8\\x&=1\end{aligned}$$

Since $x=1$ checks in the original equation, the solution is $x=1$.

© 2012 Cengage Learning. All Rights Reserved. May not be scanned, copied or duplicated, or posted to a publicly accessible website, in whole or in part.

13. Multiplying both sides of the equation by $x^2-x-2=(x-2)(x+1)$:

$$\begin{aligned}(x-2)(x+1)\left(\frac{3}{x-2}-\frac{4}{x+1}\right)&=(x-2)(x+1)\bullet\frac{5}{(x-2)(x+1)}\\3(x+1)-4(x-2)&=5\\3x+3-4x+8&=5\\-x+11&=5\\-x&=-6\\x&=6\end{aligned}$$

Since $x=6$ checks in the original equation, the solution is $x=6$.

14. Let x represent the speed of the boat in still water. Completing the table:

	d	r	t
Upstream	26	$x-2$	$\frac{26}{x-2}$
Downstream	34	$x+2$	$\frac{34}{x+2}$

The equation is:

$$\begin{aligned}\frac{26}{x-2}&=\frac{34}{x+2}\\(x+2)(x-2)\bullet\frac{26}{x-2}&=(x+2)(x-2)\bullet\frac{34}{x+2}\\26(x+2)&=34(x-2)\\26x+52&=34x-68\\-8x+52&=-68\\-8x&=-120\\x&=15\end{aligned}$$

The speed of the boat in still water is 15 mph.

15. Let t represent the time to empty the pool with both pipes open. The equation is:

$$\begin{aligned}\frac{1}{12}-\frac{1}{15}&=\frac{1}{t}\\60t\left(\frac{1}{12}-\frac{1}{15}\right)&=60t\bullet\frac{1}{t}\\5t-4t&=60\\t&=60\end{aligned}$$

It will take 60 hours to empty the pool with both pipes open.

16. The ratio of alcohol to water is given by: $\frac{27\text{ ml}}{54\text{ ml}}=\frac{1}{2}$

The ratio of alcohol to total volume is given by: $\frac{27\text{ ml}}{81\text{ ml}}=\frac{1}{3}$

17. Comparing defective parts to total parts, the proportion is:

$$\begin{aligned}\frac{8}{100}&=\frac{x}{1650}\\100x&=13200\\x&=132\end{aligned}$$

The machine can be expected to produce 132 defective parts.

18. Simplifying the complex fraction: $\frac{1+\frac{1}{x}}{1-\frac{1}{x}}=\frac{\left(1+\frac{1}{x}\right)\bullet x}{\left(1-\frac{1}{x}\right)\bullet x}=\frac{x+1}{x-1}$

© 2012 Cengage Learning. All Rights Reserved. May not be scanned, copied or duplicated, or posted to a publicly accessible website, in whole or in part.

19. Simplifying the complex fraction: $\dfrac{1-\dfrac{16}{x^2}}{1-\dfrac{2}{x}-\dfrac{8}{x^2}}=\dfrac{\left(1-\dfrac{16}{x^2}\right)\bullet x^2}{\left(1-\dfrac{2}{x}-\dfrac{8}{x^2}\right)\bullet x^2}=\dfrac{x^2-16}{x^2-2x-8}=\dfrac{(x+4)(x-4)}{(x-4)(x+2)}=\dfrac{x+4}{x+2}$

20. The variation equation is $y=Kx^2$. Finding K:

$$36=K\bullet 3^2$$
$$36=9K$$
$$K=4$$

So $y=4x^2$. Substituting $x=5$: $y=4\bullet 5^2=4\bullet 25=100$

21. The variation equation is $y=\dfrac{K}{x}$. Finding K:

$$6=\frac{K}{3}$$
$$K=18$$

So $y=\dfrac{18}{x}$. Substituting $x=9$: $y=\dfrac{18}{9}=2$

22. Finding the ratio: $\dfrac{188\text{ dB}}{60\text{ dB}}\approx 3.1$

23. Converting the speeds of each car:

Koenigsegg CCX: $\dfrac{245\text{ miles}}{1\text{ hour}}\bullet\dfrac{5,280\text{ feet}}{1\text{ mile}}\bullet\dfrac{1\text{ hour}}{3,600\text{ seconds}}\approx 359.3\text{ feet/second}$

Saleen S7 Twin Turbo: $\dfrac{248\text{ miles}}{1\text{ hour}}\bullet\dfrac{5,280\text{ feet}}{1\text{ mile}}\bullet\dfrac{1\text{ hour}}{3,600\text{ seconds}}=363.7\text{ feet/second}$

Bugatti Veyron: $\dfrac{253\text{ miles}}{1\text{ hour}}\bullet\dfrac{5,280\text{ feet}}{1\text{ mile}}\bullet\dfrac{1\text{ hour}}{3,600\text{ seconds}}\approx 371.1\text{ feet/second}$

SSC Ultimate Aero: $\dfrac{257\text{ miles}}{1\text{ hour}}\bullet\dfrac{5,280\text{ feet}}{1\text{ mile}}\bullet\dfrac{1\text{ hour}}{3,600\text{ seconds}}=376.9\text{ feet/second}$

© 2012 Cengage Learning. All Rights Reserved. May not be scanned, copied or duplicated, or posted to a publicly accessible website, in whole or in part.

Chapter 8
Roots and Radicals

Getting Ready for Chapter 8

1. Expanding: $1^2 = 1 \bullet 1 = 1$

2. Expanding: $2^2 = 2 \bullet 2 = 4$

3. Expanding: $3^2 = 3 \bullet 3 = 9$

4. Expanding: $4^2 = 4 \bullet 4 = 16$

5. Expanding: $5^2 = 5 \bullet 5 = 25$

6. Expanding: $6^2 = 6 \bullet 6 = 36$

7. Expanding: $7^2 = 7 \bullet 7 = 49$

8. Expanding: $8^2 = 8 \bullet 8 = 64$

9. Expanding: $9^2 = 9 \bullet 9 = 81$

10. Expanding: $10^2 = 10 \bullet 10 = 100$

11. Expanding: $11^2 = 11 \bullet 11 = 121$

12. Expanding: $12^2 = 12 \bullet 12 = 144$

13. Expanding: $15^2 = 15 \bullet 15 = 225$

14. Expanding: $16^2 = 16 \bullet 16 = 256$

15. Expanding: $2^3 = 2 \bullet 2 \bullet 2 = 8$

16. Expanding: $3^3 = 3 \bullet 3 \bullet 3 = 27$

17. Expanding: $4^3 = 4 \bullet 4 \bullet 4 = 64$

18. Expanding: $5^3 = 5 \bullet 5 \bullet 5 = 125$

8.1 Definitions and Common Roots

1. Finding the root: $\sqrt{9} = 3$

3. Finding the root: $-\sqrt{9} = -3$

5. Finding the root: $\sqrt{-25}$ is not a real number

7. Finding the root: $-\sqrt{144} = -12$

9. Finding the root: $\sqrt{625} = 25$

11. Finding the root: $\sqrt{-49}$ is not a real number

13. Finding the root: $-\sqrt{64} = -8$

15. Finding the root: $-\sqrt{100} = -10$

17. Finding the root: $\sqrt{1225} = 35$

19. Finding the root: $\sqrt[4]{1} = 1$

21. Finding the root: $\sqrt[3]{-8} = -2$

23. Finding the root: $-\sqrt[3]{125} = -5$

25. Finding the root: $\sqrt[3]{-1} = -1$

27. Finding the root: $\sqrt[3]{-27} = -3$

29. Finding the root: $-\sqrt[4]{16} = -2$

31. Simplifying the expression: $\sqrt{x^2} = x$

33. Simplifying the expression: $\sqrt{9x^2} = 3x$

35. Simplifying the expression: $\sqrt{x^2y^2} = xy$

37. Simplifying the expression: $\sqrt{(a+b)^2} = a+b$

39. Simplifying the expression: $\sqrt{49x^2y^2} = 7xy$

41. Simplifying the expression: $\sqrt[3]{x^3} = x$

43. Simplifying the expression: $\sqrt[3]{8x^3} = 2x$

45. Simplifying the expression: $\sqrt{x^4} = x^2$

47. Simplifying the expression: $\sqrt{36a^6} = 6a^3$

49. Simplifying the expression: $\sqrt{25a^8b^4} = 5a^4b^2$

51. Simplifying the expression: $\sqrt[3]{x^6} = x^2$

53. Simplifying the expression: $\sqrt[3]{27a^{12}} = 3a^4$

55. Simplifying the expression: $\sqrt[4]{x^8} = x^2$

© 2012 Cengage Learning. All Rights Reserved. May not be scanned, copied or duplicated, or posted to a publicly accessible website, in whole or in part.

57. **a.** Simplifying the expression: $\sqrt[4]{16}=2$

b. Simplifying the expression: $\sqrt[4]{x^4}=x$

c. Simplifying the expression: $\sqrt[4]{x^8}=x^2$

d. Simplifying the expression: $\sqrt[4]{16x^8y^{12}}=2x^2y^3$

59. Simplifying the expression: $\sqrt{9\bullet 16}=\sqrt{144}=12$

61. Simplifying the expression: $\sqrt{25}\sqrt{16}=5\bullet 4=20$

63. Simplifying the expression: $\sqrt{9}+\sqrt{16}=3+4=7$

65. Simplifying the expression: $\sqrt{9+16}=\sqrt{25}=5$

67. Simplifying the expression: $\sqrt{144}+\sqrt{25}=12+5=17$

69. Simplifying the expression: $\sqrt{144+25}=\sqrt{169}=13$

71. **a.** Approximating the expression: $\dfrac{1+\sqrt{5}}{2}\approx\dfrac{1+2.236}{2}=\dfrac{3.236}{2}=1.618$

b. Approximating the expression: $\dfrac{1-\sqrt{5}}{2}\approx\dfrac{1-2.236}{2}=\dfrac{-1.236}{2}=-0.618$

c. Approximating the expression: $\dfrac{1+\sqrt{5}}{2}+\dfrac{1-\sqrt{5}}{2}\approx 1.618-0.618=1$

73. **a.** Evaluating the root: $\sqrt{9}=3$

b. Evaluating the root: $\sqrt{900}=30$

c. Evaluating the root: $\sqrt{0.09}=0.3$

75. Simplifying each expression:

$$\frac{5+\sqrt{49}}{2}=\frac{5+7}{2}=\frac{12}{2}=6 \qquad \frac{5-\sqrt{49}}{2}=\frac{5-7}{2}=\frac{-2}{2}=-1$$

77. Simplifying each expression:

$$\frac{2+\sqrt{16}}{2}=\frac{2+4}{2}=\frac{6}{2}=3 \qquad \frac{2-\sqrt{16}}{2}=\frac{2-4}{2}=\frac{-2}{2}=-1$$

79. Simplifying the expression: $\sqrt{x^2+6x+9}=\sqrt{(x+3)^2}=x+3$

81. Let x represent the distance between Pomona and Upland. Using the Pythagorean theorem:

$$x^2=5.7^2+3.6^2=45.45$$
$$x=\sqrt{45.45}\approx 6.7\text{ miles}$$

83. Using the Pythagorean theorem:

$$x^2=3^2+4^2=9+16=25$$
$$x=\sqrt{25}=5$$

85. Using the Pythagorean theorem:

$$x^2=5^2+10^2=25+100=125$$
$$x=\sqrt{125}\approx 11.2$$

87. Let l represent the length of the wire. Using the Pythagorean theorem:

$$l^2=24^2+18^2=576+324=900$$
$$l=\sqrt{900}=30$$

The length of the wire is 30 feet.

89. Let x represent the length of the log. Using the Pythagorean theorem:

$$x^2=5^2+12^2=25+144=169$$
$$x=\sqrt{169}=13$$

The log must be 13 feet to just barely reach.

91. Simplifying: $3\bullet\sqrt{16}=3\bullet 4=12$

93. Factoring: $75=25\bullet 3$

95. Factoring: $50=25\bullet 2$

97. Factoring: $40=4\bullet 10$

99. Factoring: $x^5=x^4\bullet x$

101. Factoring: $12x^2=4x^2\bullet 3$

103. Factoring: $50x^3y^2=25x^2y^2\bullet 2x$

105. Reducing the rational expression: $\dfrac{x^2-16}{x+4}=\dfrac{(x+4)(x-4)}{x+4}=x-4$

107. Reducing the rational expression: $\dfrac{10a+20}{5a^2-20}=\dfrac{10(a+2)}{5(a^2-4)}=\dfrac{10(a+2)}{5(a+2)(a-2)}=\dfrac{2}{a-2}$

© 2012 Cengage Learning. All Rights Reserved. May not be scanned, copied or duplicated, or posted to a publicly accessible website, in whole or in part.

109. Reducing the rational expression: $\frac{2x^2-5x-3}{x^2-3x}=\frac{(2x+1)(x-3)}{x(x-3)}=\frac{2x+1}{x}$

111. Reducing the rational expression: $\frac{xy+3x+2y+6}{xy+3x+ay+3a}=\frac{x(y+3)+2(y+3)}{x(y+3)+a(y+3)}=\frac{(y+3)(x+2)}{(y+3)(x+a)}=\frac{x+2}{x+a}$

8.2 Properties of Radicals

1. Simplifying the radical expression: $\sqrt{8}=\sqrt{4\bullet 2}=\sqrt{4}\sqrt{2}=2\sqrt{2}$

3. Simplifying the radical expression: $\sqrt{12}=\sqrt{4\bullet 3}=\sqrt{4}\sqrt{3}=2\sqrt{3}$

5. Simplifying the radical expression: $\sqrt[3]{24}=\sqrt[3]{8\bullet 3}=\sqrt[3]{8}\sqrt[3]{3}=2\sqrt[3]{3}$

7. Simplifying the radical expression: $\sqrt{50x^2}=\sqrt{25x^2\bullet 2}=\sqrt{25x^2}\sqrt{2}=5x\sqrt{2}$

9. Simplifying the radical expression: $\sqrt{45a^2b^2}=\sqrt{9a^2b^2\bullet 5}=\sqrt{9a^2b^2}\sqrt{5}=3ab\sqrt{5}$

11. Simplifying the radical expression: $\sqrt[3]{54x^3}=\sqrt[3]{27x^3\bullet 2}=\sqrt[3]{27x^3}\sqrt[3]{2}=3x\sqrt[3]{2}$

13. Simplifying the radical expression: $\sqrt{32x^4}=\sqrt{16x^4\bullet 2}=\sqrt{16x^4}\sqrt{2}=4x^2\sqrt{2}$

15. Simplifying the radical expression: $5\sqrt{80}=5\sqrt{16\bullet 5}=5\sqrt{16}\sqrt{5}=5\bullet 4\sqrt{5}=20\sqrt{5}$

17. Simplifying the radical expression: $\frac{1}{2}\sqrt{28x^3}=\frac{1}{2}\sqrt{4x^2\bullet 7x}=\frac{1}{2}\sqrt{4x^2}\sqrt{7x}=\frac{1}{2}\bullet 2x\sqrt{7x}=x\sqrt{7x}$

19. Simplifying the radical expression: $x\sqrt[3]{8x^4}=x\sqrt[3]{8x^3\bullet x}=x\sqrt[3]{8x^3}\sqrt[3]{x}=x\bullet 2x\sqrt[3]{x}=2x^2\sqrt[3]{x}$

21. Simplifying the radical expression: $2a\sqrt[3]{27a^5}=2a\sqrt[3]{27a^3\bullet a^2}=2a\sqrt[3]{27a^3}\sqrt[3]{a^2}=2a\bullet 3a\sqrt[3]{a^2}=6a^2\sqrt[3]{a^2}$

23. Simplifying the radical expression: $\frac{4}{3}\sqrt{45a^3}=\frac{4}{3}\sqrt{9a^2\bullet 5a}=\frac{4}{3}\sqrt{9a^2}\sqrt{5a}=\frac{4}{3}\bullet 3a\sqrt{5a}=4a\sqrt{5a}$

25. Simplifying the radical expression: $3\sqrt{50xy^2}=3\sqrt{25y^2\bullet 2x}=3\sqrt{25y^2}\sqrt{2x}=3\bullet 5y\sqrt{2x}=15y\sqrt{2x}$

27. Simplifying the radical expression: $7\sqrt{12x^2y}=7\sqrt{4x^2\bullet 3y}=7\sqrt{4x^2}\sqrt{3y}=7\bullet 2x\sqrt{3y}=14x\sqrt{3y}$

29. Simplifying the radical expression: $\sqrt{\frac{16}{25}}=\frac{\sqrt{16}}{\sqrt{25}}=\frac{4}{5}$

31. Simplifying the radical expression: $\sqrt{\frac{4}{9}}=\frac{\sqrt{4}}{\sqrt{9}}=\frac{2}{3}$

33. Simplifying the radical expression: $\sqrt[3]{\frac{8}{27}}=\frac{\sqrt[3]{8}}{\sqrt[3]{27}}=\frac{2}{3}$

35. Simplifying the radical expression: $\sqrt[4]{\frac{16}{81}}=\frac{\sqrt[4]{16}}{\sqrt[4]{81}}=\frac{2}{3}$

37. Simplifying the radical expression: $\sqrt{\frac{100x^2}{25}}=\sqrt{4x^2}=2x$

39. Simplifying the radical expression: $\sqrt{\frac{81a^2b^2}{9}}=\sqrt{9a^2b^2}=3ab$

41. Simplifying the radical expression: $\sqrt[3]{\frac{27x^3}{8y^3}}=\frac{\sqrt[3]{27x^3}}{\sqrt[3]{8y^3}}=\frac{3x}{2y}$

43. Simplifying the radical expression: $\sqrt{\frac{50}{9}}=\frac{\sqrt{50}}{\sqrt{9}}=\frac{\sqrt{25\bullet 2}}{3}=\frac{5\sqrt{2}}{3}$

45. Simplifying the radical expression: $\sqrt{\frac{75}{25}}=\sqrt{3}$

47. Simplifying the radical expression: $\sqrt{\frac{128}{49}}=\frac{\sqrt{128}}{\sqrt{49}}=\frac{\sqrt{64\bullet 2}}{7}=\frac{8\sqrt{2}}{7}$

49. Simplifying the radical expression: $\sqrt{\frac{288x}{25}}=\frac{\sqrt{288x}}{\sqrt{25}}=\frac{\sqrt{144\bullet 2x}}{5}=\frac{12\sqrt{2x}}{5}$

© 2012 Cengage Learning. All Rights Reserved. May not be scanned, copied or duplicated, or posted to a publicly accessible website, in whole or in part.

51. Simplifying the radical expression: $\sqrt{\frac{54a^2}{25}}=\frac{\sqrt{54a^2}}{\sqrt{25}}=\frac{\sqrt{9a^2\cdot 6}}{5}=\frac{3a\sqrt{6}}{5}$

53. Simplifying the radical expression: $\frac{3\sqrt{50}}{2}=\frac{3\sqrt{25\cdot 2}}{2}=\frac{3\cdot 5\sqrt{2}}{2}=\frac{15\sqrt{2}}{2}$

55. Simplifying the radical expression: $\frac{7\sqrt{28y^2}}{3}=\frac{7\sqrt{4y^2\cdot 7}}{3}=\frac{7\cdot 2y\sqrt{7}}{3}=\frac{14y\sqrt{7}}{3}$

57. Simplifying the radical expression: $\frac{5\sqrt{72a^2b^2}}{\sqrt{36}}=\frac{5\sqrt{36a^2b^2\cdot 2}}{6}=\frac{5\cdot 6ab\sqrt{2}}{6}=\frac{30ab\sqrt{2}}{6}=5ab\sqrt{2}$

59. Simplifying the radical expression: $\frac{6\sqrt{8x^2y}}{\sqrt{4}}=\frac{6\sqrt{4x^2\cdot 2y}}{2}=\frac{6\cdot 2x\sqrt{2y}}{2}=\frac{12x\sqrt{2y}}{2}=6x\sqrt{2y}$

61. a. Substituting the values: $\sqrt{b^2-4ac}=\sqrt{(4)^2-4(2)(-3)}=\sqrt{16+24}=\sqrt{40}=\sqrt{4\cdot 10}=2\sqrt{10}$

b. Substituting the values: $\sqrt{b^2-4ac}=\sqrt{(1)^2-4(1)(-6)}=\sqrt{1+24}=\sqrt{25}=5$

c. Substituting the values: $\sqrt{b^2-4ac}=\sqrt{(1)^2-4(1)(-11)}=\sqrt{1+44}=\sqrt{45}=\sqrt{9\cdot 5}=3\sqrt{5}$

d. Substituting the values: $\sqrt{b^2-4ac}=\sqrt{(6)^2-4(3)(2)}=\sqrt{36-24}=\sqrt{12}=\sqrt{4\cdot 3}=2\sqrt{3}$

63. a. Simplifying: $\sqrt{32x^{10}y^5}=\sqrt{16x^{10}y^4\cdot 2y}=\sqrt{16x^{10}y^4}\sqrt{2y}=4x^5y^2\sqrt{2y}$

b. Simplifying: $\sqrt[3]{32x^{10}y^5}=\sqrt[3]{8x^9y^3\cdot 4xy^2}=\sqrt[3]{8x^9y^3}\sqrt[3]{4xy^2}=2x^3y\sqrt[3]{4xy^2}$

c. Simplifying: $\sqrt[4]{32x^{10}y^5}=\sqrt[4]{16x^8y^4\cdot 2x^2y}=\sqrt[4]{16x^8y^4}\sqrt[4]{2x^2y}=2x^2y\sqrt[4]{2x^2y}$

d. Simplifying: $\sqrt[5]{32x^{10}y^5}=2x^2y$

65. a. Simplifying: $\sqrt{4}=2$

b. Simplifying: $\sqrt{0.04}=0.2$

c. Simplifying: $\sqrt{400}=20$

d. Simplifying: $\sqrt{0.0004}=0.02$

67. Completing the table:

x	$\sqrt{x}$	$2\sqrt{x}$	$\sqrt{4x}$
1	1	2	2
2	1.414	2.828	2.828
3	1.732	3.464	3.464
4	2	4	4

69. Completing the table:

x	$\sqrt{x}$	$3\sqrt{x}$	$\sqrt{9x}$
1	1	3	3
2	1.414	4.243	4.243
3	1.732	5.196	5.196
4	2	6	6

71. Simplifying: $\sqrt{4x^3y^2}=\sqrt{4x^2y^2\cdot x}=2xy\sqrt{x}$

73. Simplifying: $\frac{6}{2}\sqrt{16}=3\cdot 4=12$

75. Simplifying: $\frac{\sqrt{2}}{\sqrt{4}}=\frac{\sqrt{2}}{2}$

77. Simplifying: $\frac{\sqrt[3]{18}}{\sqrt[3]{27}}=\frac{\sqrt[3]{18}}{3}$

79. Multiplying: $\frac{\sqrt{2}}{\sqrt{3}}\cdot\frac{\sqrt{3}}{\sqrt{3}}=\frac{\sqrt{6}}{\sqrt{9}}=\frac{\sqrt{6}}{3}$

81. Multiplying: $\sqrt[3]{3}\cdot\sqrt[3]{9}=\sqrt[3]{27}=3$

83. Performing the operations: $\frac{8x}{x^2-5x}\cdot\frac{x^2-25}{4x^2+4x}=\frac{8x}{x(x-5)}\cdot\frac{(x+5)(x-5)}{4x(x+1)}=\frac{8x(x+5)(x-5)}{4x^2(x-5)(x+1)}=\frac{2(x+5)}{x(x+1)}$

© 2012 Cengage Learning. All Rights Reserved. May not be scanned, copied or duplicated, or posted to a publicly accessible website, in whole or in part.

85. Performing the operations:

$$\frac{x^2+3x-4}{3x^2+7x-20} \div \frac{x^2-2x+1}{3x^2-2x-5} = \frac{x^2+3x-4}{3x^2+7x-20} \bullet \frac{3x^2-2x-5}{x^2-2x+1}$$
$$= \frac{(x+4)(x-1)}{(3x-5)(x+4)} \bullet \frac{(3x-5)(x+1)}{(x-1)^2}$$
$$= \frac{(x+4)(x-1)(3x-5)(x+1)}{(3x-5)(x+4)(x-1)^2}$$
$$= \frac{x+1}{x-1}$$

87. Performing the operations: $(x^2-36)\left(\frac{x+3}{x-6}\right) = \frac{(x+6)(x-6)}{1} \bullet \frac{x+3}{x-6} = \frac{(x+6)(x-6)(x+3)}{x-6} = (x+6)(x+3)$

8.3 Simplified Form for Radicals

1. Simplifying the radical expression: $\sqrt{\frac{1}{2}} = \frac{\sqrt{1}}{\sqrt{2}} \bullet \frac{\sqrt{2}}{\sqrt{2}} = \frac{\sqrt{2}}{\sqrt{4}} = \frac{\sqrt{2}}{2}$

3. Simplifying the radical expression: $\sqrt{\frac{1}{3}} = \frac{\sqrt{1}}{\sqrt{3}} \bullet \frac{\sqrt{3}}{\sqrt{3}} = \frac{\sqrt{3}}{\sqrt{9}} = \frac{\sqrt{3}}{3}$

5. Simplifying the radical expression: $\sqrt{\frac{2}{5}} = \frac{\sqrt{2}}{\sqrt{5}} \bullet \frac{\sqrt{5}}{\sqrt{5}} = \frac{\sqrt{10}}{\sqrt{25}} = \frac{\sqrt{10}}{5}$

7. Simplifying the radical expression: $\sqrt{\frac{3}{2}} = \frac{\sqrt{3}}{\sqrt{2}} \bullet \frac{\sqrt{2}}{\sqrt{2}} = \frac{\sqrt{6}}{\sqrt{4}} = \frac{\sqrt{6}}{2}$

9. Simplifying the radical expression: $\sqrt{\frac{20}{3}} = \frac{\sqrt{20}}{\sqrt{3}} \bullet \frac{\sqrt{3}}{\sqrt{3}} = \frac{\sqrt{60}}{\sqrt{9}} = \frac{\sqrt{4 \bullet 15}}{3} = \frac{2\sqrt{15}}{3}$

11. Simplifying the radical expression: $\sqrt{\frac{45}{6}} = \sqrt{\frac{15}{2}} = \frac{\sqrt{15}}{\sqrt{2}} \bullet \frac{\sqrt{2}}{\sqrt{2}} = \frac{\sqrt{30}}{\sqrt{4}} = \frac{\sqrt{30}}{2}$

13. Simplifying the radical expression: $\sqrt{\frac{20}{5}} = \sqrt{4} = 2$

15. Simplifying the radical expression: $\frac{\sqrt{21}}{\sqrt{3}} = \sqrt{\frac{21}{3}} = \sqrt{7}$

17. Simplifying the radical expression: $\frac{\sqrt{35}}{\sqrt{7}} = \sqrt{\frac{35}{7}} = \sqrt{5}$

19. Simplifying the radical expression: $\frac{10\sqrt{15}}{5\sqrt{3}} = \frac{10}{5} \bullet \sqrt{\frac{15}{3}} = 2\sqrt{5}$

21. Simplifying the radical expression: $\frac{6\sqrt{21}}{3\sqrt{7}} = \frac{6}{3} \bullet \sqrt{\frac{21}{7}} = 2\sqrt{3}$

23. Simplifying the radical expression: $\frac{6\sqrt{35}}{12\sqrt{5}} = \frac{6}{12} \bullet \sqrt{\frac{35}{5}} = \frac{1}{2}\sqrt{7} = \frac{\sqrt{7}}{2}$

25. Simplifying the radical expression: $\sqrt{\frac{4x^2y^2}{2}} = \sqrt{2x^2y^2} = xy\sqrt{2}$

27. Simplifying the radical expression: $\sqrt{\frac{5x^2y}{3}} = \frac{\sqrt{5x^2y}}{\sqrt{3}} \bullet \frac{\sqrt{3}}{\sqrt{3}} = \frac{\sqrt{15x^2y}}{\sqrt{9}} = \frac{x\sqrt{15y}}{3}$

29. Simplifying the radical expression: $\sqrt{\frac{16a^4}{5}} = \frac{\sqrt{16a^4}}{\sqrt{5}} \bullet \frac{\sqrt{5}}{\sqrt{5}} = \frac{4a^2\sqrt{5}}{\sqrt{25}} = \frac{4a^2\sqrt{5}}{5}$

© 2012 Cengage Learning. All Rights Reserved. May not be scanned, copied or duplicated, or posted to a publicly accessible website, in whole or in part.

31. Simplifying the radical expression: $\sqrt{\frac{72a^5}{5}} = \frac{\sqrt{72a^5}}{\sqrt{5}} \bullet \frac{\sqrt{5}}{\sqrt{5}} = \frac{\sqrt{360a^5}}{\sqrt{25}} = \frac{\sqrt{36a^4 \bullet 10a}}{5} = \frac{6a^2\sqrt{10a}}{5}$

33. Simplifying the radical expression: $\sqrt{\frac{20x^2y^3}{3}} = \frac{\sqrt{20x^2y^3}}{\sqrt{3}} \bullet \frac{\sqrt{3}}{\sqrt{3}} = \frac{\sqrt{60x^2y^3}}{\sqrt{9}} = \frac{\sqrt{4x^2y^2 \bullet 15y}}{3} = \frac{2xy\sqrt{15y}}{3}$

35. Simplifying the radical expression: $\frac{2\sqrt{20x^2y^3}}{3} = \frac{2\sqrt{4x^2y^2 \bullet 5y}}{3} = \frac{2 \bullet 2xy\sqrt{5y}}{3} = \frac{4xy\sqrt{5y}}{3}$

37. Simplifying the radical expression: $\frac{6\sqrt{54a^2b^3}}{5} = \frac{6\sqrt{9a^2b^2 \bullet 6b}}{5} = \frac{6 \bullet 3ab\sqrt{6b}}{5} = \frac{18ab\sqrt{6b}}{5}$

39. Simplifying the radical expression: $\frac{3\sqrt{72x^4}}{\sqrt{2x}} = 3\sqrt{\frac{72x^4}{2x}} = 3\sqrt{36x^3} = 3\sqrt{36x^2 \bullet x} = 3 \bullet 6x\sqrt{x} = 18x\sqrt{x}$

41. Simplifying the radical expression: $\sqrt[3]{\frac{1}{2}} = \frac{\sqrt[3]{1}}{\sqrt[3]{2}} \bullet \frac{\sqrt[3]{4}}{\sqrt[3]{4}} = \frac{\sqrt[3]{4}}{\sqrt[3]{8}} = \frac{\sqrt[3]{4}}{2}$

43. Simplifying the radical expression: $\sqrt[3]{\frac{1}{9}} = \frac{\sqrt[3]{1}}{\sqrt[3]{9}} \bullet \frac{\sqrt[3]{3}}{\sqrt[3]{3}} = \frac{\sqrt[3]{3}}{\sqrt[3]{27}} = \frac{\sqrt[3]{3}}{3}$

45. Simplifying the radical expression: $\sqrt[3]{\frac{3}{2}} = \frac{\sqrt[3]{3}}{\sqrt[3]{2}} \bullet \frac{\sqrt[3]{4}}{\sqrt[3]{4}} = \frac{\sqrt[3]{12}}{\sqrt[3]{8}} = \frac{\sqrt[3]{12}}{2}$

47. **a.** Simplifying the radical expression: $\frac{6}{\sqrt{\pi}} = \frac{6}{\sqrt{\pi}} \bullet \frac{\sqrt{\pi}}{\sqrt{\pi}} = \frac{6\sqrt{\pi}}{\pi}$

b. Simplifying the radical expression: $\sqrt{\frac{A}{\pi}} = \frac{\sqrt{A}}{\sqrt{\pi}} \bullet \frac{\sqrt{\pi}}{\sqrt{\pi}} = \frac{\sqrt{A\pi}}{\pi}$

c. Simplifying the radical expression: $\sqrt[3]{\frac{3V}{4\pi}} = \frac{\sqrt[3]{3V}}{\sqrt[3]{4\pi}} \bullet \frac{\sqrt[3]{2\pi^2}}{\sqrt[3]{2\pi^2}} = \frac{\sqrt[3]{6V\pi^2}}{\sqrt[3]{8\pi^3}} = \frac{\sqrt[3]{6V\pi^2}}{2\pi}$

d. Simplifying the radical expression: $\frac{2}{\sqrt[3]{2\pi}} = \frac{2}{\sqrt[3]{2\pi}} \bullet \frac{\sqrt[3]{4\pi^2}}{\sqrt[3]{4\pi^2}} = \frac{2\sqrt[3]{4\pi^2}}{\sqrt[3]{8\pi^3}} = \frac{2\sqrt[3]{4\pi^2}}{2\pi} = \frac{\sqrt[3]{4\pi^2}}{\pi}$

49. Completing the table:

x	$\sqrt{x}$	$\frac{1}{\sqrt{x}}$	$\frac{\sqrt{x}}{x}$
1	1	1	1
2	1.414	0.707	0.707
3	1.732	0.577	0.577
4	2	0.5	0.5
5	2.236	0.447	0.447
6	2.449	0.408	0.408

51. Completing the table:

x	$\sqrt{x^2}$	$\sqrt{x^3}$	$x\sqrt{x}$
1	1	1	1
2	2	2.828	2.828
3	3	5.196	5.196
4	4	8	8
5	5	11.180	11.180
6	6	14.697	14.697

53. Let x represent the base. Using the Pythagorean theorem:

$$x^2 + 3.6^2 = 6.7^2$$
$$x^2 = 6.7^2 - 3,6^2 = 31.93$$
$$x = \sqrt{31.93} \approx 5.65 \text{ miles}$$

Thus the area is: $A = \frac{1}{2}(5.65 \text{ miles})(3.6 \text{ miles}) \approx 10.2$ square miles

55. Combining the terms: $15x + 8x = (15+8)x = 23x$

57. Combining the terms: $25y + 3y - y = (25+3-1)y = 27y$

© 2012 Cengage Learning. All Rights Reserved. May not be scanned, copied or duplicated, or posted to a publicly accessible website, in whole or in part.

59. Combining the terms: $2ab+5ab=(2+5)ab=7ab$

61. Combining the terms: $2xy-9xy+50x=-7xy+50x$

63. Combining the terms: $3x+7x=(3+7)x=10x$

65. Combining the terms: $15x+8x=(15+8)x=23x$

67. Combining the terms: $7a-3a+6a=(7-3+6)a=10a$

69. Combining the rational expressions: $\frac{x^2}{x+5}+\frac{10x+25}{x+5}=\frac{x^2+10x+25}{x+5}=\frac{(x+5)^2}{x+5}=x+5$

71. Combining the rational expressions: $\frac{a}{3}+\frac{2}{5}=\frac{a\bullet 5}{3\bullet 5}+\frac{2\bullet 3}{5\bullet 3}=\frac{5a}{15}+\frac{6}{15}=\frac{5a+6}{15}$

73. Combining the rational expressions:

$$\begin{aligned}\frac{6}{a^2-9}-\frac{5}{a^2-a-6}&=\frac{6}{(a+3)(a-3)}-\frac{5}{(a-3)(a+2)}\\&=\frac{6\bullet(a+2)}{(a+3)(a-3)(a+2)}-\frac{5\bullet(a+3)}{(a+3)(a-3)(a+2)}\\&=\frac{6a+12}{(a+3)(a-3)(a+2)}-\frac{5a+15}{(a+3)(a-3)(a+2)}\\&=\frac{6a+12-5a-15}{(a+3)(a-3)(a+2)}\\&=\frac{a-3}{(a+3)(a-3)(a+2)}\\&=\frac{1}{(a+3)(a+2)}\end{aligned}$$

75. Drawing the figure:

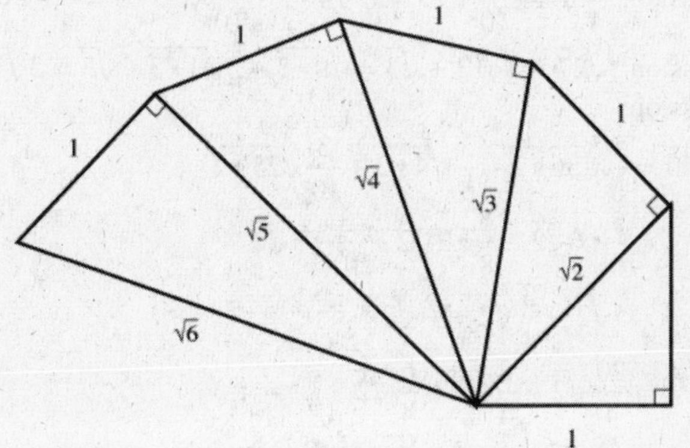

77. Simplifying the terms in the sequence:

$$\sqrt{1^2+1}=\sqrt{1+1}=\sqrt{2}$$
$$\sqrt{\left(\sqrt{2}\right)^2+1}=\sqrt{2+1}=\sqrt{3}$$
$$\sqrt{\left(\sqrt{3}\right)^2+1}=\sqrt{3+1}=\sqrt{4}=2$$

© 2012 Cengage Learning. All Rights Reserved. May not be scanned, copied or duplicated, or posted to a publicly accessible website, in whole or in part.

8.4 Addition and Subtraction of Radical Expressions

1. **a.** Simplifying: $3x+4x=7x$
b. Simplifying: $3y+4y=7y$
c. Simplifying: $3\sqrt{5}+4\sqrt{5}=7\sqrt{5}$

3. **a.** Simplifying: $x+6x=7x$
b. Simplifying: $t+6t=7t$
c. Simplifying: $\sqrt{x}+6\sqrt{x}=7\sqrt{x}$

5. Simplifying the radical expression: $3\sqrt{2}+4\sqrt{2}=7\sqrt{2}$

7. Simplifying the radical expression: $9\sqrt{5}-7\sqrt{5}=2\sqrt{5}$

9. Simplifying the radical expression: $\sqrt{3}+6\sqrt{3}=7\sqrt{3}$

11. Simplifying the radical expression: $\frac{5}{8}\sqrt{5}-\frac{3}{7}\sqrt{5}=\frac{35}{56}\sqrt{5}-\frac{24}{56}\sqrt{5}=\frac{11}{56}\sqrt{5}$

13. Simplifying the radical expression: $14\sqrt{13}-\sqrt{13}=13\sqrt{13}$

15. Simplifying the radical expression: $-3\sqrt{10}+9\sqrt{10}=6\sqrt{10}$

17. Simplifying the radical expression: $5\sqrt{5}+\sqrt{5}=6\sqrt{5}$

19. Simplifying the radical expression: $\sqrt{8}+2\sqrt{2}=\sqrt{4\bullet 2}+2\sqrt{2}=2\sqrt{2}+2\sqrt{2}=4\sqrt{2}$

21. Simplifying the radical expression: $3\sqrt{3}-\sqrt{27}=3\sqrt{3}-\sqrt{9\bullet 3}=3\sqrt{3}-3\sqrt{3}=0$

23. Simplifying the radical expression:
$$5\sqrt{12}-10\sqrt{48}=5\sqrt{4\bullet 3}-10\sqrt{16\bullet 3}=5\bullet 2\sqrt{3}-10\bullet 4\sqrt{3}=10\sqrt{3}-40\sqrt{3}=-30\sqrt{3}$$

25. Simplifying the radical expression: $-\sqrt{75}-\sqrt{3}=-\sqrt{25\bullet 3}-\sqrt{3}=-5\sqrt{3}-\sqrt{3}=-6\sqrt{3}$

27. Simplifying the radical expression: $\frac{1}{5}\sqrt{75}-\frac{1}{2}\sqrt{12}=\frac{1}{5}\sqrt{25\bullet 3}-\frac{1}{2}\sqrt{4\bullet 3}=\frac{1}{5}\bullet 5\sqrt{3}-\frac{1}{2}\bullet 2\sqrt{3}=\sqrt{3}-\sqrt{3}=0$

29. Simplifying the radical expression: $\frac{3}{4}\sqrt{8}+\frac{3}{10}\sqrt{75}=\frac{3}{4}\sqrt{4\bullet 2}+\frac{3}{10}\sqrt{25\bullet 3}=\frac{3}{4}\bullet 2\sqrt{2}+\frac{3}{10}\bullet 5\sqrt{3}=\frac{3}{2}\sqrt{2}+\frac{3}{2}\sqrt{3}$

31. Simplifying the radical expression: $\sqrt{27}-2\sqrt{12}+\sqrt{3}=\sqrt{9\bullet 3}-2\sqrt{4\bullet 3}+\sqrt{3}=3\sqrt{3}-4\sqrt{3}+\sqrt{3}=0$

33. Simplifying the radical expression:
$$\begin{aligned}\frac{5}{6}\sqrt{72}-\frac{3}{8}\sqrt{8}+\frac{3}{10}\sqrt{50}&=\frac{5}{6}\sqrt{36\bullet 2}-\frac{3}{8}\sqrt{4\bullet 2}+\frac{3}{10}\sqrt{25\bullet 2}\\&=\frac{5}{6}\bullet 6\sqrt{2}-\frac{3}{8}\bullet 2\sqrt{2}+\frac{3}{10}\bullet 5\sqrt{2}\\&=5\sqrt{2}-\frac{3}{4}\sqrt{2}+\frac{3}{2}\sqrt{2}\\&=\frac{20}{4}\sqrt{2}-\frac{3}{4}\sqrt{2}+\frac{6}{4}\sqrt{2}\\&=\frac{23}{4}\sqrt{2}\end{aligned}$$

35. Simplifying the radical expression:
$$5\sqrt{7}+2\sqrt{28}-4\sqrt{63}=5\sqrt{7}+2\sqrt{4\bullet 7}-4\sqrt{9\bullet 7}=5\sqrt{7}+2\bullet 2\sqrt{7}-4\bullet 3\sqrt{7}=5\sqrt{7}+4\sqrt{7}-12\sqrt{7}=-3\sqrt{7}$$

37. Simplifying the radical expression:
$$\begin{aligned}6\sqrt{48}-2\sqrt{12}+5\sqrt{27}&=6\sqrt{16\bullet 3}-2\sqrt{4\bullet 3}+5\sqrt{9\bullet 3}\\&=6\bullet 4\sqrt{3}-2\bullet 2\sqrt{3}+5\bullet 3\sqrt{3}\\&=24\sqrt{3}-4\sqrt{3}+15\sqrt{3}\\&=35\sqrt{3}\end{aligned}$$

39. Simplifying the radical expression:
$$\begin{aligned}6\sqrt{48}-\sqrt{72}-3\sqrt{300}&=6\sqrt{16\bullet 3}-\sqrt{36\bullet 2}-3\sqrt{100\bullet 3}\\&=6\bullet 4\sqrt{3}-6\sqrt{2}-3\bullet 10\sqrt{3}\\&=24\sqrt{3}-6\sqrt{2}-30\sqrt{3}\\&=-6\sqrt{3}-6\sqrt{2}\end{aligned}$$

41. Simplifying the radical expression: $9+6\sqrt{2}+2-18-6\sqrt{2}=(9+2-18)+\left(6\sqrt{2}-6\sqrt{2}\right)=-7$

© 2012 Cengage Learning. All Rights Reserved. May not be scanned, copied or duplicated, or posted to a publicly accessible website, in whole or in part.

43. Simplifying the radical expression: $4+4\sqrt{5}+5-8-4\sqrt{5}-1=(4+5-8-1)+\left(4\sqrt{5}-4\sqrt{5}\right)=0$

45. Simplifying the radical expression: $\sqrt{x^3}+x\sqrt{x}=\sqrt{x^2 \bullet x}+x\sqrt{x}=x\sqrt{x}+x\sqrt{x}=2x\sqrt{x}$

47. Simplifying the radical expression: $5\sqrt{3a^2}-a\sqrt{3}=5a\sqrt{3}-a\sqrt{3}=4a\sqrt{3}$

49. Simplifying the radical expression:

$$5\sqrt{8x^3}+x\sqrt{50x}=5\sqrt{4x^2 \bullet 2x}+x\sqrt{25 \bullet 2x}=5 \bullet 2x\sqrt{2x}+x \bullet 5\sqrt{2x}=10x\sqrt{2x}+5x\sqrt{2x}=15x\sqrt{2x}$$

51. Simplifying the radical expression:

$$3\sqrt{75x^3y}-2x\sqrt{3xy}=3\sqrt{25x^2 \bullet 3xy}-2x\sqrt{3xy}=3 \bullet 5x\sqrt{3xy}-2x\sqrt{3xy}=15x\sqrt{3xy}-2x\sqrt{3xy}=13x\sqrt{3xy}$$

53. Simplifying the radical expression: $\sqrt{20ab^2}-b\sqrt{45a}=\sqrt{4b^2 \bullet 5a}-b\sqrt{9 \bullet 5a}=2b\sqrt{5a}-3b\sqrt{5a}=-b\sqrt{5a}$

55. Simplifying the radical expression:

$$9\sqrt{18x^3}-2x\sqrt{48x}=9\sqrt{9x^2 \bullet 2x}-2x\sqrt{16 \bullet 3x}=9 \bullet 3x\sqrt{2x}-2x \bullet 4\sqrt{3x}=27x\sqrt{2x}-8x\sqrt{3x}$$

57. Simplifying the radical expression:

$$\begin{aligned}7\sqrt{50x^2y}+8x\sqrt{8y}-7\sqrt{32x^2y}&=7\sqrt{25x^2 \bullet 2y}+8x\sqrt{4 \bullet 2y}-7\sqrt{16x^2 \bullet 2y}\\&=7 \bullet 5x\sqrt{2y}+8x \bullet 2\sqrt{2y}-7 \bullet 4x\sqrt{2y}\\&=35x\sqrt{2y}+16x\sqrt{2y}-28x\sqrt{2y}\\&=23x\sqrt{2y}\end{aligned}$$

59. Simplifying the expression: $\dfrac{6+2\sqrt{2}}{2}=\dfrac{2\left(3+\sqrt{2}\right)}{2}=3+\sqrt{2}$

61. Simplifying the expression: $\dfrac{9-3\sqrt{3}}{3}=\dfrac{3\left(3-\sqrt{3}\right)}{3}=3-\sqrt{3}$

63. Simplifying the expression: $\dfrac{8-\sqrt{24}}{6}=\dfrac{8-\sqrt{4 \bullet 6}}{6}=\dfrac{8-2\sqrt{6}}{6}=\dfrac{2\left(4-\sqrt{6}\right)}{6}=\dfrac{4-\sqrt{6}}{3}$

65. Simplifying the expression: $\dfrac{6+\sqrt{8}}{2}=\dfrac{6+\sqrt{4 \bullet 2}}{2}=\dfrac{6+2\sqrt{2}}{2}=\dfrac{2\left(3+\sqrt{2}\right)}{2}=3+\sqrt{2}$

67. Simplifying the expression: $\dfrac{-10+\sqrt{50}}{10}=\dfrac{-10+\sqrt{25 \bullet 2}}{10}=\dfrac{-10+5\sqrt{2}}{10}=\dfrac{5\left(-2+\sqrt{2}\right)}{10}=\dfrac{-2+\sqrt{2}}{2}$

69. **a.** Approximating: $\dfrac{3+\sqrt{12}}{2}=\dfrac{3+\sqrt{4 \bullet 3}}{2}=\dfrac{3+2\sqrt{3}}{2}\approx\dfrac{3+2(1.732)}{2}=\dfrac{6.464}{2}=3.232$

b. Approximating: $\dfrac{3-\sqrt{12}}{2}=\dfrac{3-\sqrt{4 \bullet 3}}{2}=\dfrac{3-2\sqrt{3}}{2}\approx\dfrac{3-2(1.732)}{2}=\dfrac{-0.464}{2}=-0.232$

c. Approximating: $\dfrac{3+\sqrt{12}}{2}+\dfrac{3-\sqrt{12}}{2}=3.232-0.232=3$

71. Completing the table:

x	$\sqrt{x^2+9}$	$x+3$
1	3.162	4
2	3.606	5
3	4.243	6
4	5	7
5	5.831	8
6	6.708	9

73. Completing the table:

x	$\sqrt{x+3}$	$\sqrt{x}+\sqrt{3}$
1	2	2.732
2	2.236	3.146
3	2.449	3.464
4	2.646	3.732
5	2.828	3.968
6	3	4.182

75. The correct statement is: $4\sqrt{3}+5\sqrt{3}=9\sqrt{3}$

77. Multiplying: $\sqrt{5}\sqrt{2}=\sqrt{10}$

© 2012 Cengage Learning. All Rights Reserved. May not be scanned, copied or duplicated, or posted to a publicly accessible website, in whole or in part.

79. Multiplying: $\sqrt{5}\sqrt{5}=\sqrt{25}=5$

81. Multiplying: $\sqrt{x}\sqrt{x}=\sqrt{x^2}=x$

83. Multiplying: $\sqrt{5}\sqrt{7}=\sqrt{35}$

85. Combining like terms: $5+7\sqrt{5}+2\sqrt{5}+14=19+9\sqrt{5}$

87. Combining like terms: $x-7\sqrt{x}+3\sqrt{x}-21=x-4\sqrt{x}-21$

89. Multiplying the expressions: $(3x+y)^2=(3x)^2+2(3x)(y)+y^2=9x^2+6xy+y^2$

91. Multiplying the expressions: $(3x-4y)(3x+4y)=(3x)^2-(4y)^2=9x^2-16y^2$

93. Multiplying each side of the equation by 6:

$$\begin{aligned} 6\left(\frac{x}{3}-\frac{1}{2}\right)&=6\left(\frac{5}{2}\right)\\ 2x-3&=15\\ 2x&=18\\ x&=9 \end{aligned}$$

Since $x=9$ checks in the original equation, the solution is $x=9$.

95. Multiplying each side of the equation by x^2:

$$\begin{aligned} x^2\left(1-\frac{5}{x}\right)&=x^2\left(\frac{-6}{x^2}\right)\\ x^2-5x&=-6\\ x^2-5x+6&=0\\ (x-2)(x-3)&=0\\ x&=2,3 \end{aligned}$$

Both $x=2$ and $x=3$ check in the original equation.

97. Multiplying each side of the equation by $2(a-4)$:

$$\begin{aligned} 2(a-4)\left(\frac{a}{a-4}-\frac{a}{2}\right)&=2(a-4)\bullet\frac{4}{a-4}\\ 2a-a(a-4)&=8\\ 2a-a^2+4a&=8\\ -a^2+6a-8&=0\\ a^2-6a+8&=0\\ (a-4)(a-2)&=0\\ a&=2,4 \end{aligned}$$

Since $a=4$ does not check in the original equation, the solution is $a=2$.

8.5 Multiplication and Division of Radicals

1. Multiplying the radicals: $\sqrt{3}\sqrt{2}=\sqrt{6}$
3. Multiplying the radicals: $\sqrt{6}\sqrt{2}=\sqrt{12}=\sqrt{4\bullet 3}=2\sqrt{3}$
5. Multiplying the radicals: $(2\sqrt{3})(5\sqrt{7})=10\sqrt{21}$
7. Multiplying the radicals: $(4\sqrt{3})(2\sqrt{6})=8\sqrt{18}=8\sqrt{9\bullet 2}=8\bullet 3\sqrt{2}=24\sqrt{2}$
9. Multiplying the radicals: $(2\sqrt{2})^2=(2\sqrt{2})(2\sqrt{2})=4\sqrt{4}=4\bullet 2=8$
11. Multiplying the radicals: $(-2\sqrt{6})^2=(-2\sqrt{6})(-2\sqrt{6})=4\sqrt{36}=4\bullet 6=24$
13. Multiplying the radicals: $(-1+5\sqrt{2}+1)^2=(5\sqrt{2})^2=(5\sqrt{2})(5\sqrt{2})=25\sqrt{4}=25\bullet 2=50$
15. Multiplying the radicals: $\left[2(-3+\sqrt{2})+6\right]^2=(-6+2\sqrt{2}+6)=(2\sqrt{2})^2=(2\sqrt{2})(2\sqrt{2})=4\sqrt{4}=4\bullet 2=8$

© 2012 Cengage Learning. All Rights Reserved. May not be scanned, copied or duplicated, or posted to a publicly accessible website, in whole or in part.

17. Multiplying the radicals: $\sqrt{2}\left(\sqrt{3}-1\right)=\sqrt{6}-\sqrt{2}$

19. Multiplying the radicals: $\sqrt{2}\left(\sqrt{3}+\sqrt{2}\right)=\sqrt{6}+\sqrt{4}=\sqrt{6}+2$

21. Multiplying the radicals: $\sqrt{3}\left(2\sqrt{2}+\sqrt{3}\right)=2\sqrt{6}+\sqrt{9}=2\sqrt{6}+3$

23. Multiplying the radicals: $\sqrt{3}\left(2\sqrt{3}-\sqrt{5}\right)=2\sqrt{9}-\sqrt{15}=2\bullet 3-\sqrt{15}=6-\sqrt{15}$

25. Multiplying the radicals: $2\sqrt{3}\left(\sqrt{2}+\sqrt{5}\right)=2\sqrt{6}+2\sqrt{15}$

27. Simplifying the expression: $\left(\sqrt{2}+1\right)^2=\left(\sqrt{2}\right)^2+2\left(\sqrt{2}\right)(1)+(1)^2=2+2\sqrt{2}+1=3+2\sqrt{2}$

29. Simplifying the expression: $\left(\sqrt{x}+3\right)^2=\left(\sqrt{x}\right)^2+2\left(\sqrt{x}\right)(3)+(3)^2=x+6\sqrt{x}+9$

31. Simplifying the expression: $\left(5-\sqrt{2}\right)^2=(5)^2-2(5)\left(\sqrt{2}\right)+\left(\sqrt{2}\right)^2=25-10\sqrt{2}+2=27-10\sqrt{2}$

33. Simplifying the expression: $\left(\sqrt{a}-\frac{1}{2}\right)^2=\left(\sqrt{a}\right)^2-2\left(\sqrt{a}\right)\left(\frac{1}{2}\right)+\left(\frac{1}{2}\right)^2=a-\sqrt{a}+\frac{1}{4}$

35. Simplifying the expression: $\left(3+\sqrt{7}\right)^2=(3)^2+2(3)\left(\sqrt{7}\right)+\left(\sqrt{7}\right)^2=9+6\sqrt{7}+7=16+6\sqrt{7}$

37. Simplifying the expression: $\left(\sqrt{5}+3\right)\left(\sqrt{5}+2\right)=5+3\sqrt{5}+2\sqrt{5}+6=11+5\sqrt{5}$

39. Simplifying the expression: $\left(\sqrt{2}-5\right)\left(\sqrt{2}+6\right)=2-5\sqrt{2}+6\sqrt{2}-30=-28+\sqrt{2}$

41. Simplifying the expression: $\left(\sqrt{3}+\frac{1}{2}\right)\left(\sqrt{2}+\frac{1}{3}\right)=\sqrt{6}+\frac{1}{2}\sqrt{2}+\frac{1}{3}\sqrt{3}+\frac{1}{6}$

43. Simplifying the expression: $\left(\sqrt{x}+6\right)\left(\sqrt{x}-6\right)=\left(\sqrt{x}\right)^2-(6)^2=x-36$

45. Simplifying the expression: $\left(\sqrt{a}+\frac{1}{3}\right)\left(\sqrt{a}+\frac{2}{3}\right)=a+\frac{1}{3}\sqrt{a}+\frac{2}{3}\sqrt{a}+\frac{2}{9}=a+\sqrt{a}+\frac{2}{9}$

47. Simplifying the expression: $\left(\sqrt{5}-2\right)\left(\sqrt{5}+2\right)=\left(\sqrt{5}\right)^2-(2)^2=5-4=1$

49. Simplifying the expression: $\left(2\sqrt{7}+3\right)\left(3\sqrt{7}-4\right)=6\sqrt{49}+9\sqrt{7}-8\sqrt{7}-12=42+\sqrt{7}-12=30+\sqrt{7}$

51. Simplifying the expression: $\left(2\sqrt{x}+4\right)\left(3\sqrt{x}+2\right)=6x+12\sqrt{x}+4\sqrt{x}+8=6x+16\sqrt{x}+8$

53. Simplifying the expression: $\left(7\sqrt{a}+2\sqrt{b}\right)\left(7\sqrt{a}-2\sqrt{b}\right)=\left(7\sqrt{a}\right)^2-\left(2\sqrt{b}\right)^2=49a-4b$

55. Simplifying the expression:

$$\left(3+\sqrt{2}\right)^2-6\left(3+\sqrt{2}\right)=\left(3+\sqrt{2}\right)\left(3+\sqrt{2}\right)-6\left(3+\sqrt{2}\right)=9+3\sqrt{2}+3\sqrt{2}+2-18-6\sqrt{2}=-7$$

57. Simplifying the expression:

$$\left(2+\sqrt{5}\right)^2-4\left(2+\sqrt{5}\right)=\left(2+\sqrt{5}\right)\left(2+\sqrt{5}\right)-4\left(2+\sqrt{5}\right)=4+2\sqrt{5}+2\sqrt{5}+5-8-4\sqrt{5}=1$$

59. Simplifying the expression:

$$\begin{aligned}\left(7-\sqrt{5}\right)^2-14\left(7-\sqrt{5}\right)+44&=\left(7-\sqrt{5}\right)\left(7-\sqrt{5}\right)-14\left(7-\sqrt{5}\right)+44\\&=49-7\sqrt{5}-7\sqrt{5}+5-98+14\sqrt{5}+44\\&=0\end{aligned}$$

61. Rationalizing the denominator: $\frac{\sqrt{3}}{\sqrt{5}-\sqrt{2}}\bullet\frac{\sqrt{5}+\sqrt{2}}{\sqrt{5}+\sqrt{2}}=\frac{\sqrt{15}+\sqrt{6}}{5-2}=\frac{\sqrt{15}+\sqrt{6}}{3}$

63. Rationalizing the denominator: $\frac{\sqrt{5}}{\sqrt{5}+\sqrt{2}}\bullet\frac{\sqrt{5}-\sqrt{2}}{\sqrt{5}-\sqrt{2}}=\frac{\sqrt{25}-\sqrt{10}}{5-2}=\frac{5-\sqrt{10}}{3}$

© 2012 Cengage Learning. All Rights Reserved. May not be scanned, copied or duplicated, or posted to a publicly accessible website, in whole or in part.

65. Rationalizing the denominator: $\frac{8}{3-\sqrt{5}} \bullet \frac{3+\sqrt{5}}{3+\sqrt{5}} = \frac{8(3+\sqrt{5})}{9-5} = \frac{8(3+\sqrt{5})}{4} = 2(3+\sqrt{5}) = 6+2\sqrt{5}$

67. Rationalizing the denominator: $\frac{\sqrt{3}+\sqrt{2}}{\sqrt{3}-\sqrt{2}} \bullet \frac{\sqrt{3}+\sqrt{2}}{\sqrt{3}+\sqrt{2}} = \frac{3+\sqrt{6}+\sqrt{6}+2}{3-2} = \frac{5+2\sqrt{6}}{1} = 5+2\sqrt{6}$

69. Rationalizing the denominator: $\frac{\sqrt{7}-\sqrt{3}}{\sqrt{7}+\sqrt{3}} \bullet \frac{\sqrt{7}-\sqrt{3}}{\sqrt{7}-\sqrt{3}} = \frac{7-\sqrt{21}-\sqrt{21}+3}{7-3} = \frac{10-2\sqrt{21}}{4} = \frac{2(5-\sqrt{21})}{4} = \frac{5-\sqrt{21}}{2}$

71. Rationalizing the denominator: $\frac{\sqrt{x}+2}{\sqrt{x}-2} \bullet \frac{\sqrt{x}+2}{\sqrt{x}+2} = \frac{x+2\sqrt{x}+2\sqrt{x}+4}{x-4} = \frac{x+4\sqrt{x}+4}{x-4}$

73. Rationalizing the denominator: $\frac{\sqrt{5}-\sqrt{2}}{\sqrt{5}+\sqrt{3}} \bullet \frac{\sqrt{5}-\sqrt{3}}{\sqrt{5}-\sqrt{3}} = \frac{5-\sqrt{10}-\sqrt{15}+\sqrt{6}}{5-3} = \frac{5-\sqrt{10}-\sqrt{15}+\sqrt{6}}{2}$

75. a. Subtracting: $(\sqrt{7}+\sqrt{3})-(\sqrt{7}-\sqrt{3}) = \sqrt{7}+\sqrt{3}-\sqrt{7}+\sqrt{3} = 2\sqrt{3}$

b. Multiplying: $(\sqrt{7}+\sqrt{3})(\sqrt{7}-\sqrt{3}) = 7-\sqrt{21}+\sqrt{21}-3 = 4$

c. Squaring: $(\sqrt{7}-\sqrt{3})^2 = (\sqrt{7}-\sqrt{3})(\sqrt{7}-\sqrt{3}) = 7-\sqrt{21}-\sqrt{21}+3 = 10-2\sqrt{21}$

d. Dividing: $\frac{\sqrt{7}-\sqrt{3}}{\sqrt{7}+\sqrt{3}} \bullet \frac{\sqrt{7}-\sqrt{3}}{\sqrt{7}-\sqrt{3}} = \frac{7-\sqrt{21}-\sqrt{21}+3}{7-3} = \frac{10-2\sqrt{21}}{4} = \frac{2(5-\sqrt{21})}{4} = \frac{5-\sqrt{21}}{2}$

77. a. Adding: $(\sqrt{x}+2)+(\sqrt{x}-2) = \sqrt{x}+2+\sqrt{x}-2 = 2\sqrt{x}$

b. Multiplying: $(\sqrt{x}+2)(\sqrt{x}-2) = x+2\sqrt{x}-2\sqrt{x}-4 = x-4$

c. Squaring: $(\sqrt{x}+2)^2 = (\sqrt{x}+2)(\sqrt{x}+2) = x+2\sqrt{x}+2\sqrt{x}+4 = x+4\sqrt{x}+4$

d. Dividing: $\frac{\sqrt{x}+2}{\sqrt{x}-2} \bullet \frac{\sqrt{x}+2}{\sqrt{x}+2} = \frac{x+2\sqrt{x}+2\sqrt{x}+4}{x-4} = \frac{x+4\sqrt{x}+4}{x-4}$

79. a. Adding: $(5+\sqrt{2})+(5-\sqrt{2}) = 5+\sqrt{2}+5-\sqrt{2} = 10$

b. Multiplying: $(5+\sqrt{2})(5-\sqrt{2}) = 25+5\sqrt{2}-5\sqrt{2}-2 = 23$

c. Squaring: $(5+\sqrt{2})^2 = (5+\sqrt{2})(5+\sqrt{2}) = 25+5\sqrt{2}+5\sqrt{2}+2 = 27+10\sqrt{2}$

d. Dividing: $\frac{5+\sqrt{2}}{5-\sqrt{2}} \bullet \frac{5+\sqrt{2}}{5+\sqrt{2}} = \frac{25+5\sqrt{2}+5\sqrt{2}+2}{25-2} = \frac{27+10\sqrt{2}}{23}$

81. a. Adding: $\sqrt{2}+(\sqrt{6}+\sqrt{2}) = \sqrt{2}+\sqrt{6}+\sqrt{2} = 2\sqrt{2}+\sqrt{6}$

b. Multiplying: $\sqrt{2}(\sqrt{6}+\sqrt{2}) = \sqrt{12}+\sqrt{4} = 2\sqrt{3}+2$

c. Dividing: $\frac{\sqrt{6}+\sqrt{2}}{\sqrt{2}} \bullet \frac{\sqrt{2}}{\sqrt{2}} = \frac{\sqrt{12}+\sqrt{4}}{2} = \frac{2\sqrt{3}+2}{2} = 1+\sqrt{3}$

d. Dividing: $\frac{\sqrt{2}}{\sqrt{6}+\sqrt{2}} \bullet \frac{\sqrt{6}-\sqrt{2}}{\sqrt{6}-\sqrt{2}} = \frac{\sqrt{12}-\sqrt{4}}{6-2} = \frac{2\sqrt{3}-2}{4} = \frac{\sqrt{3}-1}{2}$

83. a. Adding: $\frac{1+\sqrt{5}}{2} + \frac{1-\sqrt{5}}{2} = \frac{2}{2} = 1$

b. Multiplying: $\left(\frac{1+\sqrt{5}}{2}\right)\left(\frac{1-\sqrt{5}}{2}\right) = \frac{1+\sqrt{5}-\sqrt{5}-5}{4} = \frac{-4}{4} = -1$

© 2012 Cengage Learning. All Rights Reserved. May not be scanned, copied or duplicated, or posted to a publicly accessible website, in whole or in part.

85. The correct statement is: $2(3\sqrt{5})=6\sqrt{5}$

87. The correct statement is: $(\sqrt{3}+7)^2=(\sqrt{3})^2+2(\sqrt{3})(7)+(7)^2=3+14\sqrt{3}+49=52+14\sqrt{3}$

89. Simplifying: $7^2=7\bullet 7=49$

91. Simplifying: $(-9)^2=(-9)\bullet(-9)=81$

93. Simplifying: $(\sqrt{x+1})^2=x+1$

95. Simplifying: $(\sqrt{2x-3})^2=2x-3$

97. Simplifying: $(x+3)^2=x^2+2(x)(3)+3^2=x^2+6x+9$

99. Solving the equation:

$$\begin{aligned}3a-2&=4\\3a&=6\\a&=2\end{aligned}$$

101. Solving the equation:

$$\begin{aligned}x+15&=x^2+6x+9\\0&=x^2+5x-6\\0&=(x+6)(x-1)\\x&=-6,1\end{aligned}$$

103. **a.** Substituting $x=-6$:

$\sqrt{-6+15}=\sqrt{9}=3$

$-6+3=-3$

No, $x=-6$ is not a solution.

b. Substituting $x=1$:

$\sqrt{1+15}=\sqrt{16}=4$

$1+3=4$

Yes, $x=1$ is a solution.

105. The values are given in the table:

x	$y=3\sqrt{x}$
–4	undefined
–1	undefined
0	$3\sqrt{0}=0$
1	$3\sqrt{1}=3$
4	$3\sqrt{4}=6$
9	$3\sqrt{9}=9$
16	$3\sqrt{16}=12$

107. Solving the equation:

$$\begin{aligned}x^2+5x-6&=0\\(x+6)(x-1)&=0\\x&=-6,1\end{aligned}$$

109. Solving the equation:

$$\begin{aligned}x^2-3x&=0\\x(x-3)&=0\\x&=0,3\end{aligned}$$

111. Solving the proportion:

$$\begin{aligned}\frac{x}{3}&=\frac{27}{x}\\x^2&=81\\x^2-81&=0\\(x+9)(x-9)&=0\\x&=-9,9\end{aligned}$$

113. Solving the proportion:

$$\begin{aligned}\frac{x}{5}&=\frac{3}{x+2}\\x^2+2x&=15\\x^2+2x-15&=0\\(x+5)(x-3)&=0\\x&=-5,3\end{aligned}$$

© 2012 Cengage Learning. All Rights Reserved. May not be scanned, copied or duplicated, or posted to a publicly accessible website, in whole or in part.

8.6 Equations Involving Radicals

1. Solving the equation by squaring:

$$\begin{aligned}\sqrt{x+1}&=2\\ \left(\sqrt{x+1}\right)^2&=(2)^2\\ x+1&=4\\ x&=3\end{aligned}$$

This value checks in the original equation.

3. Solving the equation by squaring:

$$\begin{aligned}\sqrt{x+5}&=7\\ \left(\sqrt{x+5}\right)^2&=(7)^2\\ x+5&=49\\ x&=44\end{aligned}$$

This value checks in the original equation.

5. Solving the equation by squaring:

$$\begin{aligned}\sqrt{x-9}&=-6\\ \left(\sqrt{x-9}\right)^2&=(-6)^2\\ x-9&=36\\ x&=45\end{aligned}$$

Since $\sqrt{45-9}=\sqrt{36}=6\neq-6$, there is no solution to the equation.

7. Solving the equation by squaring:

$$\begin{aligned}\sqrt{x-5}&=-4\\ \left(\sqrt{x-5}\right)^2&=(-4)^2\\ x-5&=16\\ x&=21\end{aligned}$$

Since $\sqrt{21-5}=\sqrt{16}=4\neq-4$, there is no solution to the equation.

9. Solving the equation by squaring:

$$\begin{aligned}\sqrt{x-8}&=0\\ \left(\sqrt{x-8}\right)^2&=(0)^2\\ x-8&=0\\ x&=8\end{aligned}$$

This value checks in the original equation.

11. Solving the equation by squaring:

$$\begin{aligned}\sqrt{2x+1}&=3\\ \left(\sqrt{2x+1}\right)^2&=(3)^2\\ 2x+1&=9\\ 2x&=8\\ x&=4\end{aligned}$$

This value checks in the original equation.

13. Solving the equation by squaring:

$$\begin{aligned}\sqrt{2x-3}&=-5\\ \left(\sqrt{2x-3}\right)^2&=(-5)^2\\ 2x-3&=25\\ 2x&=28\\ x&=14\end{aligned}$$

Since $\sqrt{2\cdot14-3}=\sqrt{28-3}=\sqrt{25}=5\neq-5$, there is no solution to the equation.

15. Solving the equation by squaring:

$$\begin{aligned}\sqrt{3x+6}&=2\\ \left(\sqrt{3x+6}\right)^2&=(2)^2\\ 3x+6&=4\\ 3x&=-2\\ x&=-\frac{2}{3}\end{aligned}$$

This value checks in the original equation.

17. Solving the equation by squaring:

$$\begin{aligned}2\sqrt{x}&=10\\ \sqrt{x}&=5\\ \left(\sqrt{x}\right)^2&=(5)^2\\ x&=25\end{aligned}$$

This value checks in the original equation.

© 2012 Cengage Learning. All Rights Reserved. May not be scanned, copied or duplicated, or posted to a publicly accessible website, in whole or in part.

19. Solving the equation by squaring:

$$3\sqrt{a} = 6$$
$$\sqrt{a} = 2$$
$$\left(\sqrt{a}\right)^2 = (2)^2$$
$$a = 4$$

This value checks in the original equation.

21. Solving the equation by squaring:

$$\sqrt{3x+4} - 3 = 2$$
$$\sqrt{3x+4} = 5$$
$$\left(\sqrt{3x+4}\right)^2 = (5)^2$$
$$3x + 4 = 25$$
$$3x = 21$$
$$x = 7$$

This value checks in the original equation.

23. Solving the equation by squaring:

$$\sqrt{5y-4} - 2 = 4$$
$$\sqrt{5y-4} = 6$$
$$\left(\sqrt{5y-4}\right)^2 = (6)^2$$
$$5y - 4 = 36$$
$$5y = 40$$
$$y = 8$$

This value checks in the original equation.

25. Solving the equation by squaring:

$$\sqrt{2x+1} + 5 = 2$$
$$\sqrt{2x+1} = -3$$
$$\left(\sqrt{2x+1}\right)^2 = (-3)^2$$
$$2x + 1 = 9$$
$$2x = 8$$
$$x = 4$$

Since $\sqrt{2 \bullet 4 + 1} + 5 = \sqrt{9} + 5 = 3 + 5 = 8 \neq 2$, there is no solution to the equation.

27. Solving the equation by squaring:

$$\sqrt{x+3} = x - 3$$
$$\left(\sqrt{x+3}\right)^2 = (x-3)^2$$
$$x + 3 = x^2 - 6x + 9$$
$$0 = x^2 - 7x + 6$$
$$0 = (x-6)(x-1)$$
$$x = 6, 1$$

Since $x = 1$ does not check in the original equation, the solution is $x = 6$.

29. Solving the equation by squaring:

$$\sqrt{a+2} = a + 2$$
$$\left(\sqrt{a+2}\right)^2 = (a+2)^2$$
$$a + 2 = a^2 + 4a + 4$$
$$0 = a^2 + 3a + 2$$
$$0 = (a+2)(a+1)$$
$$a = -2, -1$$

Both values check in the original equation.

31. Solving the equation by squaring:

$$\sqrt{2x+9} = x + 5$$
$$\left(\sqrt{2x+9}\right)^2 = (x+5)^2$$
$$2x + 9 = x^2 + 10x + 25$$
$$0 = x^2 + 8x + 16$$
$$0 = (x+4)^2$$
$$x = -4$$

This value checks in the original equation.

© 2012 Cengage Learning. All Rights Reserved. May not be scanned, copied or duplicated, or posted to a publicly accessible website, in whole or in part.

33. Solving the equation by squaring:

$$\begin{aligned}\sqrt{y-4} &= y-6\\ \left(\sqrt{y-4}\right)^2 &= (y-6)^2\\ y-4 &= y^2-12y+36\\ 0 &= y^2-13y+40\\ 0 &= (y-8)(y-5)\\ y &= 5,8\end{aligned}$$

Since $y=5$ does not check in the original equation, the solution is $y=8$.

35. a. Solving the equation:

$$\begin{aligned}\sqrt{y}-4 &= 6\\ \sqrt{y} &= 10\\ \left(\sqrt{y}\right)^2 &= (10)^2\\ y &= 100\end{aligned}$$

This value checks in the original equation.

b. Solving the equation:

$$\begin{aligned}\sqrt{y-4} &= 6\\ \left(\sqrt{y-4}\right)^2 &= (6)^2\\ y-4 &= 36\\ y &= 40\end{aligned}$$

This value checks in the original equation.

c. Solving the equation:

$$\begin{aligned}\sqrt{y-4} &= -6\\ \left(\sqrt{y-4}\right)^2 &= (-6)^2\\ y-4 &= 36\\ y &= 40\end{aligned}$$

This value does not check in the original equation, so there is no solution $(\varnothing)$.

d. Solving the equation:

$$\begin{aligned}\sqrt{y-4} &= y-6\\ \left(\sqrt{y-4}\right)^2 &= (y-6)^2\\ y-4 &= y^2-12y+36\\ 0 &= y^2-13y+40\\ 0 &= (y-8)(y-5)\\ y &= 5,8\end{aligned}$$

Since $y=5$ does not check in the original equation, the solution is $y=8$.

37. a. Solving the equation:

$$\begin{aligned}x-3 &= 0\\ x &= 3\end{aligned}$$

b. Solving the equation:

$$\begin{aligned}\sqrt{x}-3 &= 0\\ \sqrt{x} &= 3\\ \left(\sqrt{x}\right)^2 &= (3)^2\\ x &= 9\end{aligned}$$

This value checks in the original equation.

c. Solving the equation:

$$\begin{aligned}\sqrt{x-3} &= 0\\ \left(\sqrt{x-3}\right)^2 &= (0)^2\\ x-3 &= 0\\ x &= 3\end{aligned}$$

This value checks in the original equation.

d. Solving the equation:

$$\begin{aligned}\sqrt{x}+3 &= 0\\ \sqrt{x} &= -3\\ \left(\sqrt{x}\right)^2 &= (-3)^2\\ x &= 9\end{aligned}$$

This value does not check in the original equation, so there is no solution $(\varnothing)$.

© 2012 Cengage Learning. All Rights Reserved. May not be scanned, copied or duplicated, or posted to a publicly accessible website, in whole or in part.

e. Solving the equation:

$$\sqrt{x}+3=5$$
$$\sqrt{x}=2$$
$$\left(\sqrt{x}\right)^2=(2)^2$$
$$x=4$$

This value checks in the original equation.

f. Solving the equation:

$$\sqrt{x}+3=-5$$
$$\sqrt{x}=-8$$
$$\left(\sqrt{x}\right)^2=(-8)^2$$
$$x=64$$

This value does not check in the original equation, so there is no solution $(\varnothing)$.

g. Solving the equation:

$$x-3=\sqrt{5-x}$$
$$(x-3)^2=\left(\sqrt{5-x}\right)^2$$
$$x^2-6x+9=5-x$$
$$x^2-5x+4=0$$
$$(x-1)(x-4)=0$$
$$x=1,4$$

Since $x=1$ does not check in the original equation, the solution is $x=4$.

39. Complete the table:

x	y
0	0
1	1
2	1.4
3	1.7
4	2

Sketching the graph:

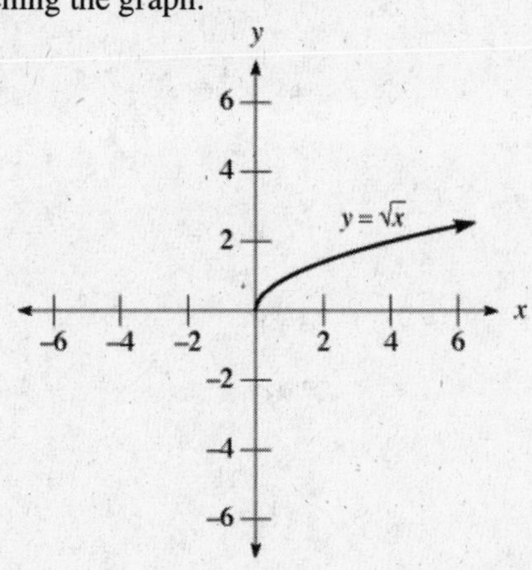

41. Complete the table:

x	y
0	0
1	2
4	4
9	6

Sketching the graph:

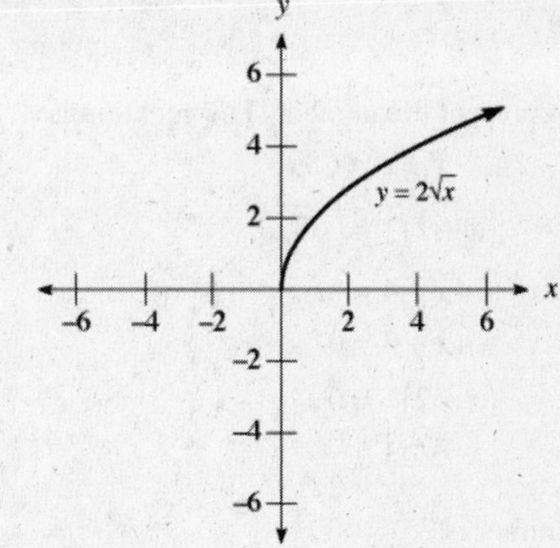

© 2012 Cengage Learning. All Rights Reserved. May not be scanned, copied or duplicated, or posted to a publicly accessible website, in whole or in part.

43. Complete the table:

x	y
0	2
1	3
2	3.4
4	4
9	5

Sketching the graph:

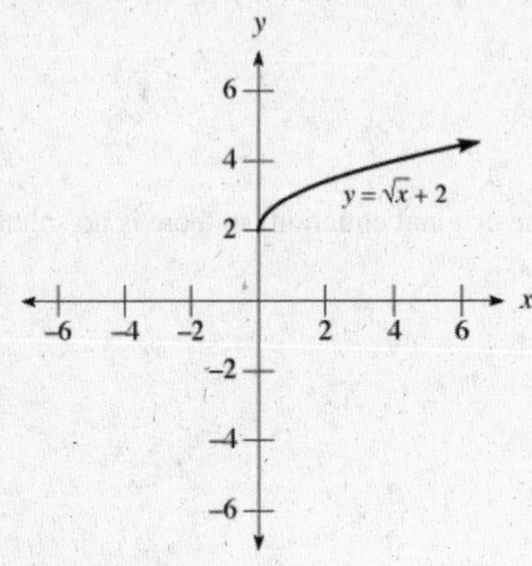

45. Finding the time: $t = \frac{1}{4}\sqrt{29,035} \approx 42.60$ seconds

47. Drawing a line graph:

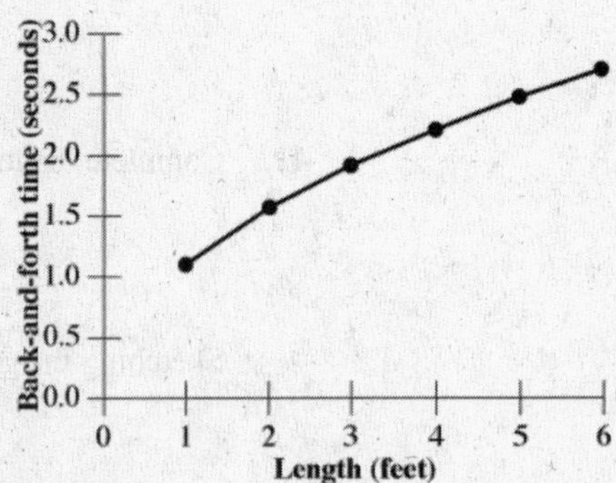

49. Let x represent the number. The equation is:

$$
\begin{aligned}
x+2 &= \sqrt{8x} \\
(x+2)^2 &= \left(\sqrt{8x}\right)^2 \\
x^2+4x+4 &= 8x \\
x^2-4x+4 &= 0 \\
(x-2)^2 &= 0 \\
x-2 &= 0 \\
x &= 2
\end{aligned}
$$

The number is 2.

© 2012 Cengage Learning. All Rights Reserved. May not be scanned, copied or duplicated, or posted to a publicly accessible website, in whole or in part.

51. Let x represent the number. The equation is:

$$\begin{aligned} x-3&=2\sqrt{x} \\ (x-3)^2&=\left(2\sqrt{x}\right)^2 \\ x^2-6x+9&=4x \\ x^2-10x+9&=0 \\ (x-9)(x-1)&=0 \\ x&=9 \quad (x=1 \text{ does not check in the original equation}) \end{aligned}$$

The number is 9.

53. Substituting $T=2$:

$$\begin{aligned} 2&=\frac{11}{7}\sqrt{\frac{L}{2}} \\ 14&=11\sqrt{\frac{L}{2}} \\ \frac{14}{11}&=\sqrt{\frac{L}{2}} \\ \left(\frac{14}{11}\right)^2&=\left(\sqrt{\frac{L}{2}}\right)^2 \\ \frac{196}{121}&=\frac{L}{2} \\ L&=\frac{392}{121}\approx 3.2 \text{ feet} \end{aligned}$$

55. Reducing the rational expression: $\frac{x^2-x-6}{x^2-9}=\frac{(x-3)(x+2)}{(x+3)(x-3)}=\frac{x+2}{x+3}$

57. Performing the operations:

$$\frac{x^2-25}{x+4}\bullet\frac{2x+8}{x^2-9x+20}=\frac{(x+5)(x-5)}{x+4}\bullet\frac{2(x+4)}{(x-4)(x-5)}=\frac{2(x+5)(x-5)(x+4)}{(x+4)(x-4)(x-5)}=\frac{2(x+5)}{x-4}$$

59. Performing the operations: $\frac{x}{x^2-16}+\frac{4}{x^2-16}=\frac{x+4}{x^2-16}=\frac{x+4}{(x+4)(x-4)}=\frac{1}{x-4}$

61. Simplifying the complex fraction: $\frac{1-\frac{25}{x^2}}{1-\frac{8}{x}+\frac{15}{x^2}}\bullet\frac{x^2}{x^2}=\frac{x^2-25}{x^2-8x+15}=\frac{(x+5)(x-5)}{(x-5)(x-3)}=\frac{x+5}{x-3}$

63. Multiplying each side of the equation by $x^2-9=(x+3)(x-3)$:

$$\begin{aligned} (x+3)(x-3)\left(\frac{x}{x^2-9}-\frac{3}{x-3}\right)&=(x+3)(x-3)\bullet\frac{1}{x+3} \\ x-3(x+3)&=x-3 \\ x-3x-9&=x-3 \\ -2x-9&=x-3 \\ -3x&=6 \\ x&=-2 \end{aligned}$$

Since $x=-2$ checks in the original equation, the solution is $x=-2$.

© 2012 Cengage Learning. All Rights Reserved. May not be scanned, copied or duplicated, or posted to a publicly accessible website, in whole or in part.

65. Let t represent the time to fill the pool with both pipes open. The equation is:

$$\frac{1}{8}-\frac{1}{12}=\frac{1}{t}$$

$$24t\left(\frac{1}{8}-\frac{1}{12}\right)=24t\bullet\frac{1}{t}$$

$$3t-2t=24$$

$$t=24$$

It will take 24 hours to fill the pool with both pipes left open.

67. The variation equation is $y=Kx$. Finding K:

$$8=K\bullet 12$$

$$K=\frac{2}{3}$$

So $y=\frac{2}{3}x$. Substituting $x=36$: $y=\frac{2}{3}(36)=24$

Chapter 8 Review

1. Finding the root: $\sqrt{25}=5$

3. Finding the root: $\sqrt[3]{-1}=-1$

5. Finding the root: $\sqrt{100x^2y^4}=10xy^2$

7. Simplifying the expression: $\sqrt{24}=\sqrt{4\bullet 6}=2\sqrt{6}$

9. Simplifying the expression: $\sqrt{90x^3y^4}=\sqrt{9x^2y^4\bullet 10x}=3xy^2\sqrt{10x}$

11. Simplifying the expression: $3\sqrt{20x^3y}=3\sqrt{4x^2\bullet 5xy}=3\bullet 2x\sqrt{5xy}=6x\sqrt{5xy}$

13. Simplifying the expression: $\sqrt{\frac{8}{81}}=\frac{\sqrt{8}}{\sqrt{81}}=\frac{\sqrt{4\bullet 2}}{9}=\frac{2\sqrt{2}}{9}$

15. Simplifying the expression: $\sqrt{\frac{49a^2b^2}{16}}=\frac{\sqrt{49a^2b^2}}{\sqrt{16}}=\frac{7ab}{4}$

17. Simplifying the expression: $\sqrt{\frac{40a^2}{121}}=\frac{\sqrt{40a^2}}{\sqrt{121}}=\frac{\sqrt{4a^2\bullet 10}}{11}=\frac{2a\sqrt{10}}{11}$

19. Simplifying the expression: $\frac{3\sqrt{120a^2b^2}}{\sqrt{25}}=\frac{3\sqrt{4a^2b^2\bullet 30}}{5}=\frac{3\bullet 2ab\sqrt{30}}{5}=\frac{6ab\sqrt{30}}{5}$

21. Simplifying the radical expression: $\frac{2}{\sqrt{7}}\bullet\frac{\sqrt{7}}{\sqrt{7}}=\frac{2\sqrt{7}}{7}$

23. Simplifying the radical expression: $\sqrt{\frac{5}{48}}=\frac{\sqrt{5}}{\sqrt{48}}\bullet\frac{\sqrt{3}}{\sqrt{3}}=\frac{\sqrt{15}}{\sqrt{144}}=\frac{\sqrt{15}}{12}$

25. Simplifying the radical expression: $\sqrt{\frac{32ab^2}{3}}=\frac{\sqrt{32ab^2}}{\sqrt{3}}\bullet\frac{\sqrt{3}}{\sqrt{3}}=\frac{\sqrt{96ab^2}}{\sqrt{9}}=\frac{\sqrt{16b^2\bullet 6a}}{3}=\frac{4b\sqrt{6a}}{3}$

27. Rationalizing the denominator: $\frac{3}{\sqrt{3}-4}\bullet\frac{\sqrt{3}+4}{\sqrt{3}+4}=\frac{3\left(\sqrt{3}+4\right)}{3-16}=\frac{3\sqrt{3}+12}{-13}=\frac{-3\sqrt{3}-12}{13}$

29. Rationalizing the denominator: $\frac{3}{\sqrt{5}-\sqrt{2}}\bullet\frac{\sqrt{5}+\sqrt{2}}{\sqrt{5}+\sqrt{2}}=\frac{3\left(\sqrt{5}+\sqrt{2}\right)}{5-2}=\frac{3\left(\sqrt{5}+\sqrt{2}\right)}{3}=\sqrt{5}+\sqrt{2}$

31. Rationalizing the denominator: $\frac{\sqrt{5}-\sqrt{2}}{\sqrt{5}+\sqrt{2}}\bullet\frac{\sqrt{5}-\sqrt{2}}{\sqrt{5}-\sqrt{2}}=\frac{5-\sqrt{10}-\sqrt{10}+2}{5-2}=\frac{7-2\sqrt{10}}{3}$

33. Combining the expressions: $3\sqrt{5}-7\sqrt{5}=-4\sqrt{5}$

© 2012 Cengage Learning. All Rights Reserved. May not be scanned, copied or duplicated, or posted to a publicly accessible website, in whole or in part.

35. Combining the expressions:

$$\begin{aligned}-2\sqrt{45}-5\sqrt{80}+2\sqrt{20}&=-2\sqrt{9\bullet 5}-5\sqrt{16\bullet 5}+2\sqrt{4\bullet 5}\\&=-2\bullet 3\sqrt{5}-5\bullet 4\sqrt{5}+2\bullet 2\sqrt{5}\\&=-6\sqrt{5}-20\sqrt{5}+4\sqrt{5}\\&=-22\sqrt{5}\end{aligned}$$

37. Combining the expressions:

$$\sqrt{40a^3b^2}-a\sqrt{90ab^2}=\sqrt{4a^2b^2\bullet 10a}-a\sqrt{9b^2\bullet 10a}=2ab\sqrt{10a}-3ab\sqrt{10a}=-ab\sqrt{10a}$$

39. Multiplying the expressions: $4\sqrt{2}\left(\sqrt{3}+\sqrt{5}\right)=4\sqrt{6}+4\sqrt{10}$

41. Multiplying the expressions: $\left(2\sqrt{5}-4\right)\left(\sqrt{5}+3\right)=2\sqrt{25}-4\sqrt{5}+6\sqrt{5}-12=10+2\sqrt{5}-12=2\sqrt{5}-2$

43. Solving the equation by squaring:

$$\begin{aligned}\sqrt{x-3}&=3\\\left(\sqrt{x-3}\right)^2&=(3)^2\\x-3&=9\\x&=12\end{aligned}$$

This value checks in the original equation.

45. Solving the equation by squaring:

$$\begin{aligned}5\sqrt{a}&=20\\\sqrt{a}&=4\\\left(\sqrt{a}\right)^2&=(4)^2\\a&=16\end{aligned}$$

This value checks in the original equation.

47. Solving the equation by squaring:

$$\begin{aligned}\sqrt{2x+1}+10&=8\\\sqrt{2x+1}&=-2\\\left(\sqrt{2x+1}\right)^2&=(-2)^2\\2x+1&=4\\2x&=3\\x&=\frac{3}{2}\end{aligned}$$

Since $\sqrt{2\bullet\frac{3}{2}+1}+10=\sqrt{3+1}+10=\sqrt{4}+10=2+10=12\neq 8$, there is no solution to the equation.

49. Using the Pythagorean theorem:

$$\begin{aligned}x^2&=\left(\sqrt{2}\right)^2+(1)^2=2+1=3\\x&=\sqrt{3}\end{aligned}$$

Chapter 8 Cumulative Review

1. Simplifying the expression: $\left(\frac{4}{5}\right)\left(\frac{5}{4}\right)=\frac{20}{20}=1$

3. Simplifying the expression: $-\left|-\frac{1}{2}\right|=-\frac{1}{2}$

5. Simplifying the expression: $10-8-11=-9$

7. Simplifying the expression: $\frac{a^{10}}{a^2}=a^{10-2}=a^8$

9. Simplifying the expression: $\frac{24a^{12}}{6a^3}+\frac{30a^{24}}{10a^{15}}=4a^9+3a^9=7a^9$

11. Simplifying the expression: $\left(\frac{1}{2}y+2\right)\left(\frac{1}{2}y-2\right)=\left(\frac{1}{2}y\right)^2-(2)^2=\frac{1}{4}y^2-4$

13. Simplifying the expression: $\dfrac{\frac{1}{a+6}+3}{\frac{1}{a+6}+2}\bullet\frac{a+6}{a+6}=\frac{1+3(a+6)}{1+2(a+6)}=\frac{1+3a+18}{1+2a+12}=\frac{3a+19}{2a+13}$

© 2012 Cengage Learning. All Rights Reserved. May not be scanned, copied or duplicated, or posted to a publicly accessible website, in whole or in part.

15. Simplifying the expression:

$$\begin{aligned}\frac{7a}{a^2-3a-54}+\frac{5}{a-9}&=\frac{7a}{(a-9)(a+6)}+\frac{5}{a-9}\bullet\frac{a+6}{a+6}\\&=\frac{7a}{(a-9)(a+6)}+\frac{5a+30}{(a-9)(a+6)}\\&=\frac{7a+5a+30}{(a-9)(a+6)}\\&=\frac{12a+30}{(a-9)(a+6)}\\&=\frac{6(2a+5)}{(a-9)(a+6)}\end{aligned}$$

17. Simplifying the expression: $\sqrt{120x^4y^3}=\sqrt{4x^4y^2\bullet 30y}=2x^2y\sqrt{30y}$

19. Solving the equation:

$$\begin{aligned}3(5x-1)&=6(2x+3)-21\\15x-3&=12x+18-21\\15x-3&=12x-3\\3x-3&=-3\\3x&=0\\x&=0\end{aligned}$$

21. Setting each factor equal to 0 results in $x=0$, $x=-\frac{2}{3}$, or $x=4$.

23. Solving the equation by squaring each side:

$$\begin{aligned}\sqrt{x+4}&=5\\\left(\sqrt{x+4}\right)^2&=(5)^2\\x+4&=25\\x&=21\end{aligned}$$

This value checks in the original equation.

25. Adding the two equations:

$$\begin{aligned}2x&=4\\x&=2\end{aligned}$$

Substituting into the first equation:

$$\begin{aligned}2+y&=1\\y&=-1\end{aligned}$$

The solution is $(2,-1)$.

© 2012 Cengage Learning. All Rights Reserved. May not be scanned, copied or duplicated, or posted to a publicly accessible website, in whole or in part.

27. Substituting into the first equation:

$$x-(3x-1)=5$$
$$x-3x+1=5$$
$$-2x+1=5$$
$$-2x=4$$
$$x=-2$$

Substituting into the second equation: $y=3(-2)-1=-6-1=-7$

The solution is $(-2,-7)$. Note this system can also be solved by graphing:

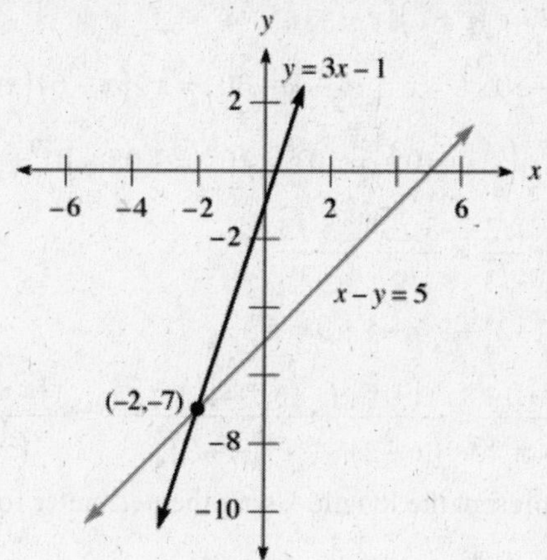

29. Graphing the line:

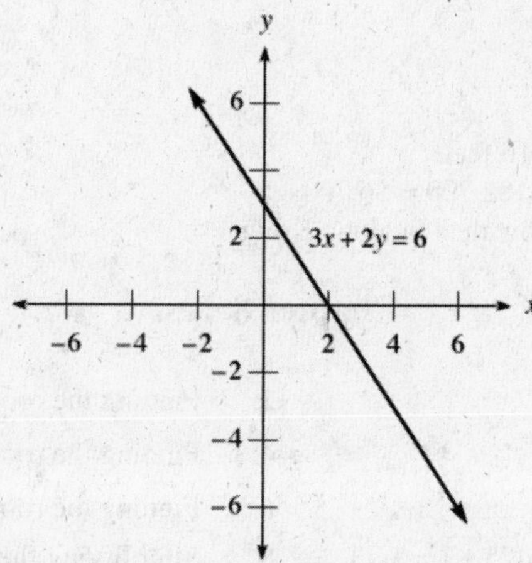

31. Checking the point $(0,0)$: $3(0)-4=-4<0$

Graphing the linear inequality:

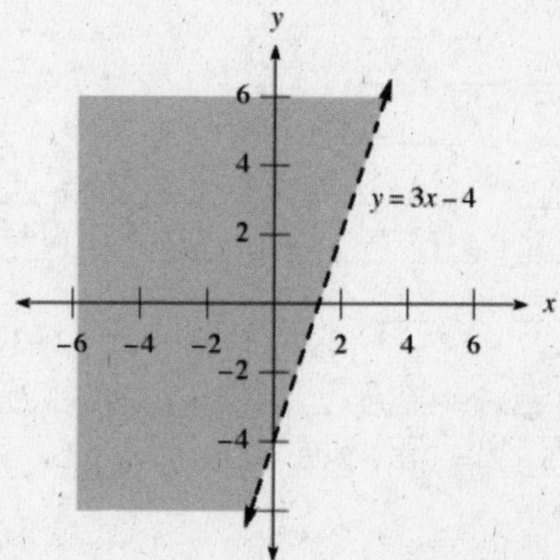

© 2012 Cengage Learning. All Rights Reserved. May not be scanned, copied or duplicated, or posted to a publicly accessible website, in whole or in part.

33. To find the x-intercept, let $y=0$:

$$0=-x+7$$
$$x=7$$

To find the y-intercept, let $x=0$: $y=-(0)+7=7$

35. Using the point-slope formula:

$$y-(-3)=-1(x-3)$$
$$y+3=-x+3$$
$$y=-x$$

37. Factoring the polynomial: $r^2+r-20=(r+5)(r-4)$

39. Factoring the polynomial: $x^5-x^4-30x^3=x^3\left(x^2-x-30\right)=x^3(x-6)(x+5)$

41. Simplifying the expression: $\left(5\times10^5\right)\left(2.1\times10^3\right)=10.5\times10^8=1.05\times10^9$

43. Rationalizing the denominator: $\dfrac{5}{\sqrt{3}}\bullet\dfrac{\sqrt{3}}{\sqrt{3}}=\dfrac{5\sqrt{3}}{\sqrt{9}}=\dfrac{5\sqrt{3}}{3}$

45. Evaluating when $x=-2$: $2x+9=2(-2)+9=-4+9=5$

47. Multiplying: $\dfrac{x^2+3x}{x^2+4x+4}\bullet\dfrac{x^2-5x-14}{x^2+6x+9}=\dfrac{x(x+3)}{(x+2)^2}\bullet\dfrac{(x+2)(x-7)}{(x+3)^2}=\dfrac{x(x+3)(x+2)(x-7)}{(x+2)^2(x+3)^2}=\dfrac{x(x-7)}{(x+2)(x+3)}$

49. Let w represent the width and $2w$ represent the length. Using the perimeter formula:

$$2(w)+2(2w)=48$$
$$2w+4w=48$$
$$6w=48$$
$$w=8$$
$$2w=16$$

The width is 8 feet and the length is 16 feet.

51. Finding the difference: \$115,400 – \$108,700 = \$6,700
It costs \$6,700 more to live in Moscow than in Hong Kong.

Chapter 8 Test

1. Finding the root: $\sqrt{16}=4$

2. Finding the root: $-\sqrt{36}=-6$

3. The root is $\sqrt{49}=7$.

4. Finding the root: $\sqrt[3]{27}=3$

5. Finding the root: $\sqrt[3]{-8}=-2$

6. Finding the root: $-\sqrt[4]{81}=-3$

7. Simplifying the expression: $\sqrt{75}=\sqrt{25\bullet3}=5\sqrt{3}$

8. Simplifying the expression: $\sqrt{32}=\sqrt{16\bullet2}=4\sqrt{2}$

9. Simplifying the expression: $\sqrt{\dfrac{2}{3}}=\dfrac{\sqrt{2}}{\sqrt{3}}\bullet\dfrac{\sqrt{3}}{\sqrt{3}}=\dfrac{\sqrt{6}}{\sqrt{9}}=\dfrac{\sqrt{6}}{3}$

10. Simplifying the expression: $\dfrac{1}{\sqrt[3]{4}}\bullet\dfrac{\sqrt[3]{2}}{\sqrt[3]{2}}=\dfrac{\sqrt[3]{2}}{\sqrt[3]{8}}=\dfrac{\sqrt[3]{2}}{2}$

11. Simplifying the expression: $3\sqrt{50x^2}=3\sqrt{25x^2\bullet2}=3\bullet5x\sqrt{2}=15x\sqrt{2}$

12. Simplifying the expression: $\sqrt{\dfrac{12x^2y^3}{5}}=\dfrac{\sqrt{12x^2y^3}}{\sqrt{5}}\bullet\dfrac{\sqrt{5}}{\sqrt{5}}=\dfrac{\sqrt{60x^2y^3}}{\sqrt{25}}=\dfrac{\sqrt{4x^2y^2\bullet15y}}{5}=\dfrac{2xy\sqrt{15y}}{5}$

13. Combining the radicals: $5\sqrt{12}-2\sqrt{27}=5\sqrt{4\bullet3}-2\sqrt{9\bullet3}=5\bullet2\sqrt{3}-2\bullet3\sqrt{3}=10\sqrt{3}-6\sqrt{3}=4\sqrt{3}$

14. Combining the radicals: $2x\sqrt{18}+5\sqrt{2x^2}=2x\sqrt{9\bullet2}+5\sqrt{x^2\bullet2}=2x\bullet3\sqrt{2}+5\bullet x\sqrt{2}=6x\sqrt{2}+5x\sqrt{2}=11x\sqrt{2}$

15. Multiplying the expressions: $\sqrt{3}\left(\sqrt{5}-2\right)=\sqrt{15}-2\sqrt{3}$

© 2012 Cengage Learning. All Rights Reserved. May not be scanned, copied or duplicated, or posted to a publicly accessible website, in whole or in part.

16. Multiplying the expressions: $\left(\sqrt{5}+7\right)\left(\sqrt{5}-8\right)=\sqrt{25}+7\sqrt{5}-8\sqrt{5}-56=5-\sqrt{5}-56=-51-\sqrt{5}$

17. Multiplying the expressions: $\left(\sqrt{x}+6\right)\left(\sqrt{x}-6\right)=\left(\sqrt{x}\right)^2-(6)^2=x-36$

18. Multiplying the expressions: $\left(\sqrt{5}-\sqrt{3}\right)^2=\left(\sqrt{5}\right)^2-2\left(\sqrt{5}\right)\left(\sqrt{3}\right)+\left(\sqrt{3}\right)^2=5-2\sqrt{15}+3=8-2\sqrt{15}$

19. Rationalizing the denominator: $\dfrac{\sqrt{7}-\sqrt{3}}{\sqrt{7}+\sqrt{3}}\bullet\dfrac{\sqrt{7}-\sqrt{3}}{\sqrt{7}-\sqrt{3}}=\dfrac{7-\sqrt{21}-\sqrt{21}+3}{7-3}=\dfrac{10-2\sqrt{21}}{4}=\dfrac{2\left(5-\sqrt{21}\right)}{4}=\dfrac{5-\sqrt{21}}{2}$

20. Rationalizing the denominator: $\dfrac{\sqrt{x}}{\sqrt{x}+5}\bullet\dfrac{\sqrt{x}-5}{\sqrt{x}-5}=\dfrac{x-5\sqrt{x}}{x-25}$

21. Solving the equation by squaring:

$$\begin{aligned}\sqrt{2x+1}+2&=7\\ \sqrt{2x+1}&=5\\ \left(\sqrt{2x+1}\right)^2&=(5)^2\\ 2x+1&=25\\ 2x&=24\\ x&=12\end{aligned}$$

This value checks in the original equation.

22. Solving the equation by squaring:

$$\begin{aligned}\sqrt{3x+1}+6&=2\\ \sqrt{3x+1}&=-4\\ \left(\sqrt{3x+1}\right)^2&=(-4)^2\\ 3x+1&=16\\ 3x&=15\\ x&=5\end{aligned}$$

Since $\sqrt{3\bullet5+1}+6=\sqrt{15+1}+6=\sqrt{16}+6=4+6=10\neq2$, there is no solution to the equation.

23. Solving the equation by squaring:

$$\begin{aligned}\sqrt{2x-3}&=x-3\\ \left(\sqrt{2x-3}\right)^2&=(x-3)^2\\ 2x-3&=x^2-6x+9\\ 0&=x^2-8x+12\\ 0&=(x-6)(x-2)\\ x&=2,6\end{aligned}$$

Since $x=2$ does not check in the original equation, the solution is $x=6$.

24. Let x represent the number. The equation is:

$$\begin{aligned}x-4&=3\sqrt{x}\\ (x-4)^2&=\left(3\sqrt{x}\right)^2\\ x^2-8x+16&=9x\\ x^2-17x+16&=0\\ (x-16)(x-1)&=0\\ x&=1,16\end{aligned}$$

Since $x=1$ does not check in the original equation, $x=16$. The number is 16.

25. Using the Pythagorean theorem:

$$\begin{aligned}x^2&=\left(\sqrt{5}\right)^2+(1)^2=5+1=6\\ x&=\sqrt{6}\end{aligned}$$

© 2012 Cengage Learning. All Rights Reserved. May not be scanned, copied or duplicated, or posted to a publicly accessible website, in whole or in part.

26. Completing the table:

x	y
0	–1
1	0
4	1
9	2

This is graph D.

27. Completing the table:

x	y
–1	0
0	1
3	2
8	3

This is graph B.

28. Completing the table:

x	y
–8	–3
–1	–2
0	–1
1	0
8	1

This is graph A.

29. Completing the table:

x	y
–9	–2
–2	–1
–1	0
0	1
7	2

This is graph C.

© 2012 Cengage Learning. All Rights Reserved. May not be scanned, copied or duplicated, or posted to a publicly accessible website, in whole or in part.

Chapter 9
Quadratic Equations

Getting Ready for Chapter 9

1. Simplifying: $\left[\frac{1}{2}(-10)\right]^2 = (-5)^2 = 25$

2. Simplifying: $\left(\frac{1}{2}\bullet 5\right)^2 = \left(\frac{5}{2}\right)^2 = \frac{25}{4}$

3. Simplifying: $\frac{5-\sqrt{49}}{2} = \frac{5-7}{2} = \frac{-2}{2} = -1$

4. Simplifying: $\frac{-4+\sqrt{40}}{4} = \frac{-4+2\sqrt{10}}{4} = \frac{2\left(-2+\sqrt{10}\right)}{4} = \frac{-2+\sqrt{10}}{2}$

5. Simplifying: $\left(2+7\sqrt{2}\right)+\left(3-5\sqrt{2}\right) = 2+7\sqrt{2}+3-5\sqrt{2} = 5+2\sqrt{2}$

6. Simplifying: $\left(3-8\sqrt{2}\right)-\left(2+4\sqrt{2}\right)+\left(3-\sqrt{2}\right) = 3-8\sqrt{2}-2-4\sqrt{2}+3-\sqrt{2} = 4-13\sqrt{2}$

7. Simplifying: $\frac{6x+5}{5x-25} - \frac{x+2}{x-5} = \frac{6x+5}{5(x-5)} - \frac{x+2}{x-5}\bullet\frac{5}{5} = \frac{6x+5}{5(x-5)} - \frac{5x+10}{5(x-5)} = \frac{6x+5-5x-10}{5(x-5)} = \frac{x-5}{5(x-5)} = \frac{1}{5}$

8. Simplifying:

$$\begin{aligned}\frac{x+1}{2x-2} - \frac{2}{x^2-1} &= \frac{x+1}{2(x-1)} - \frac{2}{(x+1)(x-1)} \\ &= \frac{x+1}{2(x-1)}\bullet\frac{x+1}{x+1} - \frac{2}{(x+1)(x-1)}\bullet\frac{2}{2} \\ &= \frac{x^2+2x+1}{2(x-1)(x+1)} - \frac{4}{2(x+1)(x-1)} \\ &= \frac{x^2+2x-3}{2(x-1)(x+1)} \\ &= \frac{(x+3)(x-1)}{2(x-1)(x+1)} \\ &= \frac{x+3}{2(x+1)}\end{aligned}$$

9. Simplifying: $\sqrt{49} = 7$

10. Simplifying: $\sqrt{12} = \sqrt{4\bullet 3} = 2\sqrt{3}$

11. Factoring: $y^2 - 12y + 36 = (y-6)^2$

12. Factoring: $x^2 + 12x + 36 = (x+6)^2$

13. Factoring: $10x^2 + 19x - 15 = (5x-3)(2x+5)$

14. Factoring: $6x^2 + x - 2 = (2x-1)(3x+2)$

15. Factoring: $y^2 + 3y + \frac{9}{4} = \left(y+\frac{3}{2}\right)^2$

16. Factoring: $y^2 - 5y + \frac{25}{4} = \left(y-\frac{5}{2}\right)^2$

© 2012 Cengage Learning. All Rights Reserved. May not be scanned, copied or duplicated, or posted to a publicly accessible website, in whole or in part.

17. Substituting $x = 3$: $y=(3-5)^2+2=(-2)^2+2=4+2=6$

18. Substituting $x = 2$: $y=(2-5)^2+2=(-3)^2+2=9+2=11$

19. Substituting $x = 1$: $y=(1-5)^2+2=(-4)^2+2=16+2=18$

20. Substituting $x = 4$: $y=(4-5)^2+2=(-1)^2+2=1+2=3$

9.1 More Quadratic Equations

1. Solving the equation:
$$\begin{aligned} x^2 &= 9 \\ x &= \pm\sqrt{9} \\ x &= \pm 3 \end{aligned}$$

3. Solving the equation:
$$\begin{aligned} a^2 &= 25 \\ a &= \pm\sqrt{25} \\ a &= \pm 5 \end{aligned}$$

5. Solving the equation:
$$\begin{aligned} y^2 &= 8 \\ y &= \pm\sqrt{8} \\ y &= \pm 2\sqrt{2} \end{aligned}$$

7. Solving the equation:
$$\begin{aligned} 2x^2 &= 100 \\ x^2 &= 50 \\ x &= \pm\sqrt{50} \\ x &= \pm 5\sqrt{2} \end{aligned}$$

9. Solving the equation:
$$\begin{aligned} 3a^2 &= 54 \\ a^2 &= 18 \\ a &= \pm\sqrt{18} \\ a &= \pm 3\sqrt{2} \end{aligned}$$

11. Solving the equation:
$$\begin{aligned} (x+2)^2 &= 4 \\ x+2 &= \pm\sqrt{4} \\ x+2 &= \pm 2 \\ x &= -2 \pm 2 \\ x &= -4, 0 \end{aligned}$$

13. Solving the equation:
$$\begin{aligned} (x+1)^2 &= 25 \\ x+1 &= \pm\sqrt{25} \\ x+1 &= \pm 5 \\ x &= -1 \pm 5 \\ x &= -6, 4 \end{aligned}$$

15. Solving the equation:
$$\begin{aligned} (a-5)^2 &= 75 \\ a-5 &= \pm\sqrt{75} \\ a-5 &= \pm 5\sqrt{3} \\ a &= 5 \pm 5\sqrt{3} \end{aligned}$$

17. Solving the equation:
$$\begin{aligned} (y+1)^2 &= 50 \\ y+1 &= \pm\sqrt{50} \\ y+1 &= \pm 5\sqrt{2} \\ y &= -1 \pm 5\sqrt{2} \end{aligned}$$

19. Solving the equation:
$$\begin{aligned} (2x+1)^2 &= 25 \\ 2x+1 &= \pm\sqrt{25} \\ 2x+1 &= \pm 5 \\ 2x &= -1 \pm 5 \\ 2x &= -6, 4 \\ x &= -3, 2 \end{aligned}$$

21. Solving the equation:
$$\begin{aligned} (4a-5)^2 &= 36 \\ 4a-5 &= \pm\sqrt{36} \\ 4a-5 &= \pm 6 \\ 4a &= 5 \pm 6 \\ 4a &= -1, 11 \\ a &= -\frac{1}{4}, \frac{11}{4} \end{aligned}$$

23. Solving the equation:
$$\begin{aligned} (3y-1)^2 &= 12 \\ 3y-1 &= \pm\sqrt{12} \\ 3y-1 &= \pm 2\sqrt{3} \\ 3y &= 1 \pm 2\sqrt{3} \\ y &= \frac{1 \pm 2\sqrt{3}}{3} \end{aligned}$$

© 2012 Cengage Learning. All Rights Reserved. May not be scanned, copied or duplicated, or posted to a publicly accessible website, in whole or in part.

25. Solving the equation:

$$\begin{aligned}(6x+2)^2&=27\\6x+2&=\pm\sqrt{27}\\6x+2&=\pm3\sqrt{3}\\6x&=-2\pm3\sqrt{3}\\x&=\frac{-2\pm3\sqrt{3}}{6}\end{aligned}$$

27. Solving the equation:

$$\begin{aligned}(3x-9)^2&=27\\3x-9&=\pm\sqrt{27}\\3x-9&=\pm3\sqrt{3}\\3x&=9\pm3\sqrt{3}\\x&=3\pm\sqrt{3}\end{aligned}$$

29. Solving the equation:

$$\begin{aligned}(3x+6)^2&=45\\3x+6&=\pm\sqrt{45}\\3x+6&=\pm3\sqrt{5}\\3x&=-6\pm3\sqrt{5}\\x&=-2\pm\sqrt{5}\end{aligned}$$

31. Solving the equation:

$$\begin{aligned}(2y-4)^2&=8\\2y-4&=\pm\sqrt{8}\\2y-4&=\pm2\sqrt{2}\\2y&=4\pm2\sqrt{2}\\y&=2\pm\sqrt{2}\end{aligned}$$

33. Solving the equation:

$$\begin{aligned}\left(x-\frac{2}{3}\right)^2&=\frac{25}{9}\\x-\frac{2}{3}&=\pm\sqrt{\frac{25}{9}}\\x-\frac{2}{3}&=\pm\frac{5}{3}\\x&=\frac{2}{3}\pm\frac{5}{3}\\x&=-1,\frac{7}{3}\end{aligned}$$

35. Solving the equation:

$$\begin{aligned}\left(x+\frac{1}{2}\right)^2&=\frac{7}{4}\\x+\frac{1}{2}&=\pm\sqrt{\frac{7}{4}}\\x+\frac{1}{2}&=\pm\frac{\sqrt{7}}{2}\\x&=\frac{-1\pm\sqrt{7}}{2}\end{aligned}$$

37. Solving the equation:

$$\begin{aligned}\left(a-\frac{4}{5}\right)^2&=\frac{12}{25}\\a-\frac{4}{5}&=\pm\sqrt{\frac{12}{25}}\\a-\frac{4}{5}&=\pm\frac{2\sqrt{3}}{5}\\a&=\frac{4\pm2\sqrt{3}}{5}\end{aligned}$$

39. Solving the equation:

$$\begin{aligned}x^2+10x+25&=7\\(x+5)^2&=7\\x+5&=\pm\sqrt{7}\\x&=-5\pm\sqrt{7}\end{aligned}$$

41. Solving the equation:

$$\begin{aligned}x^2-2x+1&=9\\(x-1)^2&=9\\x-1&=\pm\sqrt{9}\\x-1&=\pm3\\x&=1\pm3\\x&=-2,4\end{aligned}$$

43. Solving the equation:

$$\begin{aligned}x^2+12x+36&=8\\(x+6)^2&=8\\x+6&=\pm\sqrt{8}\\x+6&=\pm2\sqrt{2}\\x&=-6\pm2\sqrt{2}\end{aligned}$$

45. **a.** No, as long as only rational numbers are used in the factoring.

b. Solving the equation:

$$\begin{aligned}x^2&=3\\x&=\pm\sqrt{3}\end{aligned}$$

© 2012 Cengage Learning. All Rights Reserved. May not be scanned, copied or duplicated, or posted to a publicly accessible website, in whole or in part.

47. **a.** Yes, the equation can be solved by factoring.

b. Solving the equation:

$$\begin{aligned}(x-3)^2&=4\\ x^2-6x+9&=4\\ x^2-6x+5&=0\\ (x-1)(x-5)&=0\\ x&=1,5\end{aligned}$$

Note that this equation can also be solved by taking square roots:

$$\begin{aligned}(x-3)^2&=4\\ x-3&=\pm\sqrt{4}\\ x-3&=\pm2\\ x&=3\pm2\\ x&=1,5\end{aligned}$$

49. Checking the solution: $(2x-6)^2=\left(2\left(3+\sqrt{3}\right)-6\right)^2=\left(6+2\sqrt{3}-6\right)^2=\left(2\sqrt{3}\right)^2=4\bullet3=12$

Yes, it is a solution to the equation.

51. **a.** Solving the equation:

$$\begin{aligned}2x-1&=0\\ 2x&=1\\ x&=\frac{1}{2}\end{aligned}$$

b. Solving the equation:

$$\begin{aligned}2x-1&=4\\ 2x&=5\\ x&=\frac{5}{2}\end{aligned}$$

c. Solving the equation:

$$\begin{aligned}(2x-1)^2&=4\\ 2x-1&=\pm\sqrt{4}\\ 2x-1&=\pm2\\ 2x&=1\pm2\\ 2x&=-1,3\\ x&=-\frac{1}{2},\frac{3}{2}\end{aligned}$$

d. Solving the equation:

$$\begin{aligned}\sqrt{2x}-1&=0\\ \sqrt{2x}&=1\\ \left(\sqrt{2x}\right)^2&=(1)^2\\ 2x&=1\\ x&=\frac{1}{2}\end{aligned}$$

e. Solving the equation:

$$\begin{aligned}\frac{1}{2x}-1&=\frac{1}{4}\\ 4x\left(\frac{1}{2x}-1\right)&=4x\left(\frac{1}{4}\right)\\ 2-4x&=x\\ 2&=5x\\ x&=\frac{2}{5}\end{aligned}$$

53. Checking the solution: $(x+1)^2=\left(-1+5\sqrt{2}+1\right)^2=\left(5\sqrt{2}\right)^2=25\bullet2=50$

55. **a.** Finding the sum: $\left(5+\sqrt{3}\right)+\left(5-\sqrt{3}\right)=5+\sqrt{3}+5-\sqrt{3}=10$

b. Finding the product: $\left(5+\sqrt{3}\right)\left(5-\sqrt{3}\right)=25+5\sqrt{3}-5\sqrt{3}-3=22$

© 2012 Cengage Learning. All Rights Reserved. May not be scanned, copied or duplicated, or posted to a publicly accessible website, in whole or in part.

57. Finding the radius r:

$$\pi r^2 = 36\pi$$
$$r^2 = 36$$
$$r = \sqrt{36}$$
$$r = 6$$

59. Solving for t:

$$16t^2 = 100$$
$$t^2 = \frac{25}{4}$$
$$t = \sqrt{\frac{25}{4}}$$
$$t = \frac{5}{2} \text{ seconds}$$

61. Let x represent the number. The equation is:

$$(x+3)^2 = 16$$
$$x+3 = \pm\sqrt{16}$$
$$x+3 = \pm 4$$
$$x = -3 \pm 4$$
$$x = -7, 1$$

The number is either –7 or 1.

63. Solving for r:

$$100(1+r)^2 = A$$
$$(1+r)^2 = \frac{A}{100}$$
$$1+r = \sqrt{\frac{A}{100}} \quad (\text{since } 1+r>0)$$
$$1+r = \frac{\sqrt{A}}{10}$$
$$r = -1 + \frac{\sqrt{A}}{10}$$

65. Let x represent the height of the triangle. Draw the figure:

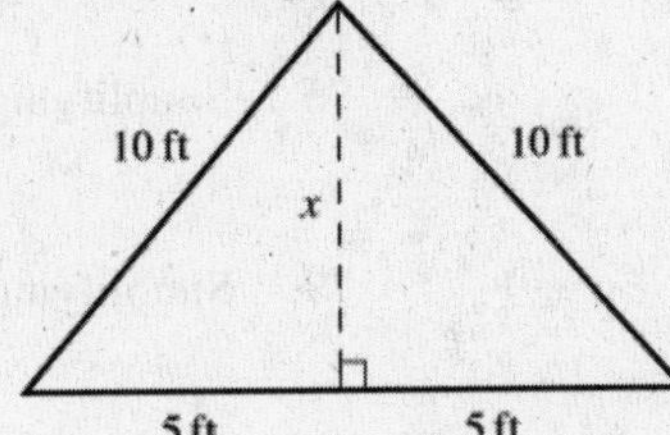

Using the Pythagorean theorem:

$$5^2 + x^2 = 10^2$$
$$25 + x^2 = 100$$
$$x^2 = 75$$
$$x = \sqrt{75}$$
$$x = 5\sqrt{3} \approx 8.66$$

The height is $5\sqrt{3} \approx 8.66$ feet.

67. Let x represent the height of the triangle. Draw the figure:

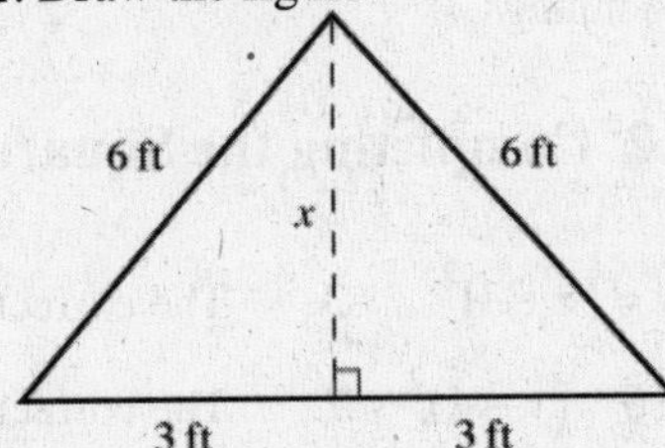

Using the Pythagorean theorem:

$$3^2 + x^2 = 6^2$$
$$9 + x^2 = 36$$
$$x^2 = 27$$
$$x = \sqrt{27}$$
$$x = 3\sqrt{3} \approx 5.2 \text{ feet}$$

No, a person 5 ft 8 in. tall cannot stand up inside the tent.

© 2012 Cengage Learning. All Rights Reserved. May not be scanned, copied or duplicated, or posted to a publicly accessible website, in whole or in part.

69. Let x represent the height of the triangle. Draw the figure:

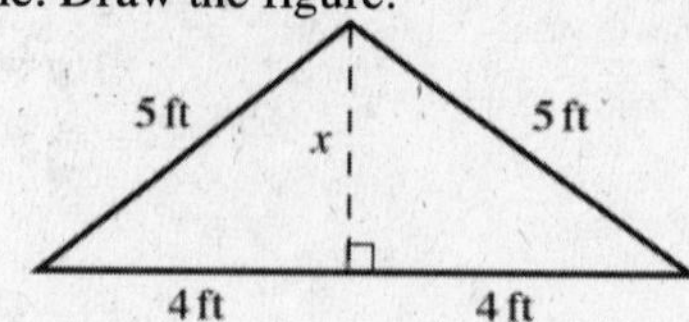

Using the Pythagorean theorem:

$$4^2 + x^2 = 5^2$$
$$16 + x^2 = 25$$
$$x^2 = 9$$
$$x = \sqrt{9} = 3$$

The height is 3 feet.

71. Substituting $c = 6$, $s = 3.53$, and $d = 192$ in the formula:

$$192 = (3.14)(3.53)(6)\left(\frac{1}{2}b\right)^2$$
$$\left(\frac{1}{2}b\right)^2 \approx 2.8870$$
$$\frac{1}{2}b \approx 1.6991$$
$$b \approx 3.4 \text{ inches}$$

73. Simplifying: $\left(\frac{1}{2} \cdot 18\right)^2 = (9)^2 = 81$

75. Simplifying: $\left[\frac{1}{2}(-2)\right]^2 = (-1)^2 = 1$

77. Simplifying: $\left(\frac{1}{2} \cdot 3\right)^2 = \left(\frac{3}{2}\right)^2 = \frac{9}{4}$

79. Simplifying: $\frac{2x^2 + 16}{2} = \frac{2(x^2 + 8)}{2} = x^2 + 8$

81. Factoring: $x^2 + 6x + 9 = (x+3)^2$

83. Factoring: $y^2 - 3y + \frac{9}{4} = \left(y - \frac{3}{2}\right)^2$

85. Multiplying using the square of binomial formula: $(x-5)^2 = x^2 - 2(x)(5) + (5)^2 = x^2 - 10x + 25$

87. Factoring the polynomial: $x^2 - 12x + 36 = (x-6)(x-6) = (x-6)^2$

89. Factoring the polynomial: $x^2 + 4x + 4 = (x+2)(x+2) = (x+2)^2$

91. Finding the root: $\sqrt[3]{8} = 2$

93. Finding the root: $\sqrt[4]{16} = 2$

95. Finding the strikeouts per inning: $\frac{3{,}154 \text{ strikeouts}}{2{,}734 \text{ innings}} \approx 1.15$ strikeouts per inning

9.2 Completing the Square

1. The correct term is 9, since: $x^2 + 6x + 9 = (x+3)^2$

3. The correct term is 1, since: $x^2 + 2x + 1 = (x+1)^2$

5. The correct term is 16, since: $y^2 - 8y + 16 = (y-4)^2$

7. The correct term is 1, since: $y^2 - 2y + 1 = (y-1)^2$

9. The correct term is 64, since: $x^2 + 16x + 64 = (x+8)^2$

11. The correct term is $\frac{9}{4}$, since: $a^2 - 3a + \frac{9}{4} = \left(a - \frac{3}{2}\right)^2$

13. The correct term is $\frac{49}{4}$, since: $x^2 - 7x + \frac{49}{4} = \left(x - \frac{7}{2}\right)^2$

15. The correct term is $\frac{1}{4}$, since: $y^2 + y + \frac{1}{4} = \left(y + \frac{1}{2}\right)^2$

© 2012 Cengage Learning. All Rights Reserved. May not be scanned, copied or duplicated, or posted to a publicly accessible website, in whole or in part.

17. The correct term is $\frac{9}{16}$, since: $x^2-\frac{3}{2}x+\frac{9}{16}=\left(x-\frac{3}{4}\right)^2$

19. Solve by completing the square:

$$\begin{aligned} x^2+4x&=12\\ x^2+4x+4&=12+4\\ (x+2)^2&=16\\ x+2&=\pm4\\ x&=-2\pm4\\ x&=-6,2 \end{aligned}$$

21. Solve by completing the square:

$$\begin{aligned} x^2-6x&=16\\ x^2-6x+9&=16+9\\ (x-3)^2&=25\\ x-3&=\pm5\\ x&=3\pm5\\ x&=-2,8 \end{aligned}$$

23. Solve by completing the square:

$$\begin{aligned} a^2+2a&=3\\ a^2+2a+1&=3+1\\ (a+1)^2&=4\\ a+1&=\pm2\\ a&=-1\pm2\\ a&=-3,1 \end{aligned}$$

25. Solve by completing the square:

$$\begin{aligned} x^2-10x&=0\\ x^2-10x+25&=0+25\\ (x-5)^2&=25\\ x-5&=\pm5\\ x&=5\pm5\\ x&=0,10 \end{aligned}$$

27. Solve by completing the square:

$$\begin{aligned} y^2+2y-15&=0\\ y^2+2y&=15\\ y^2+2y+1&=15+1\\ (y+1)^2&=16\\ y+1&=\pm4\\ y&=-1\pm4\\ y&=-5,3 \end{aligned}$$

29. Solve by completing the square:

$$\begin{aligned} x^2+4x-3&=0\\ x^2+4x&=3\\ x^2+4x+4&=3+4\\ (x+2)^2&=7\\ x+2&=\pm\sqrt{7}\\ x&=-2\pm\sqrt{7} \end{aligned}$$

31. Solve by completing the square:

$$\begin{aligned} x^2-4x&=4\\ x^2-4x+4&=4+4\\ (x-2)^2&=8\\ x-2&=\pm\sqrt{8}\\ x-2&=\pm2\sqrt{2}\\ x&=2\pm2\sqrt{2} \end{aligned}$$

33. Solve by completing the square:

$$\begin{aligned} a^2&=7a+8\\ a^2-7a&=8\\ a^2-7a+\frac{49}{4}&=8+\frac{49}{4}\\ \left(a-\frac{7}{2}\right)^2&=\frac{81}{4}\\ a-\frac{7}{2}&=\pm\sqrt{\frac{81}{4}}\\ a-\frac{7}{2}&=\pm\frac{9}{2}\\ a&=\frac{7}{2}\pm\frac{9}{2}\\ a&=-1,8 \end{aligned}$$

© 2012 Cengage Learning. All Rights Reserved. May not be scanned, copied or duplicated, or posted to a publicly accessible website, in whole or in part.

35. Solve by completing the square:

$$4x^2+8x-4=0$$
$$x^2+2x-1=0$$
$$x^2+2x=1$$
$$x^2+2x+1=1+1$$
$$(x+1)^2=2$$
$$x+1=\pm\sqrt{2}$$
$$x=-1\pm\sqrt{2}$$

37. Solve by completing the square:

$$2x^2+2x-4=0$$
$$x^2+x-2=0$$
$$x^2+x=2$$
$$x^2+x+\frac{1}{4}=2+\frac{1}{4}$$
$$\left(x+\frac{1}{2}\right)^2=\frac{9}{4}$$
$$x+\frac{1}{2}=\pm\sqrt{\frac{9}{4}}$$
$$x+\frac{1}{2}=\pm\frac{3}{2}$$
$$x=-\frac{1}{2}\pm\frac{3}{2}$$
$$x=-2,1$$

39. Solve by completing the square:

$$4x^2+8x+1=0$$
$$x^2+2x+\frac{1}{4}=0$$
$$x^2+2x=-\frac{1}{4}$$
$$x^2+2x+1=-\frac{1}{4}+1$$
$$(x+1)^2=\frac{3}{4}$$
$$x+1=\pm\sqrt{\frac{3}{4}}$$
$$x+1=\pm\frac{\sqrt{3}}{2}$$
$$x=-1\pm\frac{\sqrt{3}}{2}$$
$$x=\frac{-2\pm\sqrt{3}}{2}$$

41. Solve by completing the square:

$$2x^2-2x=1$$
$$x^2-x=\frac{1}{2}$$
$$x^2-x+\frac{1}{4}=\frac{1}{2}+\frac{1}{4}$$
$$\left(x-\frac{1}{2}\right)^2=\frac{3}{4}$$
$$x-\frac{1}{2}=\pm\sqrt{\frac{3}{4}}$$
$$x-\frac{1}{2}=\pm\frac{\sqrt{3}}{2}$$
$$x=\frac{1}{2}\pm\frac{\sqrt{3}}{2}$$
$$x=\frac{1\pm\sqrt{3}}{2}$$

43. Solve by completing the square:

$$4a^2-4a+1=0$$
$$a^2-a+\frac{1}{4}=0$$
$$\left(a-\frac{1}{2}\right)^2=0$$
$$a-\frac{1}{2}=0$$
$$a=\frac{1}{2}$$

45. Solve by completing the square:

$$3y^2-9y=2$$
$$y^2-3y=\frac{2}{3}$$
$$y^2-3y+\frac{9}{4}=\frac{2}{3}+\frac{9}{4}$$
$$\left(y-\frac{3}{2}\right)^2=\frac{8}{12}+\frac{27}{12}=\frac{35}{12}$$
$$y-\frac{3}{2}=\pm\sqrt{\frac{35}{12}}\bullet\frac{\sqrt{3}}{\sqrt{3}}$$
$$y-\frac{3}{2}=\pm\frac{\sqrt{105}}{6}$$
$$y=\frac{3}{2}\pm\frac{\sqrt{105}}{6}$$
$$y=\frac{9\pm\sqrt{105}}{6}$$

© 2012 Cengage Learning. All Rights Reserved. May not be scanned, copied or duplicated, or posted to a publicly accessible website, in whole or in part.

47. **a.** Solving by factoring:

$$x^2-2x=0$$
$$x(x-2)=0$$
$$x=0,2$$

b. Solving by completing the square:

$$x^2-2x=0$$
$$x^2-2x+1=1$$
$$(x-1)^2=1$$
$$x-1=\pm\sqrt{1}$$
$$x-1=-1,1$$
$$x=0,2$$

49. Substituting into the equation: $x^2+6x=(3-\sqrt{2})^2+6(3-\sqrt{2})=9-6\sqrt{2}+2+18-6\sqrt{2}=29-12\sqrt{2}$

No, it is not a solution to the equation.

51. Yes, it is faster to solve the equation by factoring:

$$x^2-6x=0$$
$$x(x-6)=0$$
$$x=0,6$$

53. Yes, another method is to solve the equation by factoring:

$$x^2-6x=7$$
$$x^2-6x-7=0$$
$$(x+1)(x-7)=0$$
$$x=-1,7$$

55. Using the Pythagorean theorem:

$$(3x)^2+(4x)^2=14^2$$
$$9x^2+16x^2=196$$
$$25x^2=196$$
$$x^2=\frac{196}{25}$$
$$x=\sqrt{\frac{196}{25}}=2.8$$
$$3x=8.4, 4x=11.2$$

The length (height) of the screen is 8.4 inches and the width of the screen is 11.2 inches.

57. Writing in standard form: $2x^2+4x-3=0$

59. Writing in standard form:

$$(x-2)(x+3)=5$$
$$x^2+3x-2x-6=5$$
$$x^2+x-11=0$$

61. The coefficients and constant term are: 1, –5, –6

63. The coefficients and constant term are: 2, 4, –3

65. Evaluating b^2-4ac: $b^2-4ac=(-5)^2-4(1)(-6)=25+24=49$

67. Evaluating b^2-4ac: $b^2-4ac=(4)^2-4(2)(-3)=16+24=40$

69. Simplifying: $\frac{5+\sqrt{49}}{2}=\frac{5+7}{2}=\frac{12}{2}=6$

71. Simplifying: $\frac{-4-\sqrt{40}}{4}=\frac{-4-2\sqrt{10}}{4}=\frac{-2-\sqrt{10}}{2}$

73. Evaluating when $a=2$: $2a=2(2)=4$

75. Evaluating when $a=2$ and $c=-3$: $4ac=4(2)(-3)=-24$

77. Evaluating when $a=2, b=4$, and $c=-3$: $\sqrt{b^2-4ac}=\sqrt{(4)^2-4(2)(-3)}=\sqrt{16+24}=\sqrt{40}=2\sqrt{10}$

79. Simplifying the radical: $\sqrt{12}=\sqrt{4\bullet 3}=\sqrt{4}\sqrt{3}=2\sqrt{3}$

81. Simplifying the radical: $\sqrt{20x^2y^3}=\sqrt{4x^2y^2\bullet 5y}=\sqrt{4x^2y^2}\sqrt{5y}=2xy\sqrt{5y}$

© 2012 Cengage Learning. All Rights Reserved. May not be scanned, copied or duplicated, or posted to a publicly accessible website, in whole or in part.

83. Simplifying the radical: $\sqrt{\frac{81}{25}}=\frac{\sqrt{81}}{\sqrt{25}}=\frac{9}{5}$

85. Since 56.1% continue to work, 43.9% stay at home. Finding the amount: $0.439(4,000,000)=1,756,000$ mothers

9.3 The Quadratic Formula

1. Using the quadratic formula with $a = 1$, $b = 3$, and $c = 2$:
$$x=\frac{-b\pm\sqrt{b^2-4ac}}{2a}=\frac{-3\pm\sqrt{(3)^2-4(1)(2)}}{2(1)}=\frac{-3\pm\sqrt{9-8}}{2}=\frac{-3\pm1}{2}=-2,-1$$

3. Using the quadratic formula with $a = 1$, $b = 5$, and $c = 6$:
$$x=\frac{-b\pm\sqrt{b^2-4ac}}{2a}=\frac{-5\pm\sqrt{(5)^2-4(1)(6)}}{2(1)}=\frac{-5\pm\sqrt{25-24}}{2}=\frac{-5\pm1}{2}=-3,-2$$

5. Using the quadratic formula with $a = 1$, $b = 6$, and $c = 9$:
$$x=\frac{-b\pm\sqrt{b^2-4ac}}{2a}=\frac{-6\pm\sqrt{(6)^2-4(1)(9)}}{2(1)}=\frac{-6\pm\sqrt{36-36}}{2}=\frac{-6\pm0}{2}=-3$$

7. Using the quadratic formula with $a = 1$, $b = 6$, and $c = 7$:
$$x=\frac{-b\pm\sqrt{b^2-4ac}}{2a}=\frac{-6\pm\sqrt{(6)^2-4(1)(7)}}{2(1)}=\frac{-6\pm\sqrt{36-28}}{2}=\frac{-6\pm\sqrt{8}}{2}=\frac{-6\pm2\sqrt{2}}{2}=-3\pm\sqrt{2}$$

9. Using the quadratic formula with $a = 2$, $b = 5$, and $c = 3$:
$$x=\frac{-b\pm\sqrt{b^2-4ac}}{2a}=\frac{-5\pm\sqrt{(5)^2-4(2)(3)}}{2(2)}=\frac{-5\pm\sqrt{25-24}}{4}=\frac{-5\pm1}{4}=-\frac{3}{2},-1$$

11. Using the quadratic formula with $a = 4$, $b = 8$, and $c = 1$:
$$x=\frac{-b\pm\sqrt{b^2-4ac}}{2a}=\frac{-8\pm\sqrt{(8)^2-4(4)(1)}}{2(4)}=\frac{-8\pm\sqrt{64-16}}{8}=\frac{-8\pm\sqrt{48}}{8}=\frac{-8\pm4\sqrt{3}}{8}=\frac{-2\pm\sqrt{3}}{2}$$

13. Using the quadratic formula with $a = 1$, $b = -2$, and $c = 1$:
$$x=\frac{-b\pm\sqrt{b^2-4ac}}{2a}=\frac{-(-2)\pm\sqrt{(-2)^2-4(1)(1)}}{2(1)}=\frac{2\pm\sqrt{4-4}}{2}=\frac{2\pm0}{2}=1$$

15. First write the equation as $x^2-5x-7=0$. Using $a = 1$, $b = -5$, and $c = -7$ in the quadratic formula:
$$x=\frac{-b\pm\sqrt{b^2-4ac}}{2a}=\frac{-(-5)\pm\sqrt{(-5)^2-4(1)(-7)}}{2(1)}=\frac{5\pm\sqrt{25+28}}{2}=\frac{5\pm\sqrt{53}}{2}$$

17. Using $a = 6$, $b = -1$, and $c = -2$ in the quadratic formula:
$$x=\frac{-b\pm\sqrt{b^2-4ac}}{2a}=\frac{-(-1)\pm\sqrt{(-1)^2-4(6)(-2)}}{2(6)}=\frac{1\pm\sqrt{1+48}}{12}=\frac{1\pm\sqrt{49}}{12}=\frac{1\pm7}{12}=-\frac{1}{2},\frac{2}{3}$$

19. First simplify the equation:
$$(x-2)(x+1)=3$$
$$x^2-2x+x-2=3$$
$$x^2-x-5=0$$
Using $a = 1$, $b = -1$, and $c = -5$ in the quadratic formula:
$$x=\frac{-b\pm\sqrt{b^2-4ac}}{2a}=\frac{-(-1)\pm\sqrt{(-1)^2-4(1)(-5)}}{2(1)}=\frac{1\pm\sqrt{1+20}}{2}=\frac{1\pm\sqrt{21}}{2}$$

© 2012 Cengage Learning. All Rights Reserved. May not be scanned, copied or duplicated, or posted to a publicly accessible website, in whole or in part.

21. First simplify the equation:

$$(2x-3)(x+2)=1$$
$$2x^2-3x+4x-6=1$$
$$2x^2+x-7=0$$

Using $a = 2$, $b = 1$, and $c = -7$ in the quadratic formula:

$$x=\frac{-b\pm\sqrt{b^2-4ac}}{2a}=\frac{-1\pm\sqrt{(1)^2-4(2)(-7)}}{2(2)}=\frac{-1\pm\sqrt{1+56}}{4}=\frac{-1\pm\sqrt{57}}{4}$$

23. First write the equation as $2x^2-3x-5=0$. Using $a = 2$, $b = -3$, and $c = -5$ in the quadratic formula:

$$x=\frac{-b\pm\sqrt{b^2-4ac}}{2a}=\frac{-(-3)\pm\sqrt{(-3)^2-4(2)(-5)}}{2(2)}=\frac{3\pm\sqrt{9+40}}{4}=\frac{3\pm\sqrt{49}}{4}=\frac{3\pm7}{4}=-1,\frac{5}{2}$$

25. First write the equation as $2x^2+6x-7=0$. Using $a = 2$, $b = 6$, and $c = -7$ in the quadratic formula:

$$x=\frac{-b\pm\sqrt{b^2-4ac}}{2a}=\frac{-6\pm\sqrt{(6)^2-4(2)(-7)}}{2(2)}=\frac{-6\pm\sqrt{36+56}}{4}=\frac{-6\pm\sqrt{92}}{4}=\frac{-6\pm2\sqrt{23}}{4}=\frac{-3\pm\sqrt{23}}{2}$$

27. First write the equation as $3x^2+4x-2=0$. Using $a = 3$, $b = 4$, and $c = -2$ in the quadratic formula:

$$x=\frac{-b\pm\sqrt{b^2-4ac}}{2a}=\frac{-4\pm\sqrt{(4)^2-4(3)(-2)}}{2(3)}=\frac{-4\pm\sqrt{16+24}}{6}=\frac{-4\pm\sqrt{40}}{6}=\frac{-4\pm2\sqrt{10}}{6}=\frac{-2\pm\sqrt{10}}{3}$$

29. First write the equation as $2x^2-2x-5=0$. Using $a = 2$, $b = -2$, and $c = -5$ in the quadratic formula:

$$x=\frac{-b\pm\sqrt{b^2-4ac}}{2a}=\frac{-(-2)\pm\sqrt{(-2)^2-4(2)(-5)}}{2(2)}=\frac{2\pm\sqrt{4+40}}{4}=\frac{2\pm\sqrt{44}}{4}=\frac{2\pm2\sqrt{11}}{4}=\frac{1\pm\sqrt{11}}{2}$$

31. Factoring out x results in the equation $x(2x^2+3x-4)=0$, so $x=0$ is one solution. The other two solutions are found by using $a = 2$, $b = 3$, and $c = -4$ in the quadratic formula:

$$x=\frac{-b\pm\sqrt{b^2-4ac}}{2a}=\frac{-3\pm\sqrt{(3)^2-4(2)(-4)}}{2(2)}=\frac{-3\pm\sqrt{9+32}}{4}=\frac{-3\pm\sqrt{41}}{4}$$

33. Using $a = 3$, $b = -4$, and $c = 0$ in the quadratic formula:

$$x=\frac{-b\pm\sqrt{b^2-4ac}}{2a}=\frac{-(-4)\pm\sqrt{(-4)^2-4(3)(0)}}{2(3)}=\frac{4\pm\sqrt{16-0}}{6}=\frac{4\pm4}{6}=0,\frac{4}{3}$$

35. Solving the equation:

$$0=2,716-16t^2$$
$$16t^2=2,716$$
$$t^2=\frac{2,716}{16}$$
$$t=\sqrt{\frac{2,716}{16}}\sqrt{\frac{679}{4}}=\frac{\sqrt{679}}{2}\approx13.0\text{ seconds}$$

The object will reach the ground after $\frac{\sqrt{679}}{2}\approx13.0$ seconds.

37. Solving the equation:

$$56=8+64t-16t^2$$
$$16t^2-64t+48=0$$
$$t^2-4t+3=0$$
$$(t-1)(t-3)=0$$
$$t=1,3$$

The arrow is 56 feet above the ground after 1 second and after 3 seconds.

© 2012 Cengage Learning. All Rights Reserved. May not be scanned, copied or duplicated, or posted to a publicly accessible website, in whole or in part.

39. Combining terms: $\left(3+4\sqrt{2}\right)+\left(2-6\sqrt{2}\right)=5-2\sqrt{2}$

41. Combining terms: $\left(2-5\sqrt{2}\right)-\left(3+7\sqrt{2}\right)+\left(2-\sqrt{2}\right)=2-5\sqrt{2}-3-7\sqrt{2}+2-\sqrt{2}=1-13\sqrt{2}$

43. Multiplying: $4\sqrt{2}\left(3+5\sqrt{2}\right)=12\sqrt{2}+20\sqrt{4}=12\sqrt{2}+40$

45. Multiplying: $\left(3+2\sqrt{2}\right)\left(4-3\sqrt{2}\right)=12-9\sqrt{2}+8\sqrt{2}-12=-\sqrt{2}$

47. Rationalizing the denominator: $\dfrac{2}{3+\sqrt{5}}\bullet\dfrac{3-\sqrt{5}}{3-\sqrt{5}}=\dfrac{2\left(3-\sqrt{5}\right)}{9-5}=\dfrac{2\left(3-\sqrt{5}\right)}{4}=\dfrac{3-\sqrt{5}}{2}$

49. Rationalizing the denominator: $\dfrac{2+\sqrt{3}}{2-\sqrt{3}}\bullet\dfrac{2+\sqrt{3}}{2+\sqrt{3}}=\dfrac{4+2\sqrt{3}+2\sqrt{3}+3}{4-3}=\dfrac{7+4\sqrt{3}}{1}=7+4\sqrt{3}$

51. Simplifying: $\dfrac{3}{x-2}+\dfrac{6}{2x-4}=\dfrac{3}{x-2}+\dfrac{6}{2(x-2)}=\dfrac{3}{x-2}+\dfrac{3}{x-2}=\dfrac{6}{x-2}$

53. Simplifying:

$$\begin{aligned}\frac{x}{x^2-9}+\frac{4}{4x-12}&=\frac{x}{(x+3)(x-3)}\bullet\frac{4}{4}+\frac{4}{4(x-3)}\bullet\frac{x+3}{x+3}\\&=\frac{4x}{4(x+3)(x-3)}+\frac{4x+12}{4(x+3)(x-3)}\\&=\frac{8x+12}{4(x+3)(x-3)}\\&=\frac{4(2x+3)}{4(x+3)(x-3)}\\&=\frac{2x+3}{(x+3)(x-3)}\end{aligned}$$

55. Simplifying:

$$\begin{aligned}\frac{2x}{x^2-1}+\frac{x}{x^2-3x+2}&=\frac{2x}{(x+1)(x-1)}\bullet\frac{x-2}{x-2}+\frac{x}{(x-1)(x-2)}\bullet\frac{x+1}{x+1}\\&=\frac{2x^2-4x}{(x+1)(x-1)(x-2)}+\frac{x^2+x}{(x+1)(x-1)(x-2)}\\&=\frac{3x^2-3x}{(x+1)(x-1)(x-2)}\\&=\frac{3x(x-1)}{(x+1)(x-1)(x-2)}\\&=\frac{3x}{(x+1)(x-2)}\end{aligned}$$

57. Multiplying each side of the equation by 6 results in the equation $3x^2-3x-1=0$. Using $a=3$, $b=-3$, and $c=-1$ in the quadratic formula: $x=\dfrac{-b\pm\sqrt{b^2-4ac}}{2a}=\dfrac{-(-3)\pm\sqrt{(-3)^2-4(3)(-1)}}{2(3)}=\dfrac{3\pm\sqrt{9+12}}{6}=\dfrac{3\pm\sqrt{21}}{6}$

© 2012 Cengage Learning. All Rights Reserved. May not be scanned, copied or duplicated, or posted to a publicly accessible website, in whole or in part.

9.4 Complex Numbers

1. Combining the complex numbers: $(3-2i)+3i=3+(-2i+3i)=3+i$
3. Combining the complex numbers: $(6+2i)-10i=6+(2i-10i)=6-8i$
5. Combining the complex numbers: $(11+9i)-9i=11+(9i-9i)=11$
7. Combining the complex numbers: $(3+2i)+(6-i)=(3+6)+(2i-i)=9+i$
9. Combining the complex numbers: $(5+7i)-(6+8i)=5+7i-6-8i=-1-i$
11. Combining the complex numbers: $(9-i)+(2-i)=9-i+2-i=11-2i$
13. Combining the complex numbers: $(6+i)-4i-(2-i)=6+i-4i-2+i=4-2i$
15. Combining the complex numbers: $(6-11i)+3i+(2+i)=6-11i+3i+2+i=8-7i$
17. Combining the complex numbers: $(2+3i)-(6-2i)+(3-i)=2+3i-6+2i+3-i=-1+4i$
19. Multiplying the complex numbers: $3(2-i)=6-3i$
21. Multiplying the complex numbers: $2i(8-7i)=16i-14i^2=16i-14(-1)=14+16i$
23. Multiplying the complex numbers: $(2+i)(4-i)=8+4i-2i-i^2=8+2i-(-1)=8+2i+1=9+2i$
25. Multiplying the complex numbers: $(2+i)(3-5i)=6+3i-10i-5i^2=6-7i-5(-1)=6-7i+5=11-7i$
27. Multiplying the complex numbers: $(3+5i)(3-5i)=(3)^2-(5i)^2=9-25i^2=9-25(-1)=9+25=34$
29. Multiplying the complex numbers: $(2+i)(2-i)=(2)^2-(i)^2=4-i^2=4-(-1)=4+1=5$
31. Dividing the complex numbers: $\frac{2}{3-2i}\bullet\frac{3+2i}{3+2i}=\frac{2(3+2i)}{9-4i^2}=\frac{2(3+2i)}{9+4}=\frac{6+4i}{13}$
33. Dividing the complex numbers: $\frac{-3i}{2+3i}\bullet\frac{2-3i}{2-3i}=\frac{-6i+9i^2}{4-9i^2}=\frac{-6i-9}{4+9}=\frac{-9-6i}{13}$
35. Dividing the complex numbers: $\frac{6i}{3-i}\bullet\frac{3+i}{3+i}=\frac{18i+6i^2}{9-i^2}=\frac{18i-6}{9+1}=\frac{6(-1+3i)}{10}=\frac{3(-1+3i)}{5}=\frac{-3+9i}{5}$
37. Dividing the complex numbers: $\frac{2+i}{2-i}\bullet\frac{2+i}{2+i}=\frac{4+2i+2i+i^2}{4-i^2}=\frac{4+4i-1}{4+1}=\frac{3+4i}{5}$
39. Dividing the complex numbers:
$$\frac{4+5i}{3-6i}\bullet\frac{3+6i}{3+6i}=\frac{12+15i+24i+30i^2}{9-36i^2}=\frac{12+39i-30}{9+36}=\frac{-18+39i}{45}=\frac{3(-6+13i)}{45}=\frac{-6+13i}{15}$$
41. Finding the temperature: $C=\frac{5}{9}(40-32)=\frac{5}{9}(8)\approx 4°\text{C}$
43. Simplifying: $\sqrt{36}=6$
45. Simplifying: $\sqrt{75}=\sqrt{25\bullet 3}=5\sqrt{3}$
47. Solving the equation:
$$\begin{aligned}(x+2)^2&=9\\ x+2&=\pm\sqrt{9}=\pm 3\\ x&=-5,1\end{aligned}$$
49. Solving the equation:
$$\begin{aligned}\frac{1}{10}x^2-\frac{1}{5}x&=\frac{1}{2}\\ 10\left(\frac{1}{10}x^2-\frac{1}{5}x\right)&=10\left(\frac{1}{2}\right)\\ x^2-2x&=5\\ x^2-2x+1&=5+1\\ (x-1)^2&=6\\ x-1&=\pm\sqrt{6}\\ x&=1\pm\sqrt{6}\end{aligned}$$

© 2012 Cengage Learning. All Rights Reserved. May not be scanned, copied or duplicated, or posted to a publicly accessible website, in whole or in part.

51. Solving the equation:

$$\begin{aligned}(2x-3)(2x-1)&=4\\4x^2-2x-6x+3&=4\\4x^2-8x&=1\\x^2-2x&=\frac{1}{4}\\x^2-2x+1&=\frac{1}{4}+1\\(x-1)^2&=\frac{5}{4}\\x-1&=\pm\frac{\sqrt{5}}{2}\\x&=1\pm\frac{\sqrt{5}}{2}=\frac{2\pm\sqrt{5}}{2}\end{aligned}$$

53. Solving the equation:

$$\begin{aligned}(x-3)^2&=25\\x-3&=\pm\sqrt{25}\\x-3&=\pm5\\x&=3\pm5\\x&=-2,8\end{aligned}$$

55. Solving the equation:

$$\begin{aligned}(2x-6)^2&=16\\2x-6&=\pm\sqrt{16}\\2x-6&=\pm4\\2x&=6\pm4\\2x&=2,10\\x&=1,5\end{aligned}$$

57. Solving the equation:

$$\begin{aligned}(x+3)^2&=12\\x+3&=\pm\sqrt{12}\\x+3&=\pm2\sqrt{3}\\x&=-3\pm2\sqrt{3}\end{aligned}$$

59. Simplifying the radical expression: $\sqrt{\frac{1}{2}}=\frac{\sqrt{1}}{\sqrt{2}}\bullet\frac{\sqrt{2}}{\sqrt{2}}=\frac{\sqrt{2}}{2}$

61. Simplifying the radical expression: $\sqrt{\frac{8x^2y^3}{3}}=\frac{\sqrt{8x^2y^3}}{\sqrt{3}}\bullet\frac{\sqrt{3}}{\sqrt{3}}=\frac{\sqrt{24x^2y^3}}{3}=\frac{\sqrt{4x^2y^2\bullet6y}}{3}=\frac{2xy\sqrt{6y}}{3}$

63. Simplifying the radical expression: $\sqrt[3]{\frac{1}{4}}=\frac{\sqrt[3]{1}}{\sqrt[3]{4}}\bullet\frac{\sqrt[3]{2}}{\sqrt[3]{2}}=\frac{\sqrt[3]{2}}{\sqrt[3]{8}}=\frac{\sqrt[3]{2}}{2}$

65. Multiplying the expressions: $(x+3i)(x-3i)=x^2-(3i)^2=x^2-9i^2=x^2-9(-1)=x^2+9$

67. Simplifying: $\frac{1}{i}\bullet\frac{i}{i}=\frac{i}{i^2}=\frac{i}{-1}=-i$

9.5 Complex Solutions to Quadratic Equations

1. Writing as a complex number: $\sqrt{-16}=\sqrt{16(-1)}=\sqrt{16}\sqrt{-1}=4i$
3. Writing as a complex number: $\sqrt{-49}=\sqrt{49(-1)}=\sqrt{49}\sqrt{-1}=7i$
5. Writing as a complex number: $\sqrt{-6}=\sqrt{6(-1)}=\sqrt{6}\sqrt{-1}=i\sqrt{6}$
7. Writing as a complex number: $\sqrt{-11}=\sqrt{11(-1)}=\sqrt{11}\sqrt{-1}=i\sqrt{11}$
9. Writing as a complex number: $\sqrt{-32}=\sqrt{-16\bullet2}=\sqrt{-16}\sqrt{2}=4i\sqrt{2}$
11. Writing as a complex number: $\sqrt{-50}=\sqrt{-25\bullet2}=\sqrt{-25}\sqrt{2}=5i\sqrt{2}$
13. Writing as a complex number: $\sqrt{-8}=\sqrt{-4\bullet2}=\sqrt{-4}\sqrt{2}=2i\sqrt{2}$
15. Writing as a complex number: $\sqrt{-48}=\sqrt{-16\bullet3}=\sqrt{-16}\sqrt{3}=4i\sqrt{3}$
17. First write the equation as $x^2-2x+2=0$. Using $a=1$, $b=-2$, and $c=2$ in the quadratic formula:

$$x=\frac{-b\pm\sqrt{b^2-4ac}}{2a}=\frac{-(-2)\pm\sqrt{(-2)^2-4(1)(2)}}{2(1)}=\frac{2\pm\sqrt{4-8}}{2}=\frac{2\pm\sqrt{-4}}{2}=\frac{2\pm2i}{2}=1\pm i$$

© 2012 Cengage Learning. All Rights Reserved. May not be scanned, copied or duplicated, or posted to a publicly accessible website, in whole or in part.

19. Solving the equation:

$$\begin{aligned} x^2-4x&=-4\\ x^2-4x+4&=0\\ (x-2)^2&=0\\ x-2&=0\\ x&=2 \end{aligned}$$

21. Solving the equation:

$$\begin{aligned} 2x^2+5x&=12\\ 2x^2+5x-12&=0\\ (2x-3)(x+4)&=0\\ x&=\frac{3}{2},-4 \end{aligned}$$

23. Solving the equation:

$$\begin{aligned} (x-2)^2&=-4\\ x-2&=\pm\sqrt{-4}\\ x-2&=\pm 2i\\ x&=2\pm 2i \end{aligned}$$

25. Solving the equation:

$$\begin{aligned} \left(x+\frac{1}{2}\right)^2&=-\frac{9}{4}\\ x+\frac{1}{2}&=\pm\sqrt{-\frac{9}{4}}\\ x+\frac{1}{2}&=\pm\frac{3}{2}i\\ x&=\frac{-1\pm 3i}{2} \end{aligned}$$

27. Solving the equation:

$$\begin{aligned} \left(x-\frac{1}{2}\right)^2&=-\frac{27}{36}\\ x-\frac{1}{2}&=\pm\sqrt{-\frac{27}{36}}\\ x-\frac{1}{2}&=\pm\frac{3i\sqrt{3}}{6}\\ x-\frac{1}{2}&=\pm\frac{i\sqrt{3}}{2}\\ x&=\frac{1\pm i\sqrt{3}}{2} \end{aligned}$$

29. Using $a=1$, $b=1$, and $c=1$ in the quadratic formula:

$$x=\frac{-b\pm\sqrt{b^2-4ac}}{2a}=\frac{-1\pm\sqrt{(1)^2-4(1)(1)}}{2(1)}=\frac{-1\pm\sqrt{1-4}}{2}=\frac{-1\pm\sqrt{-3}}{2}=\frac{-1\pm i\sqrt{3}}{2}$$

31. Solving the equation:

$$\begin{aligned} x^2-5x+6&=0\\ (x-2)(x-3)&=0\\ x&=2,3 \end{aligned}$$

33. First multiply by 6 to clear the equation of fractions:

$$\begin{aligned} 6\left(\frac{1}{2}x^2+\frac{1}{3}x+\frac{1}{6}\right)&=6(0)\\ 3x^2+2x+1&=0 \end{aligned}$$

Using $a=3$, $b=2$, and $c=1$ in the quadratic formula:

$$x=\frac{-b\pm\sqrt{b^2-4ac}}{2a}=\frac{-2\pm\sqrt{(2)^2-4(3)(1)}}{2(3)}=\frac{-2\pm\sqrt{4-12}}{6}=\frac{-2\pm\sqrt{-8}}{6}=\frac{-2\pm 2i\sqrt{2}}{6}=\frac{-1\pm i\sqrt{2}}{3}$$

35. First multiply by 6 to clear the equation of fractions:

$$\begin{aligned} 6\left(\frac{1}{3}x^2\right)&=6\left(-\frac{1}{2}x+\frac{1}{3}\right)\\ 2x^2&=-3x+2\\ 2x^2+3x-2&=0\\ (2x-1)(x+2)&=0\\ x&=\frac{1}{2},-2 \end{aligned}$$

© 2012 Cengage Learning. All Rights Reserved. May not be scanned, copied or duplicated, or posted to a publicly accessible website, in whole or in part.

37. Solving the equation:

$$(x+2)(x-3)=5$$
$$x^2-x-6=5$$
$$x^2-x-11=0$$

Using $a=1$, $b=-1$, and $c=-11$ in the quadratic formula:

$$x=\frac{-b\pm\sqrt{b^2-4ac}}{2a}=\frac{-(-1)\pm\sqrt{(-1)^2-4(1)(-11)}}{2(1)}=\frac{1\pm\sqrt{1+44}}{2}=\frac{1\pm\sqrt{45}}{2}=\frac{1\pm3\sqrt{5}}{2}$$

39. Solving the equation:

$$(x-5)(x-3)=-10$$
$$x^2-8x+15=-10$$
$$x^2-8x+25=0$$

Using $a=1$, $b=-8$, and $c=25$ in the quadratic formula:

$$x=\frac{-b\pm\sqrt{b^2-4ac}}{2a}=\frac{-(-8)\pm\sqrt{(-8)^2-4(1)(25)}}{2(1)}=\frac{8\pm\sqrt{64-100}}{2}=\frac{8\pm\sqrt{-36}}{2}=\frac{8\pm6i}{2}=4\pm3i$$

41. Solving the equation:

$$(2x-2)(x-3)=9$$
$$2x^2-8x+6=9$$
$$2x^2-8x-3=0$$

Using $a=2$, $b=-8$, and $c=-3$ in the quadratic formula:

$$x=\frac{-b\pm\sqrt{b^2-4ac}}{2a}=\frac{-(-8)\pm\sqrt{(-8)^2-4(2)(-3)}}{2(2)}=\frac{8\pm\sqrt{64+24}}{4}=\frac{8\pm\sqrt{88}}{4}=\frac{8\pm2\sqrt{22}}{4}=\frac{4\pm\sqrt{22}}{2}$$

43. Using $a=1$, $b=-4$, and $c=5$ in the quadratic formula:

$$x=\frac{-b\pm\sqrt{b^2-4ac}}{2a}=\frac{-(-4)\pm\sqrt{(-4)^2-4(1)(5)}}{2(1)}=\frac{4\pm\sqrt{16-20}}{2}=\frac{4\pm\sqrt{-4}}{2}=\frac{4\pm2i}{2}=2\pm i$$

Since the solutions are non-real complex numbers, the graph cannot cross the x-axis.

45. Substituting $x=2+2i$ into the equation:

$$x^2-4x+8=(2+2i)^2-4(2+2i)+8=4+8i+4i^2-8-8i+8=4+8i-4-8-8i+8=0$$

Yes, $x=2+2i$ is a solution to the equation.

47. The other solution is $3-7i$.

49. Computing the size: $\frac{7,370 \text{ square miles}}{14}\approx 526.4$ square miles

51. Evaluating when $x=-4$: $y=(-4+1)^2-3=(-3)^2-3=9-3=6$

53. Evaluating when $x=-2$: $y=(-2+1)^2-3=(-1)^2-3=1-3=-2$

55. Evaluating when $x=1$: $y=(1+1)^2-3=(2)^2-3=4-3=1$

57. Simplifying the radical expression: $3\sqrt{50}+2\sqrt{32}=3\sqrt{25\cdot2}+2\sqrt{16\cdot2}=3\cdot5\sqrt{2}+2\cdot4\sqrt{2}=15\sqrt{2}+8\sqrt{2}=23\sqrt{2}$

59. Simplifying the radical expression: $\sqrt{24}-\sqrt{54}-\sqrt{150}=\sqrt{4\cdot6}-\sqrt{9\cdot6}-\sqrt{25\cdot6}=2\sqrt{6}-3\sqrt{6}-5\sqrt{6}=-6\sqrt{6}$

61. Simplifying the radical expression:

$$2\sqrt{27x^2}-x\sqrt{48}=2\sqrt{9x^2\cdot3}-x\sqrt{16\cdot3}=2\cdot3x\sqrt{3}-x\cdot4\sqrt{3}=6x\sqrt{3}-4x\sqrt{3}=2x\sqrt{3}$$

© 2012 Cengage Learning. All Rights Reserved. May not be scanned, copied or duplicated, or posted to a publicly accessible website, in whole or in part.

9.6 Graphing Parabolas

1. Graphing the parabola:

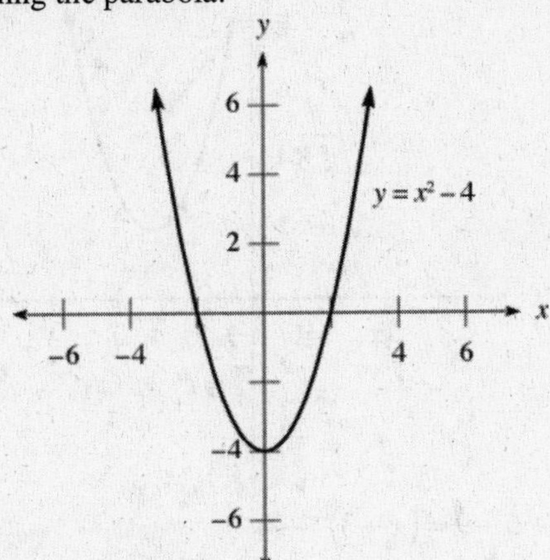

3. Graphing the parabola:

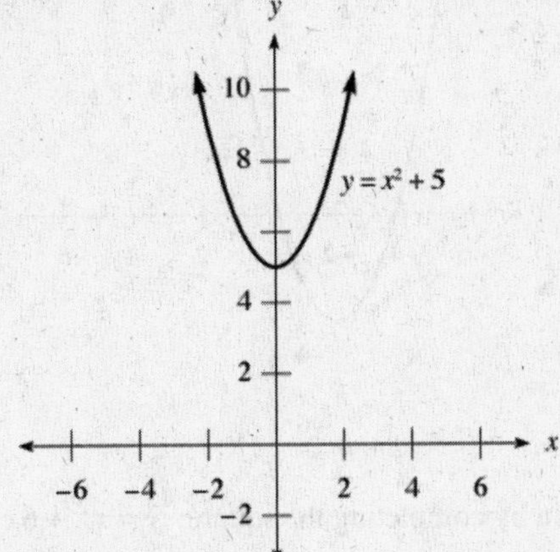

5. Graphing the parabola:

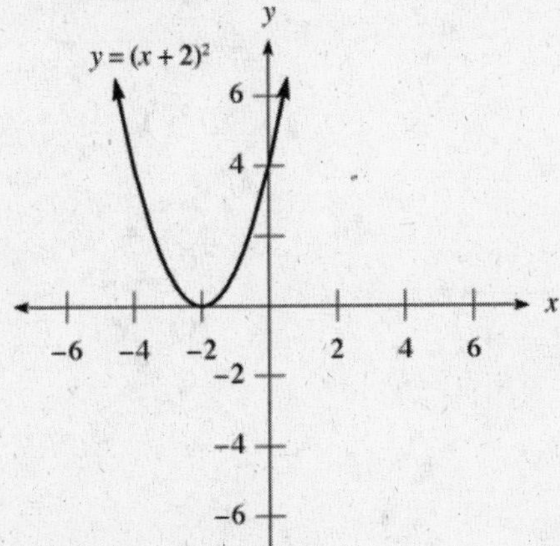

7. Graphing the parabola:

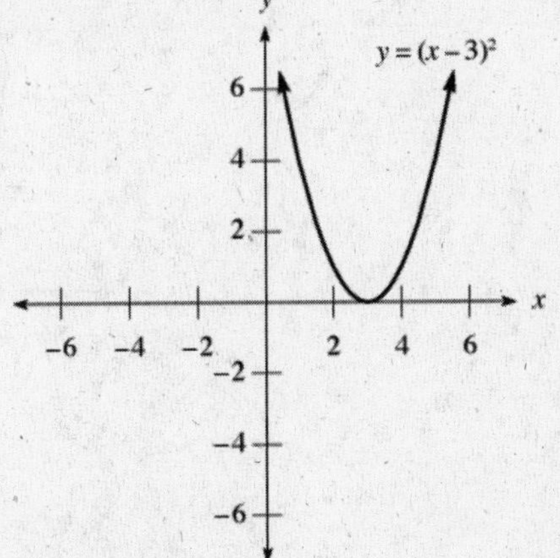

9. Graphing the parabola:

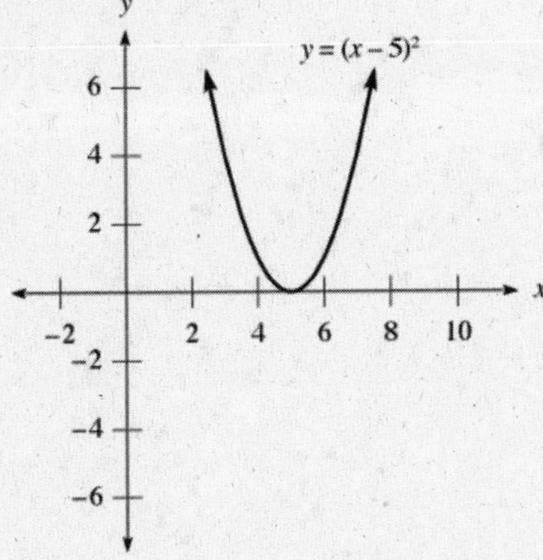

11. Graphing the parabola:

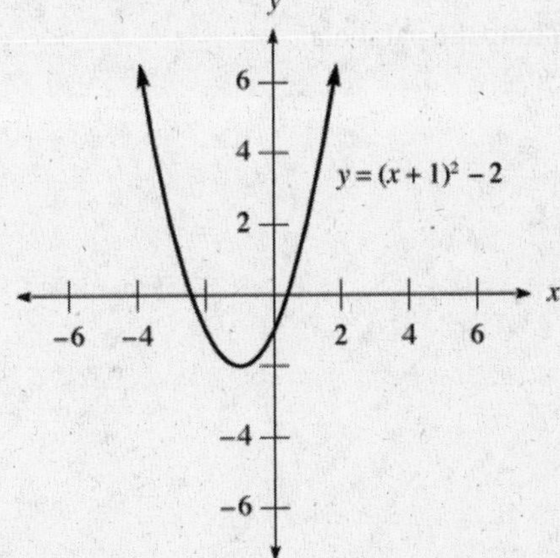

© 2012 Cengage Learning. All Rights Reserved. May not be scanned, copied or duplicated, or posted to a publicly accessible website, in whole or in part.

13. Graphing the parabola:

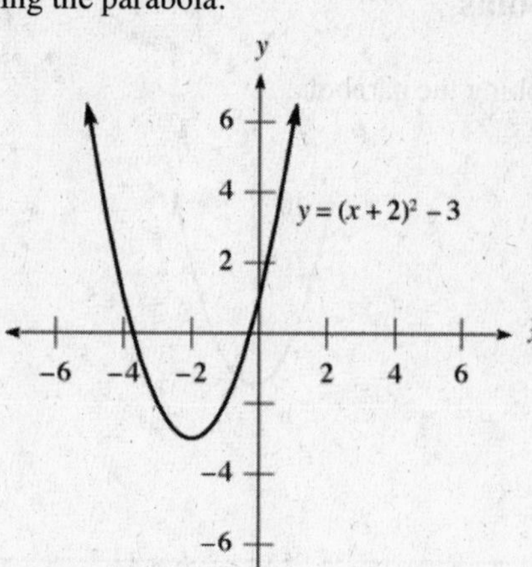

15. Graphing the parabola:

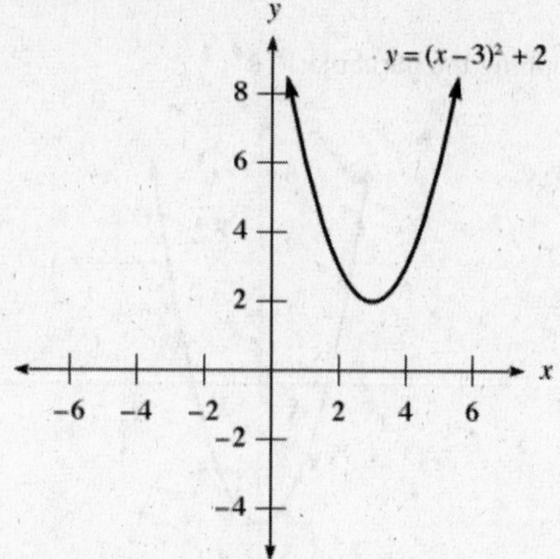

17. Begin by completing the square: $y = x^2 + 6x + 5 = \left(x^2 + 6x + 9\right) + 5 - 9 = (x + 3)^2 - 4$

Graphing the parabola:

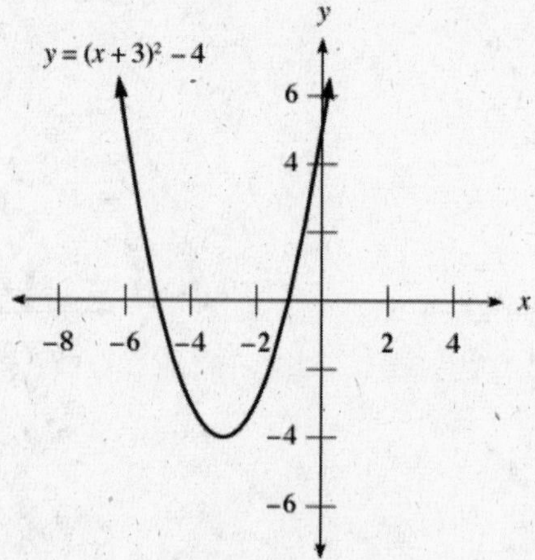

19. Begin by completing the square: $y = x^2 - 2x - 3 = \left(x^2 - 2x + 1\right) - 3 - 1 = (x - 1)^2 - 4$

Graphing the parabola:

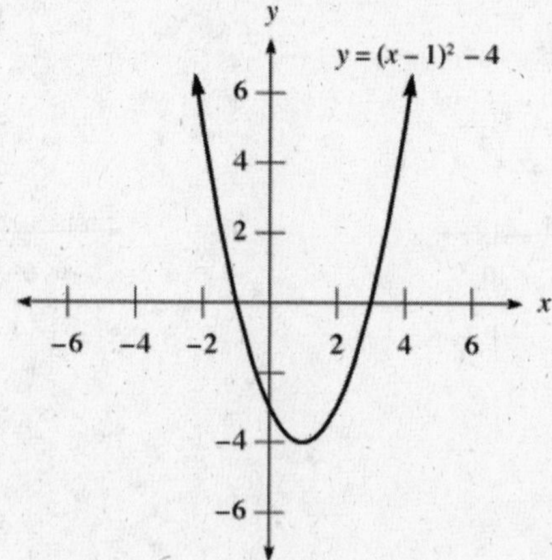

© 2012 Cengage Learning. All Rights Reserved. May not be scanned, copied or duplicated, or posted to a publicly accessible website, in whole or in part.

21. Begin by making a table of values:

x	y
–3	–5
–2	0
–1	3
0	4
1	3
2	0
3	–5

Graphing the parabola:

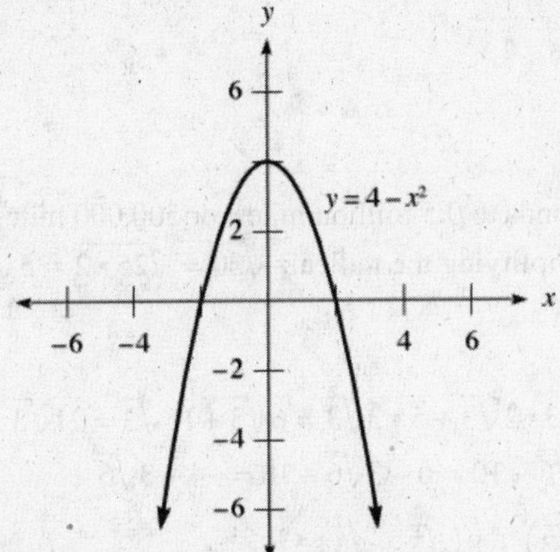

23. Begin by making a table of values:

x	y
–2	–5
–1	–2
0	–1
1	–2
2	–5

Graphing the parabola:

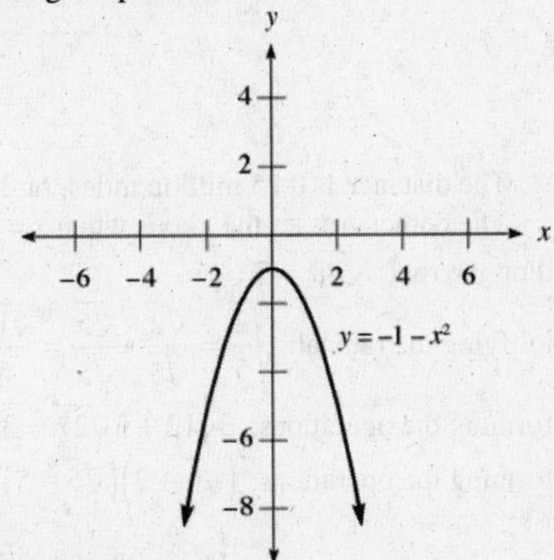

25. Graphing the line and the parabola:

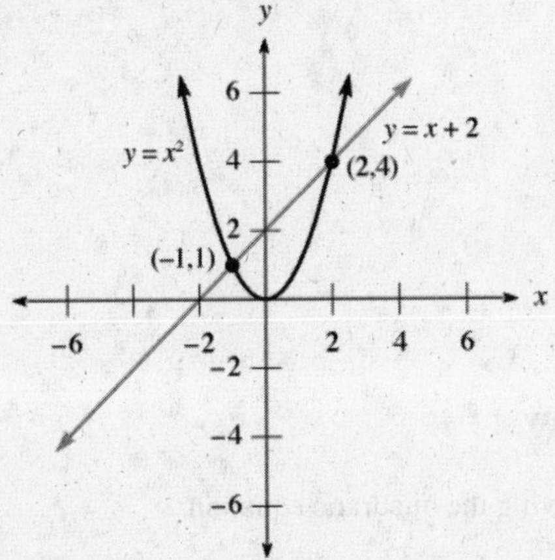

The intersection points are $(-1,1)$ and $(2,4)$.

27. Graphing the two parabolas:

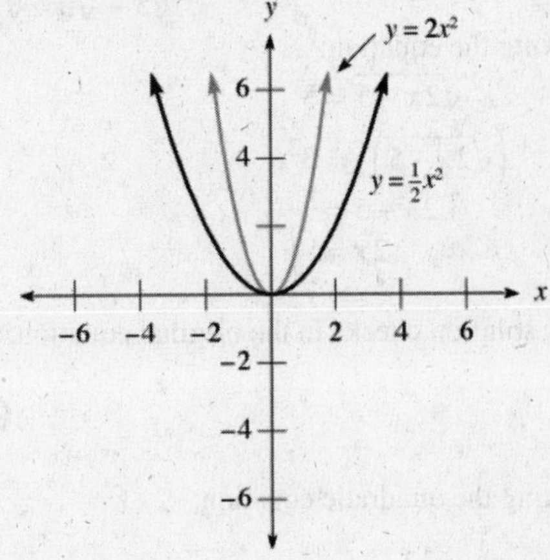

© 2012 Cengage Learning. All Rights Reserved. May not be scanned, copied or duplicated, or posted to a publicly accessible website, in whole or in part.

29. **a.** Graphing the orbit:

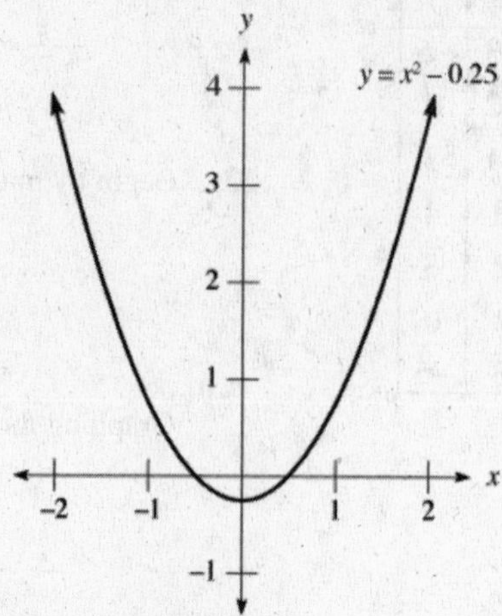

b. The distance is 0.25 million miles, or 250,000 miles.

c. The comet crosses the x-axis when $x = 0.5$, which corresponds to 0.5 million miles, or 500,000 miles.

31. Finding the root: $\sqrt{49} = 7$

33. Simplifying the radical: $\sqrt{50} = \sqrt{25 \cdot 2} = 5\sqrt{2}$

35. Simplifying the radical: $\sqrt{\frac{2}{5}} = \frac{\sqrt{2}}{\sqrt{5}} \cdot \frac{\sqrt{5}}{\sqrt{5}} = \frac{\sqrt{10}}{5}$

37. Performing the operations: $3\sqrt{12} + 5\sqrt{27} = 3\sqrt{4 \cdot 3} + 5\sqrt{9 \cdot 3} = 3 \cdot 2\sqrt{3} + 5 \cdot 3\sqrt{3} = 6\sqrt{3} + 15\sqrt{3} = 21\sqrt{3}$

39. Performing the operations: $\left(\sqrt{6}+2\right)\left(\sqrt{6}-5\right) = \sqrt{36} + 2\sqrt{6} - 5\sqrt{6} - 10 = 6 - 3\sqrt{6} - 10 = -4 - 3\sqrt{6}$

41. Rationalizing the denominator: $\frac{8}{\sqrt{5}-\sqrt{3}} \cdot \frac{\sqrt{5}+\sqrt{3}}{\sqrt{5}+\sqrt{3}} = \frac{8\left(\sqrt{5}+\sqrt{3}\right)}{5-3} = \frac{8\left(\sqrt{5}+\sqrt{3}\right)}{2} = 4\left(\sqrt{5}+\sqrt{3}\right) = 4\sqrt{5} + 4\sqrt{3}$

43. Solving the equation:

$$\begin{aligned} \sqrt{2x-5} &= 3 \\ \left(\sqrt{2x-5}\right)^2 &= 3^2 \\ 2x-5 &= 9 \\ 2x &= 14 \\ x &= 7 \end{aligned}$$

This solution checks in the original equation.

Chapter 9 Review

1. Solving the quadratic equation:

$$\begin{aligned} a^2 &= 32 \\ a &= \pm\sqrt{32} \\ a &= \pm 4\sqrt{2} \end{aligned}$$

3. Solving the quadratic equation:

$$\begin{aligned} 2x^2 &= 32 \\ x^2 &= 16 \\ x &= \pm\sqrt{16} \\ x &= \pm 4 \end{aligned}$$

5. Solving the quadratic equation:

$$\begin{aligned} (x-2)^2 &= 81 \\ x-2 &= \pm\sqrt{81} \\ x-2 &= \pm 9 \\ x &= 2 \pm 9 \\ x &= -7, 11 \end{aligned}$$

7. Solving the quadratic equation:

$$\begin{aligned} (2x+5)^2 &= 32 \\ 2x+5 &= \pm\sqrt{32} \\ 2x+5 &= \pm 4\sqrt{2} \\ 2x &= -5 \pm 4\sqrt{2} \\ x &= \frac{-5 \pm 4\sqrt{2}}{2} \end{aligned}$$

© 2012 Cengage Learning. All Rights Reserved. May not be scanned, copied or duplicated, or posted to a publicly accessible website, in whole or in part.

9. Solving the quadratic equation:

$$\left(x-\frac{2}{3}\right)^2=-\frac{25}{9}$$
$$x-\frac{2}{3}=\pm\sqrt{-\frac{25}{9}}$$
$$x-\frac{2}{3}=\pm\frac{5i}{3}$$
$$x=\frac{2}{3}\pm\frac{5}{3}i$$

11. Solving by completing the square:

$$x^2-8x=4$$
$$x^2-8x+16=4+16$$
$$(x-4)^2=20$$
$$x-4=\pm\sqrt{20}$$
$$x-4=\pm2\sqrt{5}$$
$$x=4\pm2\sqrt{5}$$

13. Solving by completing the square:

$$x^2+4x+3=0$$
$$x^2+4x=-3$$
$$x^2+4x+4=-3+4$$
$$(x+2)^2=1$$
$$x+2=\pm\sqrt{1}$$
$$x+2=\pm1$$
$$x=-2\pm1$$
$$x=-3,-1$$

15. Solving by completing the square:

$$a^2=5a+6$$
$$a^2-5a=6$$
$$a^2-5a+\frac{25}{4}=6+\frac{25}{4}$$
$$\left(a-\frac{5}{2}\right)^2=\frac{49}{4}$$
$$a-\frac{5}{2}=\pm\sqrt{\frac{49}{4}}$$
$$a-\frac{5}{2}=\pm\frac{7}{2}$$
$$a=\frac{5}{2}\pm\frac{7}{2}$$
$$a=-1,6$$

17. Solving by completing the square:

$$3x^2-6x-2=0$$
$$x^2-2x-\frac{2}{3}=0$$
$$x^2-2x=\frac{2}{3}$$
$$x^2-2x+1=\frac{2}{3}+1$$
$$(x-1)^2=\frac{5}{3}$$
$$x-1=\pm\sqrt{\frac{5}{3}}$$
$$x-1=\pm\frac{\sqrt{5}}{\sqrt{3}}\bullet\frac{\sqrt{3}}{\sqrt{3}}$$
$$x-1=\pm\frac{\sqrt{15}}{3}$$
$$x=1\pm\frac{\sqrt{15}}{3}$$
$$x=\frac{3\pm\sqrt{15}}{3}$$

19. Using $a = 1$, $b = -8$, and $c = 16$ in the quadratic formula:

$$x=\frac{-b\pm\sqrt{b^2-4ac}}{2a}=\frac{-(-8)\pm\sqrt{(-8)^2-4(1)(16)}}{2(1)}=\frac{8\pm\sqrt{64-64}}{2}=\frac{8\pm0}{2}=4$$

21. First write the equation as $2x^2+8x-5=0$. Using $a = 2$, $b = 8$, and $c = -5$ in the quadratic formula:

$$x=\frac{-b\pm\sqrt{b^2-4ac}}{2a}=\frac{-8\pm\sqrt{(8)^2-4(2)(-5)}}{2(2)}=\frac{-8\pm\sqrt{64+40}}{4}=\frac{-8\pm\sqrt{104}}{4}=\frac{-8\pm2\sqrt{26}}{4}=\frac{-4\pm\sqrt{26}}{2}$$

© 2012 Cengage Learning. All Rights Reserved. May not be scanned, copied or duplicated, or posted to a publicly accessible website, in whole or in part.

23. First multiply by 10 to clear the equation of fractions:

$$10\left(\frac{1}{5}x^2-\frac{1}{2}x\right)=10\left(\frac{3}{10}\right)$$
$$2x^2-5x=3$$
$$2x^2-5x-3=0$$

Using $a=2$, $b=-5$, and $c=-3$ in the quadratic formula:

$$x=\frac{-b\pm\sqrt{b^2-4ac}}{2a}=\frac{-(-5)\pm\sqrt{(-5)^2-4(2)(-3)}}{2(2)}=\frac{5\pm\sqrt{25+24}}{4}=\frac{5\pm\sqrt{49}}{4}=\frac{5\pm7}{4}=-\frac{1}{2},3$$

25. Combining the complex numbers: $(4-3i)+5i=4+2i$

27. Combining the complex numbers: $(5+6i)+(5-i)=10+5i$

29. Combining the complex numbers: $(3-2i)-(3-i)=3-2i-3+i=-i$

31. Combining the complex numbers: $(3+i)-5i-(4-i)=3+i-5i-4+i=-1-3i$

33. Multiplying the complex numbers: $2(3-i)=6-2i$

35. Multiplying the complex numbers: $4i(6-5i)=24i-20i^2=24i-20(-1)=20+24i$

37. Multiplying the complex numbers: $(3-4i)(5+i)=15-20i+3i-4i^2=15-17i-4(-1)=15-17i+4=19-17i$

39. Multiplying the complex numbers: $(4+i)(4-i)=16-i^2=16-(-1)=16+1=17$

41. Dividing the complex numbers: $\frac{i}{3+i}\bullet\frac{3-i}{3-i}=\frac{3i-i^2}{9-i^2}=\frac{3i+1}{9+1}=\frac{1+3i}{10}$

43. Dividing the complex numbers: $\frac{5}{2+5i}\bullet\frac{2-5i}{2-5i}=\frac{10-25i}{4-25i^2}=\frac{10-25i}{4+25}=\frac{10-25i}{29}$

45. Dividing the complex numbers: $\frac{-3i}{3-2i}\bullet\frac{3+2i}{3+2i}=\frac{-9i-6i^2}{9-4i^2}=\frac{-9i+6}{9+4}=\frac{6-9i}{13}$

47. Dividing the complex numbers: $\frac{4-5i}{4+5i}\bullet\frac{4-5i}{4-5i}=\frac{16-20i-20i+25i^2}{16-25i^2}=\frac{16-40i-25}{16+25}=\frac{-9-40i}{41}$

49. Writing as a complex number: $\sqrt{-36}=6i$

51. Writing as a complex number: $\sqrt{-17}=i\sqrt{17}$

53. Writing as a complex number: $\sqrt{-40}=\sqrt{-4\bullet10}=2i\sqrt{10}$

55. Writing as a complex number: $\sqrt{-200}=\sqrt{-100\bullet2}=10i\sqrt{2}$

57. Graphing the parabola:

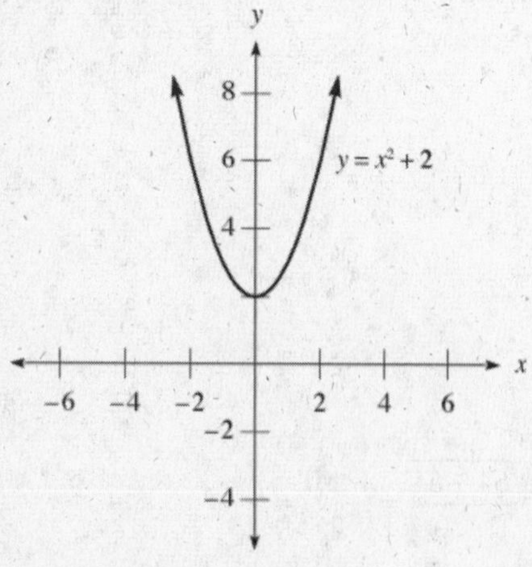

59. Graphing the parabola:

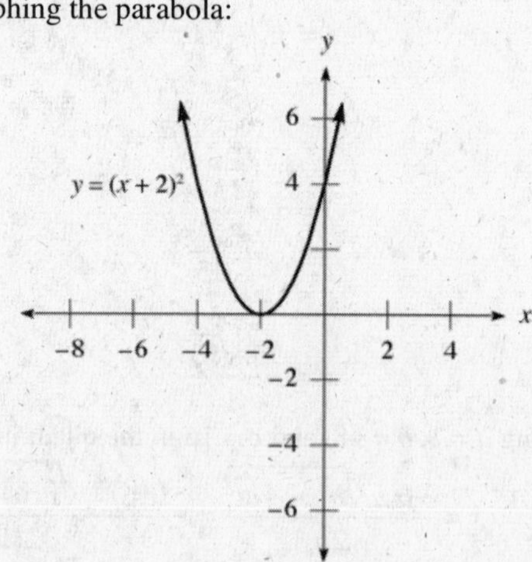

© 2012 Cengage Learning. All Rights Reserved. May not be scanned, copied or duplicated, or posted to a publicly accessible website, in whole or in part.

61. Begin by completing the square: $y = x^2 + 4x + 7 = \left(x^2 + 4x + 4\right) + 7 - 4 = (x+2)^2 + 3$

Graphing the parabola:

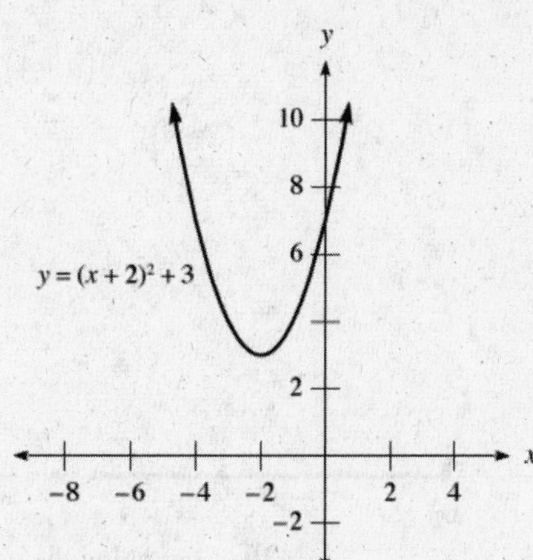

Chapter 9 Cumulative Review

1. Simplifying the expression: $\dfrac{9(-6)-10}{2(-8)} = \dfrac{-54-10}{-16} = \dfrac{-64}{-16} = 4$

3. Simplifying the expression: $\dfrac{x^4}{x^{-8}} = x^{4-(-8)} = x^{4+8} = x^{12}$

5. Simplifying the expression: $\dfrac{\frac{3x^4}{y^5}}{\frac{9x^3}{y}} = \dfrac{\frac{3x^4}{y^5}}{\frac{9x^3}{y}} \bullet \dfrac{y^5}{y^5} = \dfrac{3x^4}{9x^3y^4} = \dfrac{x}{3y^4}$

7. Simplifying the expression: $\sqrt{81} = 9$

9. Simplifying the expression: $\dfrac{10}{\sqrt{2}} = \dfrac{10}{\sqrt{2}} \bullet \dfrac{\sqrt{2}}{\sqrt{2}} = \dfrac{10\sqrt{2}}{2} = 5\sqrt{2}$

11. Simplifying the expression: $(3+3i) - 7i - (2+2i) = 3 + 3i - 7i - 2 - 2i = 1 - 6i$

13. Multiplying using the column method:

$$\begin{array}{r} a^2 - 6a + 7 \\ a - 2 \\ \hline a^3 - 6a^2 + 7a \\ -2a^2 + 12a - 14 \\ \hline a^3 - 8a^2 + 19a - 14 \end{array}$$

15. Simplifying the expression: $\sqrt{121x^4y^2} = 11x^2y$

17. Solving the equation:

$$\begin{aligned} x - \frac{3}{4} &= \frac{5}{6} \\ x &= \frac{5}{6} + \frac{3}{4} \\ x &= \frac{10}{12} + \frac{9}{12} \\ x &= \frac{19}{12} \end{aligned}$$

19. Solving the equation:

$$\begin{aligned} 7 - 4(3x+4) &= -9x \\ 7 - 12x - 16 &= -9x \\ -12x - 9 &= -9x \\ -9 &= 3x \\ x &= -3 \end{aligned}$$

© 2012 Cengage Learning. All Rights Reserved. May not be scanned, copied or duplicated, or posted to a publicly accessible website, in whole or in part.

21. Solving the equation:

$$\begin{aligned} 5x^2 &= -15x \\ 5x^2 + 15x &= 0 \\ 5x(x+3) &= 0 \\ x &= 0, -3 \end{aligned}$$

23. Solving the equation:

$$\begin{aligned} \frac{a}{a+4} &= \frac{7}{3} \\ 3(a+4) \bullet \frac{a}{a+4} &= 3(a+4) \bullet \frac{7}{3} \\ 3a &= 7(a+4) \\ 3a &= 7a + 28 \\ -4a &= 28 \\ a &= -7 \end{aligned}$$

25. Solving the inequality:

$$\begin{aligned} -5 \le 2x - 1 \le 7 \\ -4 \le 2x \le 8 \\ -2 \le x \le 4 \end{aligned}$$

Graphing the solution set:

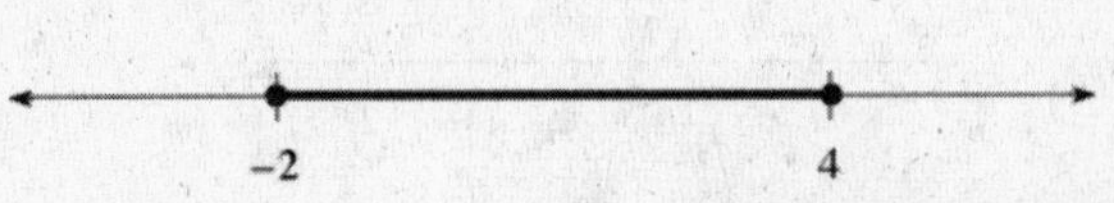

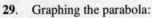

27. Graphing the line:

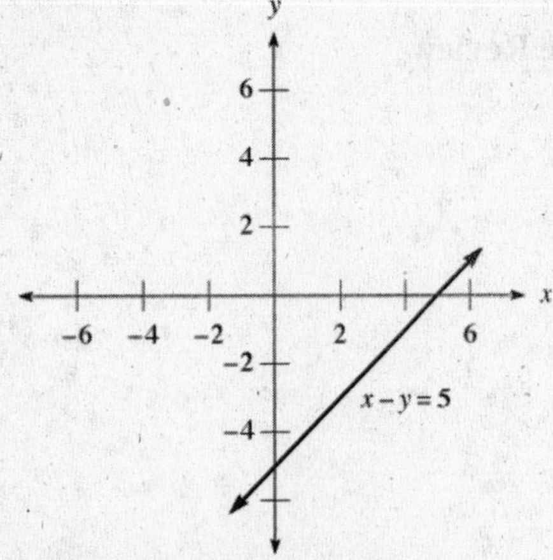

29. Graphing the parabola:

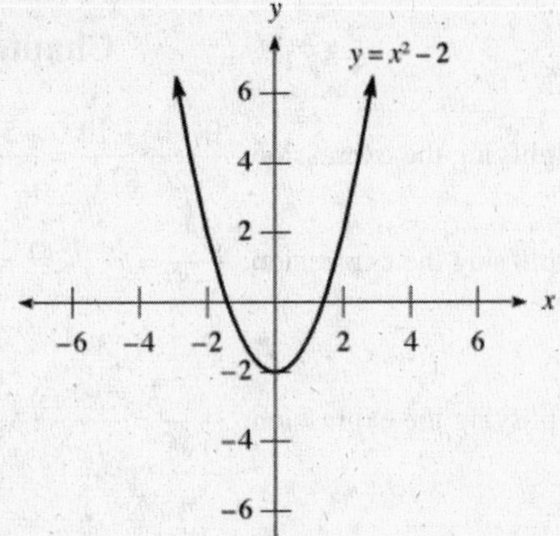

31. Substituting into the second equation:

$$\begin{aligned} 2(y-3) + 3y &= 4 \\ 2y - 6 + 3y &= 4 \\ 5y - 6 &= 4 \\ 5y &= 10 \\ y &= 2 \\ x &= 2 - 3 = -1 \end{aligned}$$

The solution is $(-1, 2)$.

33. Factoring the polynomial: $x^2 - 5x - 24 = (x-8)(x+3)$

35. Factoring the polynomial: $25 - y^2 = (5+y)(5-y)$

37. Finding the slope: $m = \frac{8-(-2)}{1-3} = \frac{8+2}{-2} = \frac{10}{-2} = -5$

39. Dividing by the monomial: $\frac{15a^3b - 10a^2b^2 - 20ab^3}{5ab} = \frac{15a^3b}{5ab} - \frac{10a^2b^2}{5ab} - \frac{20ab^3}{5ab} = 3a^2 - 2ab - 4b^2$

41. Subtracting the rational expressions: $\frac{a-3}{a-7} - \frac{a+10}{a-7} = \frac{a-3-a-10}{a-7} = -\frac{13}{a-7}$

43. Combining the radicals: $5\sqrt{200} + 9\sqrt{50} = 5\sqrt{100 \bullet 2} + 9\sqrt{25 \bullet 2} = 5 \bullet 10\sqrt{2} + 9 \bullet 5\sqrt{2} = 50\sqrt{2} + 45\sqrt{2} = 95\sqrt{2}$

45. Dividing the complex numbers: $\frac{8+i}{i} \bullet \frac{i}{i} = \frac{8i + i^2}{i^2} = \frac{8i - 1}{-1} = 1 - 8i$

© 2012 Cengage Learning. All Rights Reserved. May not be scanned, copied or duplicated, or posted to a publicly accessible website, in whole or in part.

47. Let x represent the amount invested at 5% and y represent the amount invested at 7%. The system of equations is:

$$x+y=20000$$
$$0.07y=0.05x+800$$

Re-writing the system:

$$x+y=20000$$
$$-0.05x+0.07y=800$$

Multiplying the first equation by 0.05:

$$0.05x+0.05y=1000$$
$$-0.05x+0.07y=800$$

Adding the two equations:

$$0.12y=1800$$
$$y=15000$$

Substituting into the first equation:

$$x+15000=20000$$
$$x=5000$$

Randy invested \$5,000 at 5% and \$15,000 at 7%.

49. The variation equation is $y=K\sqrt{x}$. Finding K:

$$10=K\sqrt{4}$$
$$10=2K$$
$$K=5$$

So $y=5\sqrt{x}$. Substituting $x=9$: $y=5\sqrt{9}=15$

51. Solving the proportion:

$$\frac{20.4}{1}=\frac{11,016}{x}$$
$$20.4x=11,016$$
$$x=540$$

They had 540 teachers.

Chapter 9 Test

1. Solving the equation:

$$x^2-7x-8=0$$
$$(x-8)(x+1)=0$$
$$x=-1,8$$

2. Solving the equation:

$$(x-3)^2=12$$
$$x-3=\pm\sqrt{12}$$
$$x-3=\pm2\sqrt{3}$$
$$x=3\pm2\sqrt{3}$$

3. Solving the equation:

$$\left(x-\frac{5}{2}\right)^2=-\frac{75}{4}$$
$$x-\frac{5}{2}=\pm\sqrt{-\frac{75}{4}}$$
$$x-\frac{5}{2}=\pm\frac{5i\sqrt{3}}{2}$$
$$x=\frac{5\pm5i\sqrt{3}}{2}$$

4. First multiply by 6 to clear the equation of fractions:

$$6\left(\frac{1}{3}x^2\right)=6\left(\frac{1}{2}x-\frac{5}{6}\right)$$
$$2x^2=3x-5$$
$$2x^2-3x+5=0$$

Using $a=2$, $b=-3$, and $c=5$ in the quadratic formula:

$$x=\frac{-b\pm\sqrt{b^2-4ac}}{2a}=\frac{-(-3)\pm\sqrt{(-3)^2-4(2)(5)}}{2(2)}=\frac{3\pm\sqrt{9-40}}{4}=\frac{3\pm\sqrt{-31}}{4}=\frac{3\pm i\sqrt{31}}{4}$$

© 2012 Cengage Learning. All Rights Reserved. May not be scanned, copied or duplicated, or posted to a publicly accessible website, in whole or in part.

5. Solving the equation:

$$\begin{aligned} 3x^2 &= -2x+1 \\ 3x^2+2x-1 &= 0 \\ (3x-1)(x+1) &= 0 \\ x &= \frac{1}{3},-1 \end{aligned}$$

6. Simplifying the equation:

$$\begin{aligned} (x+2)(x-1) &= 6 \\ x^2+x-2 &= 6 \\ x^2+x-8 &= 0 \end{aligned}$$

Using $a=1$, $b=1$, and $c=-8$ in the quadratic formula:

$$x=\frac{-b\pm\sqrt{b^2-4ac}}{2a}=\frac{-1\pm\sqrt{(1)^2-4(1)(-8)}}{2(1)}=\frac{-1\pm\sqrt{1+32}}{2}=\frac{-1\pm\sqrt{33}}{2}$$

7. Solving the equation:

$$\begin{aligned} 9x^2+12x+4 &= 0 \\ (3x+2)^2 &= 0 \\ 3x+2 &= 0 \\ 3x &= -2 \\ x &= -\frac{2}{3} \end{aligned}$$

8. Solving by completing the square:

$$\begin{aligned} x^2-6x-6 &= 0 \\ x^2-6x &= 6 \\ x^2-6x+9 &= 6+9 \\ (x-3)^2 &= 15 \\ x-3 &= \pm\sqrt{15} \\ x &= 3\pm\sqrt{15} \end{aligned}$$

9. Writing as a complex number: $\sqrt{-9}=3i$

10. Writing as a complex number: $\sqrt{-121}=11i$

11. Writing as a complex number: $\sqrt{-72}=\sqrt{-36\bullet 2}=6i\sqrt{2}$

12. Writing as a complex number: $\sqrt{-18}=\sqrt{-9\bullet 2}=3i\sqrt{2}$

13. Combining the complex numbers: $(3i+1)+(2+5i)=3+8i$

14. Combining the complex numbers: $(6-2i)-(7-4i)=6-2i-7+4i=-1+2i$

15. Combining the complex numbers: $(2+i)(2-i)=4-i^2=4-(-1)=4+1=5$

16. Combining the complex numbers: $(3+2i)(1+i)=3+2i+3i+2i^2=3+5i-2=1+5i$

17. Combining the complex numbers: $\frac{i}{3-i}\bullet\frac{3+i}{3+i}=\frac{3i+i^2}{9-i^2}=\frac{3i-1}{9+1}=\frac{-1+3i}{10}$

18. Combining the complex numbers: $\frac{2+i}{2-i}\bullet\frac{2+i}{2+i}=\frac{4+2i+2i+i^2}{4-i^2}=\frac{4+4i-1}{4+1}=\frac{3+4i}{5}$

19. Graphing the parabola:

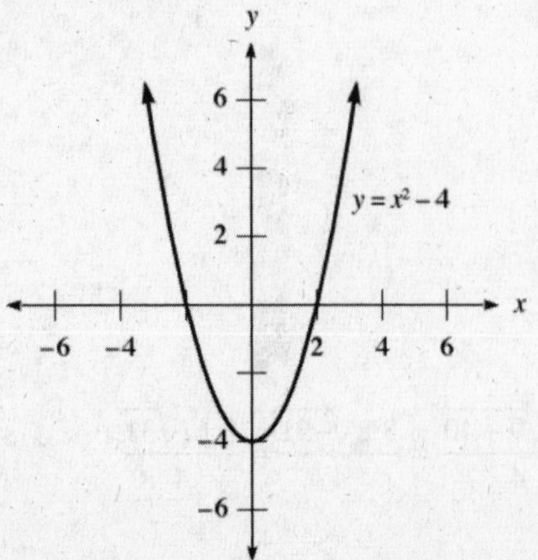

20. Graphing the parabola:

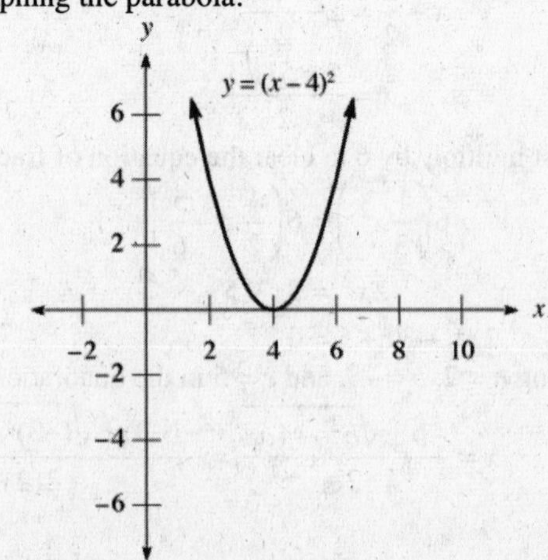

© 2012 Cengage Learning. All Rights Reserved. May not be scanned, copied or duplicated, or posted to a publicly accessible website, in whole or in part.

21. Graphing the parabola:

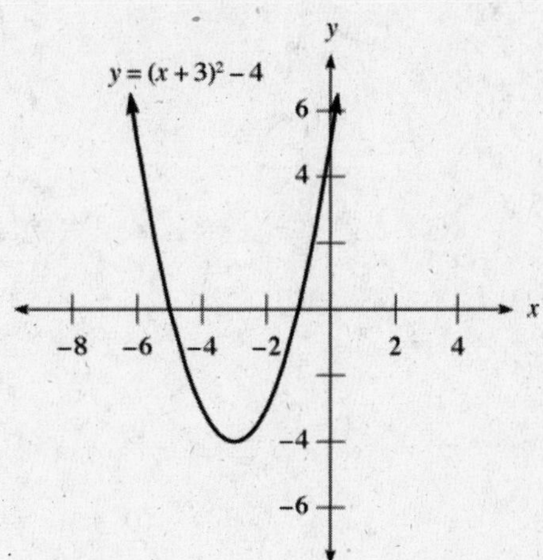

22. First complete the square: $y = x^2 - 6x + 11 = \left(x^2 - 6x + 9\right) + 11 - 9 = (x-3)^2 + 2$

Graphing the parabola:

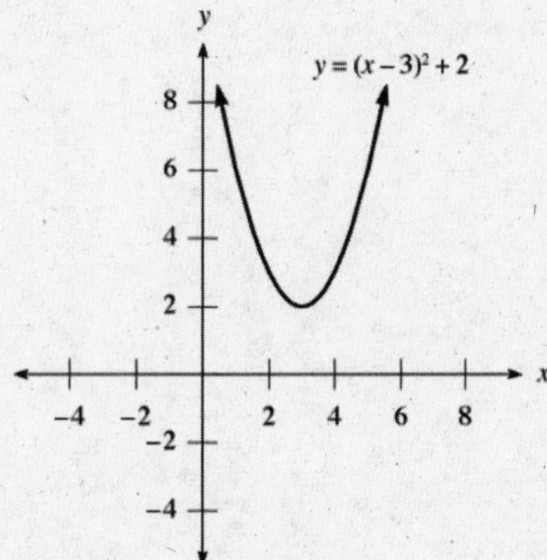

23. Let h represent the height. Using the Pythagorean theorem:

$$\begin{aligned} 6^2 + h^2 &= 10^2 \\ 36 + h^2 &= 100 \\ h^2 &= 64 \\ h &= 8 \end{aligned}$$

The height is 8 cm.

24. The correct graph is C.

25. The correct graph is B.

26. The correct graph is A.

27. The correct graph is D.

28. The correct graph is C.

29. The correct graph is D.

30. The correct graph is A.

31. The correct graph is B.

© 2012 Cengage Learning. All Rights Reserved. May not be scanned, copied or duplicated, or posted to a publicly accessible website, in whole or in part.

Appendices

Appendix A Introduction to Functions

1. a. The equation is $y = 8.5x$ for $10 \le x \le 40$.

   b. Completing the table:

Hours Worked x	Function Rule $y = 8.5x$	Gross Pay ($) y
10	$y = 8.5(10) = 85$	85
20	$y = 8.5(20) = 170$	170
30	$y = 8.5(30) = 255$	255
40	$y = 8.5(40) = 340$	340

   c. Constructing a line graph:

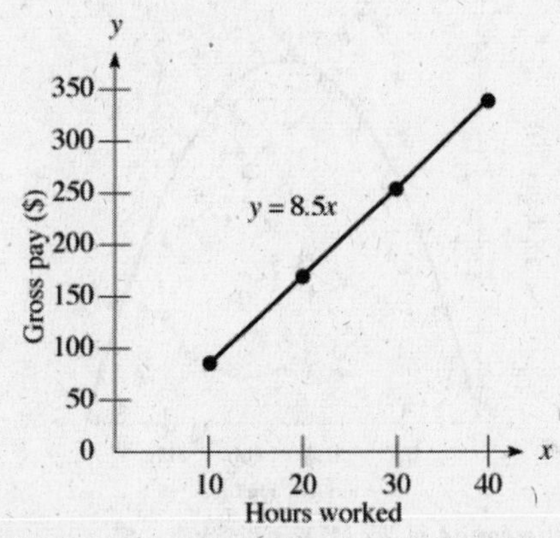

   d. The domain is $\{x \mid 10 \le x \le 40\}$ and the range is $\{y \mid 85 \le y \le 340\}$.

   e. The minimum is $85 and the maximum is $340.

3. The domain is {1,2,4} and the range is {1,3,5}. This is a function.
5. The domain is {–1,1,2} and the range is {–5,3}. This is a function.
7. The domain is {3,7} and the range is {–1,4}. This is not a function.
9. Yes, since it passes the vertical line test.
11. No, since it fails the vertical line test.
13. No, since it fails the vertical line test.
15. Yes, since it passes the vertical line test.
17. Yes, since it passes the vertical line test.

© 2012 Cengage Learning. All Rights Reserved. May not be scanned, copied or duplicated, or posted to a publicly accessible website, in whole or in part.

19. **a.** Completing the table:

Time (sec) t	Function Rule $h=16t-16t^2$	Distance (ft) h
0	$h=16(0)-16(0)^2$	0
0.1	$h=16(0.1)-16(0.1)^2$	1.44
0.2	$h=16(0.2)-16(0.2)^2$	2.56
0.3	$h=16(0.3)-16(0.3)^2$	3.36
0.4	$h=16(0.4)-16(0.4)^2$	3.84
0.5	$h=16(0.5)-16(0.5)^2$	4
0.6	$h=16(0.6)-16(0.6)^2$	3.84
0.7	$h=16(0.7)-16(0.7)^2$	3.36
0.8	$h=16(0.8)-16(0.8)^2$	2.56
0.9	$h=16(0.9)-16(0.9)^2$	1.44
1	$h=16(1)-16(1)^2$	0

b. The domain is $\{t \mid 0 \le t \le 1\}$ and the range is $\{h \mid 0 \le h \le 4\}$.

c. Graphing the function:

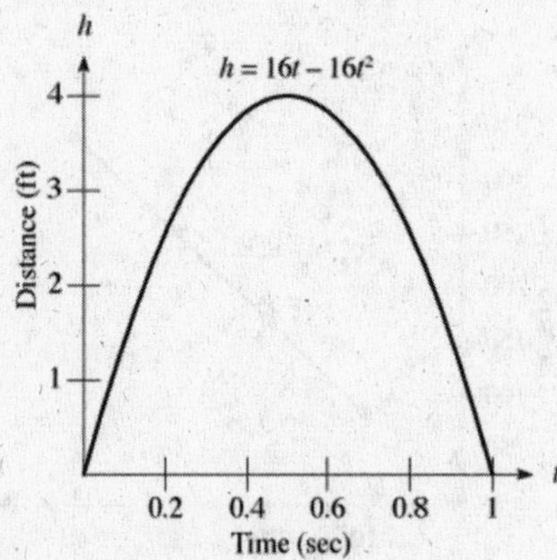

21. The domain is $\{x \mid -5 \le x \le 5\}$ and the range is $\{y \mid 0 \le y \le 5\}$.

Appendix B Functional Notation

1. Evaluating the function: $f(2)=2(2)-5=4-5=-1$

3. Evaluating the function: $f(-3)=2(-3)-5=-6-5=-11$

5. Evaluating the function: $g(-1)=(-1)^2+3(-1)+4=1-3+4=2$

7. Evaluating the function: $g(-3)=(-3)^2+3(-3)+4=9-9+4=4$

9. First evaluate each function:

$$g(4)=(4)^2+3(4)+4=16+12+4=32 \qquad f(4)=2(4)-5=8-5=3$$

Now evaluating: $g(4)+f(4)=32+3=35$

11. First evaluate each function:

$$f(3)=2(3)-5=6-5=1 \qquad g(2)=(2)^2+3(2)+4=4+6+4=14$$

Now evaluating: $f(3)-g(2)=1-14=-13$

© 2012 Cengage Learning. All Rights Reserved. May not be scanned, copied or duplicated, or posted to a publicly accessible website, in whole or in part.

13. Evaluating the function: $f(0)=3(0)^2-4(0)+1=0-0+1=1$

15. Evaluating the function: $g(-4)=2(-4)-1=-8-1=-9$

17. Evaluating the function: $f(-1)=3(-1)^2-4(-1)+1=3+4+1=8$

19. Evaluating the function: $g(10)=2(10)-1=20-1=19$

21. Evaluating the function: $f(3)=3(3)^2-4(3)+1=27-12+1=16$

23. Evaluating the function: $g\left(\frac{1}{2}\right)=2\left(\frac{1}{2}\right)-1=1-1=0$

25. Evaluating the function: $f(a)=3a^2-4a+1$

27. $f(1)=4$

29. $g\left(\frac{1}{2}\right)=0$

31. $g(-2)=2$

33. Evaluating the function: $f(0)=2(0)^2-8=0-8=-8$

35. Evaluating the function: $g(-4)=\frac{1}{2}(-4)+1=-2+1=-1$

37. Evaluating the function: $f(a)=2a^2-8$

39. Evaluating the function: $f(b)=2b^2-8$

41. Evaluating the function: $f[g(2)]=f\left[\frac{1}{2}(2)+1\right]=f(2)=2(2)^2-8=8-8=0$

43. Evaluating the function: $g[f(-1)]=g\left[2(-1)^2-8\right]=g(-6)=\frac{1}{2}(-6)+1=-3+1=-2$

45. Evaluating the function: $g[f(0)]=g\left[2(0)^2-8\right]=g(-8)=\frac{1}{2}(-8)+1=-4+1=-3$

47. Graphing the function:

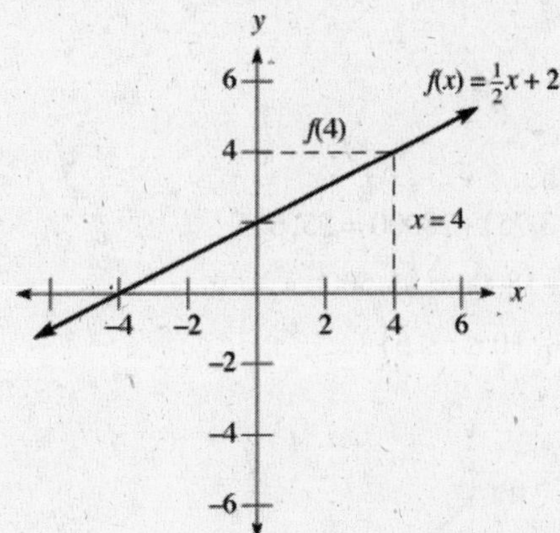

49. Finding where $f(x)=x$:

$$\frac{1}{2}x+2=x$$
$$2=\frac{1}{2}x$$
$$x=4$$

© 2012 Cengage Learning. All Rights Reserved. May not be scanned, copied or duplicated, or posted to a publicly accessible website, in whole or in part.

51. Graphing the function:

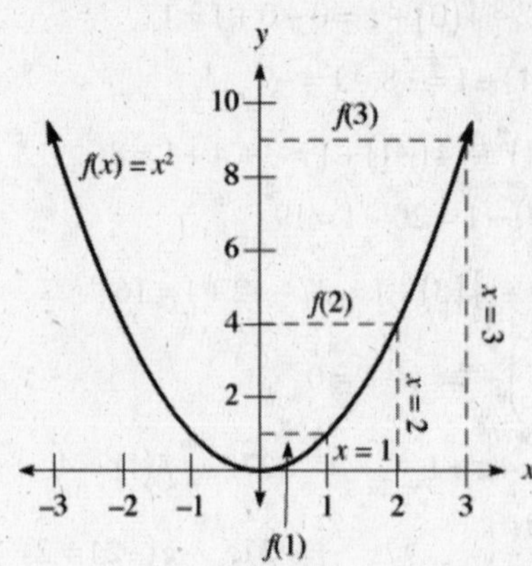

53. Evaluating: $V(3) = 150 \bullet 2^{3/3} = 150 \bullet 2 = 300$; The painting is worth \$300 in 3 years.

Evaluating: $V(6) = 150 \bullet 2^{6/3} = 150 \bullet 4 = 600$; The painting is worth \$600 in 6 years.

55. Let x represent the width and $2x + 3$ represent the length. Then the perimeter is given by:

$$P(x) = 2(x) + 2(2x+3) = 2x + 4x + 6 = 6x + 6, \text{ where } x > 0$$

57. Finding the values:

$$A(2) = 3.14(2)^2 = 3.14(4) = 12.56$$

$$A(5) = 3.14(5)^2 = 3.14(25) = 78.5$$

$$A(10) = 3.14(10)^2 = 3.14(100) = 314$$

59. **a.** Evaluating: $f(9) = 0.24(9) + 0.33 = 2.16 + 0.33 = 2.49$. The cost is \$2.49.

b. Evaluating: $f(5) = 0.24(5) + 0.33 = 1.20 + 0.33 = 1.53$. This represents the cost of making a 6 minute call.

c. Solving the equation:

$$0.24x + 0.33 = 1.29$$

$$0.24x = 0.96$$

$$x = 4$$

The call was 5 minutes in length.

61. **a.** Evaluating: $V(3.75) = -3300(3.75) + 18000 = \$5,625$

b. Evaluating: $V(5) = -3300(5) + 18000 = \$1,500$

c. The domain of this function is $\{t \mid 0 \le t \le 5\}$.

d. Sketching the graph:

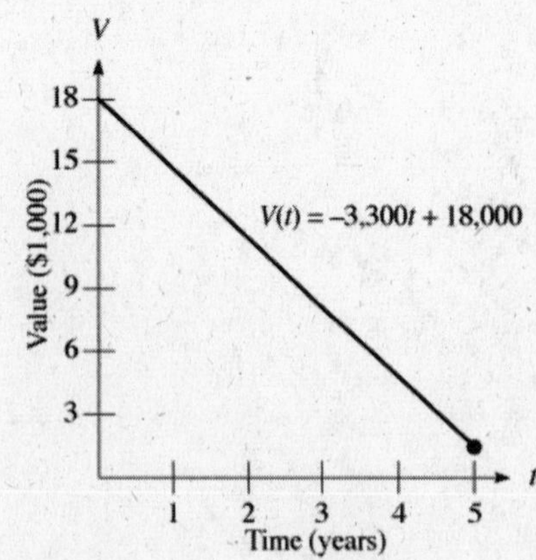

e. The range of this function is $\{V(t) \mid 1,500 \le V(t) \le 18,000\}$.

© 2012 Cengage Learning. All Rights Reserved. May not be scanned, copied or duplicated, or posted to a publicly accessible website, in whole or in part.

f. Solving $V(t)=10000$:

$$-3300t+18000=10000$$
$$-3300t=-8000$$
$$t\approx 2.42$$

The copier will be worth $10,000 after approximately 2.42 years.

Appendix C Fractional Exponents

1. Writing as a root and simplifying: $4^{1/2}=\sqrt{4}=2$

3. Writing as a root and simplifying: $16^{1/2}=\sqrt{16}=4$

5. Writing as a root and simplifying: $27^{1/3}=\sqrt[3]{27}=3$

7. Writing as a root and simplifying: $125^{1/3}=\sqrt[3]{125}=5$

9. Writing as a root and simplifying: $81^{1/4}=\sqrt[4]{81}=3$

11. Writing as a root and simplifying: $81^{1/2}=\sqrt{81}=9$

13. Writing as a root and simplifying: $8^{2/3}=\left(8^{1/3}\right)^2=\left(\sqrt[3]{8}\right)^2=2^2=4$

15. Writing as a root and simplifying: $125^{2/3}=\left(125^{1/3}\right)^2=\left(\sqrt[3]{125}\right)^2=5^2=25$

17. Writing as a root and simplifying: $16^{3/4}=\left(16^{1/4}\right)^3=\left(\sqrt[4]{16}\right)^3=2^3=8$

19. Writing as a root and simplifying: $16^{3/2}=\left(16^{1/2}\right)^3=\left(\sqrt{16}\right)^3=4^3=64$

21. Writing as a root and simplifying: $4^{3/2}=\left(4^{1/2}\right)^3=\left(\sqrt{4}\right)^3=2^3=8$

23. Writing as a root and simplifying: $(-8)^{2/3}=\left((-8)^{1/3}\right)^2=\left(\sqrt[3]{-8}\right)^2=(-2)^2=4$

25. Writing as a root and simplifying: $(-32)^{1/5}=\sqrt[5]{-32}=-2$

27. Writing as a root and simplifying: $4^{1/2}+9^{1/2}=\sqrt{4}+\sqrt{9}=2+3=5$

29. Writing as a root and simplifying: $16^{3/4}+27^{2/3}=\left(16^{1/4}\right)^3+\left(27^{1/3}\right)^2=\left(\sqrt[4]{16}\right)^3+\left(\sqrt[3]{27}\right)^2=2^3+3^2=8+9=17$

31. Writing as a root and simplifying: $4^{1/2}\bullet 27^{1/3}=\sqrt{4}\bullet\sqrt[3]{27}=2\bullet 3=6$

33. Using properties of exponents: $x^{1/4}\bullet x^{3/4}=x^{1/4+3/4}=x^1=x$

35. Using properties of exponents: $\left(x^{2/3}\right)^3=x^{2/3\bullet 3}=x^2$

37. Using properties of exponents: $\dfrac{a^{3/5}}{a^{1/5}}=a^{3/5-1/5}=a^{2/5}$

39. Using properties of exponents: $\left(27y^6\right)^{1/3}=27^{1/3}y^{6\bullet 1/3}=\sqrt[3]{27}y^2=3y^2$

41. Using properties of exponents: $\left(9a^4b^2\right)^{1/2}=9^{1/2}a^{4\bullet 1/2}b^{2\bullet 1/2}=\sqrt{9}a^2b=3a^2b$

43. Using properties of exponents: $\dfrac{x^{3/5}\bullet x^{4/5}}{x^{2/5}}=x^{3/5+4/5-2/5}=x^{5/5}=x$

45. Simplifying: $25^{-1/2}=\left(25^{1/2}\right)^{-1}=\left(\sqrt{25}\right)^{-1}=5^{-1}=\dfrac{1}{5}$

47. Simplifying: $8^{-2/3}=\left(8^{1/3}\right)^{-2}=\left(\sqrt[3]{8}\right)^{-2}=2^{-2}=\dfrac{1}{4}$

49. Simplifying: $27^{-2/3}=\left(27^{1/3}\right)^{-2}=\left(\sqrt[3]{27}\right)^{-2}=3^{-2}=\dfrac{1}{9}$

© 2012 Cengage Learning. All Rights Reserved. May not be scanned, copied or duplicated, or posted to a publicly accessible website, in whole or in part.

Appendix D Equations with Absolute Value

1. Solving the equation:
$$|x| = 4$$
$$x = -4, 4$$

3. Solving the equation:
$$2 = |a|$$
$$a = -2, 2$$

5. The equation $|x| = -3$ has no solution, or $\varnothing$.

7. Solving the equation:
$$|a| + 2 = 3$$
$$|a| = 1$$
$$a = -1, 1$$

9. Solving the equation:
$$|y| + 4 = 3$$
$$|y| = -1$$
The equation $|y| = -1$ has no solution, or $\varnothing$.

11. Solving the equation:
$$4 = |x| - 2$$
$$|x| = 6$$
$$x = -6, 6$$

13. Solving the equation:
$$|x - 2| = 5$$
$$x - 2 = -5, 5$$
$$x = -3, 7$$

15. Solving the equation:
$$|a - 4| = \frac{5}{3}$$
$$a - 4 = -\frac{5}{3}, \frac{5}{3}$$
$$a = \frac{7}{3}, \frac{17}{3}$$

17. Solving the equation:
$$1 = |3 - x|$$
$$3 - x = -1, 1$$
$$-x = -4, -2$$
$$x = 2, 4$$

19. Solving the equation:
$$\left|\frac{3}{5}a + \frac{1}{2}\right| = 1$$
$$\frac{3}{5}a + \frac{1}{2} = -1, 1$$
$$\frac{3}{5}a = -\frac{3}{2}, \frac{1}{2}$$
$$a = -\frac{5}{2}, \frac{5}{6}$$

21. Solving the equation:
$$60 = |20x - 40|$$
$$20x - 40 = -60, 60$$
$$20x = -20, 100$$
$$x = -1, 5$$

23. Since $|2x + 1| = -3$ is impossible, there is no solution, or $\varnothing$.

25. Solving the equation:
$$\left|\frac{3}{4}x - 6\right| = 9$$
$$\frac{3}{4}x - 6 = -9, 9$$
$$\frac{3}{4}x = -3, 15$$
$$3x = -12, 60$$
$$x = -4, 20$$

27. Solving the equation:
$$\left|1 - \frac{1}{2}a\right| = 3$$
$$1 - \frac{1}{2}a = -3, 3$$
$$-\frac{1}{2}a = -4, 2$$
$$a = -4, 8$$

© 2012 Cengage Learning. All Rights Reserved. May not be scanned, copied or duplicated, or posted to a publicly accessible website, in whole or in part.

29. Solving the equation:

$$\begin{aligned} |3x+4|+1&=7 \\ |3x+4|&=6 \\ 3x+4&=-6,6 \\ 3x&=-10,2 \\ x&=-\frac{10}{3},\frac{2}{3} \end{aligned}$$

31. Solving the equation:

$$\begin{aligned} |3-2y|+4&=3 \\ |3-2y|&=-1 \end{aligned}$$

Since this equation is impossible, there is no solution, or $\varnothing$.

33. Solving the equation:

$$\begin{aligned} |5x-1|+3&=7 \\ |5x-1|&=4 \\ 5x-1&=-4,4 \\ 5x&=-3,5 \\ x&=-\frac{3}{5},1 \end{aligned}$$

35. Solving the equation:

$$\begin{aligned} 2|3x-1|&=8 \\ |3x-1|&=4 \\ 3x-1&=-4,4 \\ 3x&=-3,5 \\ x&=-1,\frac{5}{3} \end{aligned}$$

37. Solving the equation:

$$\begin{aligned} 2|2x+3|-1&=5 \\ 2|2x+3|&=6 \\ |2x+3|&=3 \\ 2x+3&=-3,3 \\ 2x&=-6,0 \\ x&=-3,0 \end{aligned}$$

Appendix E Inequalities with Absolute Value

1. Solving the inequality:

$$\begin{aligned} &|x|<3 \\ &-3<x<3 \end{aligned}$$

Graphing the solution set:

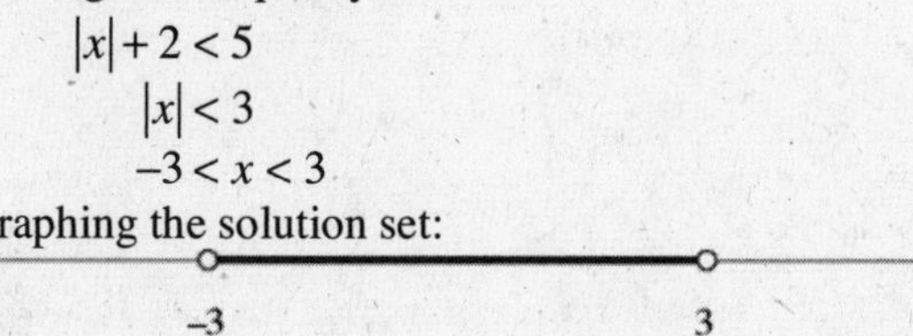

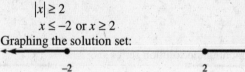

3. Solving the inequality:

$$\begin{aligned} &|x|\geq 2 \\ &x\leq -2 \text{ or } x\geq 2 \end{aligned}$$

Graphing the solution set:

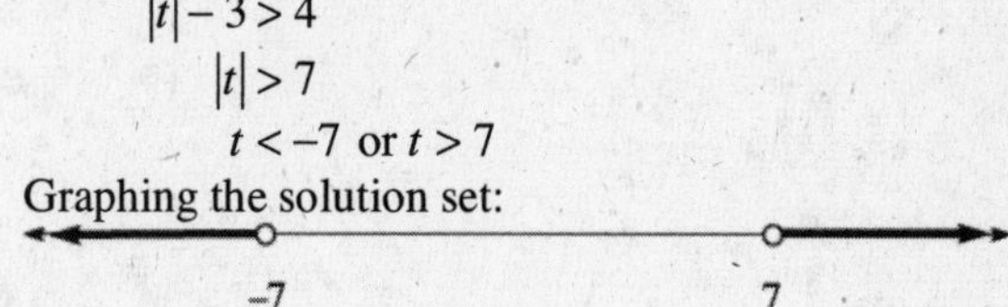

5. Solving the inequality:

$$\begin{aligned} |x|+2&<5 \\ |x|&<3 \\ -3<x&<3 \end{aligned}$$

Graphing the solution set:

7. Solving the inequality:

$$\begin{aligned} |t|-3&>4 \\ |t|&>7 \\ t<-7 &\text{ or } t>7 \end{aligned}$$

Graphing the solution set:

9. Since the inequality $|y|<-5$ is never true, there is no solution, or $\varnothing$. Graphing the solution set:

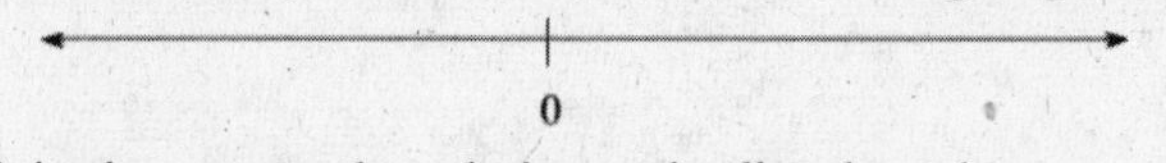

11. Since the inequality $|x|\geq -2$ is always true, the solution set is all real numbers.

Graphing the solution set:

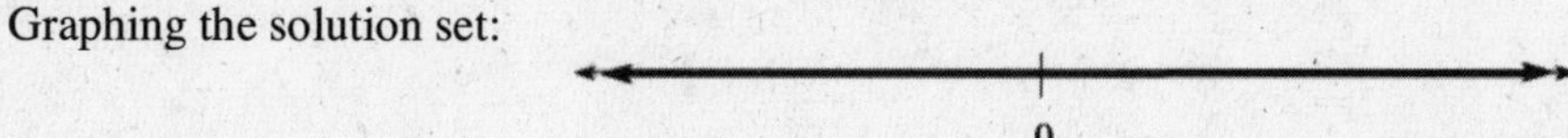

© 2012 Cengage Learning. All Rights Reserved. May not be scanned, copied or duplicated, or posted to a publicly accessible website, in whole or in part.

13. Solving the inequality:

$$|x-3|<7$$
$$-7<x-3<7$$
$$-4<x<10$$

Graphing the solution set:

-4 10

15. Solving the inequality:

$$|a+5|\ge 4$$
$$a+5\le -4 \text{ or } a+5\ge 4$$
$$a\le -9 \text{ or } \quad a\ge -1$$

Graphing the solution set:

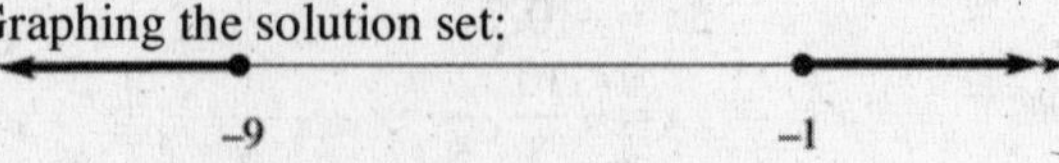

17. Since the inequality $|a-1|<-3$ is never true, there is no solution, or $\varnothing$. Graphing the solution set:

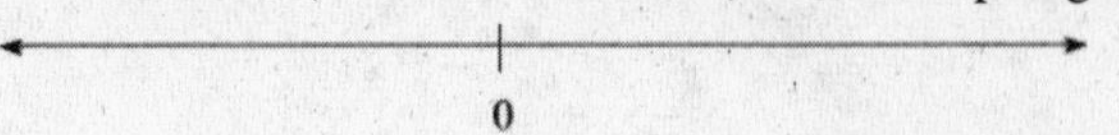

19. Solving the inequality:

$$|2x-4|<6$$
$$-6<2x-4<6$$
$$-2<2x<10$$
$$-1<x<5$$

Graphing the solution set:

-1 5

21. Solving the inequality:

$$|3y+9|\ge 6$$
$$3y+9\le -6 \quad \text{or} \quad 3y+9\ge 6$$
$$3y\le -15 \qquad 3y\ge -3$$
$$y\le -5 \qquad y\ge -1$$

Graphing the solution set:

23. Solving the inequality:

$$|2k+3|\ge 7$$
$$2k+3\le -7 \quad \text{or} \quad 2k+3\ge 7$$
$$2k\le -10 \qquad 2k\ge 4$$
$$k\le -5 \qquad k\ge 2$$

Graphing the solution set:

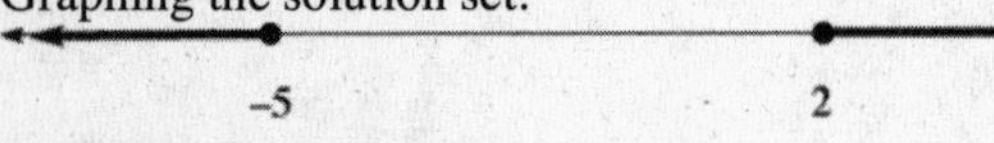

25. Solving the inequality:

$$|x-3|+2<6$$
$$|x-3|<4$$
$$-4<x-3<4$$
$$-1<x<7$$

Graphing the solution set:

27. Solving the inequality:

$$|2a+1|+4\ge 7$$
$$|2a+1|\ge 3$$
$$2a+1\le -3 \quad \text{or} \quad 2a+1\ge 3$$
$$2a\le -4 \qquad 2a\ge 2$$
$$a\le -2 \qquad a\ge 1$$

Graphing the solution set:

29. Solving the inequality:

$$|3x+5|-8<5$$
$$|3x+5|<13$$
$$-13<3x+5<13$$
$$-18<3x<8$$
$$-6<x<\frac{8}{3}$$

Graphing the solution set:

-6 8/3

© 2012 Cengage Learning. All Rights Reserved. May not be scanned, copied or duplicated, or posted to a publicly accessible website, in whole or in part.

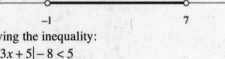